The Philosophy of Joseph Petzoldt

Also available from Bloomsbury

Advances in Experimental Philosophy of Science, edited by
Daniel A. Wilkenfeld and Richard Samuels
Early Analytic Philosophy and the German Philosophical Tradition,
by Nikolay Milkov
The History and Philosophy of Science: A Reader, edited by
Daniel J. McKaughan and Holly VandeWall
The History of Understanding in Analytic Philosophy, edited by
Adam Tamas Tuboly
The Philosophy of Anne Conway, by Jonathan Head

The Philosophy of Joseph Petzoldt

*From Mach's Positivism to
Einstein's Relativity*

Chiara Russo Krauss

BLOOMSBURY ACADEMIC
LONDON • NEW YORK • OXFORD • NEW DELHI • SYDNEY

BLOOMSBURY ACADEMIC
Bloomsbury Publishing Plc
50 Bedford Square, London, WC1B 3DP, UK
1385 Broadway, New York, NY 10018, USA
29 Earlsfort Terrace, Dublin 2, Ireland

BLOOMSBURY, BLOOMSBURY ACADEMIC and the Diana logo
are trademarks of Bloomsbury Publishing Plc

First published in Great Britain 2023
This paperback edition published 2024

Copyright © Chiara Russo Krauss, 2023

Chiara Russo Krauss has asserted her right under the Copyright,
Designs and Patents Act, 1988, to be identified as Author of this work.

English language translation © Perlocutio Ltd. 2023

For legal purposes the Acknowledgements on p. vi constitute
an extension of this copyright page.

Cover image © Ibusca / Getty Images

All rights reserved. No part of this publication may be reproduced or transmitted
in any form or by any means, electronic or mechanical, including photocopying,
recording, or any information storage or retrieval system, without
prior permission in writing from the publishers.

Bloomsbury Publishing Plc does not have any control over, or responsibility for,
any third-party websites referred to or in this book. All internet addresses given in
this book were correct at the time of going to press. The author and publisher regret
any inconvenience caused if addresses have changed or sites have ceased to exist,
but can accept no responsibility for any such changes.

A catalogue record for this book is available from the British Library.

A catalog record for this book is available from the Library of Congress.

ISBN: HB: 978-1-3503-2145-8
      PB: 978-1-3503-2149-6
      ePDF: 978-1-3503-2146-5
      eBook: 978-1-3503-2147-2

Typeset by Integra Software Services Pvt. Ltd.

To find out more about our authors and books visit www.bloomsbury.com
and sign up for our newsletters.

Contents

Acknowledgements		vi
List of abbreviations		vii
1	Introduction	1
2	Stability and Eindeutigkeit	11
3	Subjectivism and relativistic positivism	67
4	Petzoldt and Einstein's theory of relativity	107
5	Criticism of Petzoldt's interpretation of relativity	153
6	Conclusion	169
Notes		174
Bibliography		221
Index		239

Acknowledgements

I wish to thank the SIR program (Scientific Independence of young Researchers) of the Italian Ministry of University for providing the funds for the research presented in this book; and the University of Konstanz's Philosophical Archive, the Universitätsarchiv of the Berlin Technische Universität and the archive of the Deutsches Museum in Munich for granting me access to the material in their possession. I would also like to thank Professor Gereon Wolters and Professor Klaus Hentschel for their seminal works, without which my research would not have been possible, and also for kindly and helpfully discussing my research ideas. I thank Professor Riccardo Martinelli, Professor Stefano Besoli and Dr Francesco Pisano for reading the first draft of this book and helping me improve it based on their comments. Last but not least, I thank Professor Edoardo Massimilla for the unparalleled guidance in my professional and scientific career.

List of abbreviations

Works by Joseph Petzoldt:

- Avenarius1887 = 'Zu Richard Avenarius' Prinzip des kleinsten Kraftmasses und zum Begriff der Philosophie', *Vierteljahrsschrift für wissenschaftliche Philosophie*, 11 (1887), 177–203.

- Maxima1890 = 'Maxima, Minima und Ökonomie', *Vierteljahrsschrift für wissenschaftliche Philosophie*, 14 (1890), 206–39, 354–66, 417–42.

- Sittenlehre1893/1894 = 'Einiges zur Grundlegung der Sittenlehre', *Vierteljahrsschrift für wissenschaftliche Philosophie*, 17 (1893), 145–77; 18 (1894), 32–76; 196–248.

- Eindeutigkeit1895 = 'Das Gesetz der Eindeutigkeit', *Vierteljahrsschrift für wissenschaftliche Philosophie*, 19 (1895), 146–203.

- Einführung1900/1904 = *Einführung in die Philosophie der reinen Erfahrung*, 2 vols (Leipzig: Teubner, 1900–4).

- Parallelismus1902 = 'Die Notwendigkeit und Allgemeinheit des psychophysischen Parallelismus', *Archiv für systematische Philosophie*, 8 (1902), 281–337.

- Weltproblem1906 = *Das Weltproblem vom positivistischen Standpunkt aus* (Leipzig: Teubner, 1906).

- Bewegung1908 = 'Die Gebiete der absoluten und der relativen Bewegung', *Annalen der Naturphilosophie*, 7 (1908), 29–62.

- Weltproblem1911 = *Das Weltproblem vom Standpunkte des Relativistischen Positivismus aus Historisch-Kritisch dargestellt*, Zweite vermehrte Auflage (Leipzig: Teubner, 1906).

- Relativitätstheorie1914 = 'Die Relativitätstheorie der Physik', *Zeitschrift für positivistische Philosophie*, 2 (1914), 1–56.

- Weltproblem1921 = *Das Weltproblem vom Standpunkte des relativistischen Positivismus aus*. Dritte, neubearbeitete Auflage unter besonderer Berücksichtigung der Relativitätstheorie (Leipzig, Berlin: Teubner, 1921).

- Stellung1923 = *Die Stellung der Relativitätstheorie in der geistigen Entwicklung der Menschheit*, Zweite verbesserte und vermehrte Auflage (Leipzig: Barth, 1923).

Archival sources:

- MachNachlass = Ernst Mach's *Nachlass* in Deutsches Museum, Munich, Archives NL 174. Also available online at https://digital.deutsches-museum.de/archive/
- PetzoldtNachlass = Joseph Petzoldt's *Nachlass* in Universitätsarchiv der Technischen Universität Berlin.

1

Introduction

Who was Joseph Petzoldt?

Education

A volume about a 'minor' philosopher is clearly marked as such by the fact that the Introduction needs to explain who the subject was and why he is worthy of attention. We do not deny that Petzoldt is a 'minor' philosopher. It is for this very reason that we take up the challenge of demonstrating that awareness of his work is key to understanding the times in which he was active from a historical and a philosophical perspective.

Joseph Petzoldt was born on 4 November 1862 in Altenburg, Germany.[1] He opted to study natural sciences and philosophy after developing a keen interest in Darwin's theory of evolution. His studies covered physics, mathematics, geometry, zoology, psychology and the history of philosophy, at the universities of Jena, Munich, Geneva, Leipzig and Göttingen.

The German scientific and philosophical world of those times was noteworthy for its renewed interest in the work of Kant. His transcendental philosophy was regarded as a theory for the foundation of science, whose key teaching was that (1) empirical data are the basis for every piece of knowledge, and (2) empirical knowledge has inherent limits, set by the human forms of cognition, that cannot be transcended to grasp reality directly. It is therefore unsurprising that a short autobiographical piece written by Petzoldt describes how as a young man he considered himself to be 'neo-Kantian and agnostic', noting that the first fundamental works he had read included some of the foundational writings in this paradigm, such as Friedrich Albert Lange's *History of Materialism*, Du Bois-Reymond's *The Limits of Natural Science* (in which he formulates his famous *ignoramus et ignorabimus* motto) and in particular Kant's *Prolegomena*.[2]

However, Petzoldt's intellectual development took a decisive turn in 1883 and 1884 when he came across *The Science of Mechanics* (1883) by Ernst Mach (1838–1916) and *Philosophy as Thinking of the World in Accordance with the Principle of the Least Amount of Energy, Prolegomena to a Critique of Pure Experience* by Richard Avenarius (1843–96). The works of these two thinkers were to define his philosophical career. Petzoldt himself recounts how, even having read only parts of Mach's work, he was struck by the passage in the Preface that expresses the 'fundamental conception of the nature of science as Economy of Thought …'.³ Mach's doctrine of the economy of thought, together with Avenarius's concept of the 'principle of least amount of energy,' thus became the nucleus of the young Petzoldt's early reflections. As he neared the end of his studies, he chose Avenarius's principle of least amount of energy as the theme of one of the works presented for his final exams.⁴ Petzoldt decided to send the work to Avenarius, who published part of it in his journal *Vierteljahrsschrift für wissenschaftliche Philosophie* (Scientific Philosophy Quarterly), under the title 'Zu Richard Avenarius' Prinzip des kleinsten Kraftmasses und zum Begriff der Philosophie (On Richard Avenarius's Principle of Least Amount of Energy and the Concept of Philosophy, 1887). Their collaboration intensified the following year when, after a brief meeting in Dresden, Avenarius invited Petzoldt to visit him in Zurich, where he lived and worked. Petzoldt spent his summer break there, and the two engaged in daily conversations. It was this period of close contact with Avenarius that, as he put it, freed him 'from the last residues of neo-Kantian idealism', launching him definitively on the path of 'the great turning point towards a perspective based on pure experience'.⁵

Over the following years, Petzoldt continued this line of research, focusing on the work of Mach and Avenarius, and on the principles of economy and least amount of energy. In 1889, he published an analysis of the recently released first volume of Avenarius's *Kritik der reinen Erfahrung* (Critique of Pure Experience, 1888–90),⁶ and in 1891 he published 'Maxima, Minima und Ökonomie' (Maxima, Minima and Economy). This aimed to identify common ground between the principles of Mach and Avenarius, and the principles of physics such as Gauss's 'least effort' and Euler's 'least action'.

This work gained him a doctorate from the University of Göttingen, under the supervision of the experimental philosopher and psychologist Georg Elias Müller. However, Petzoldt's academic career did not take off. To earn a living, he therefore embarked on what would become his lifelong occupation: teacher of natural science at the Kant-Gymnasium in Spandau, a borough of Berlin. Over the years, Petzoldt made several unsuccessful attempts to secure a tenured position at the University of Berlin. He attributed his failure to do so to prejudice on the part of the faculty's philosophy professors – Wilhelm Dilthey, Friedrich Paulsen and Carl Stumpf – against positivist thinkers or those with socialist tendencies.⁷

With this avenue closed, Petzoldt fell back on the Technical University of Berlin, where, in 1904, he gained his professional qualification for a professorship (*Habilitation*). He produced another work on the theories of Avenarius: the two volumes of the *Einführung in die Philosophie der reinen Erfahrung* (Introduction to the Philosophy of Pure Experience, 1900–4). Avenarius, since deceased, was noted for using abstruse and idiosyncratic terminology in his main works. Petzoldt's aim was to develop the ideas contained in these works and to make them more accessible to a broader audience.

Petzoldt taught at the Technical University as a *Privatdozent* until 1925, the year in which he was promoted to a permanent position as a lecturer of philosophy (*außerordentliche Professur*), with particular reference to the natural sciences and methodology.⁸ However, notwithstanding his more permanent position, his workload at the Technical University remained limited, consisting of only four hours a week.⁹

Petzoldt's relatively lacklustre academic career did not daunt him, but rather made him more determined to make a name for himself in philosophy, becoming an interpreter and promoter of this new approach that he saw as having been shaped by Mach and Avenarius. For him, the opposition to those who sought to advance positivism was further proof of the need to fight hard to ensure that this line of thinking would eventually predominate. Despite the lack of recognition by his peers, Petzoldt's philosophical output was prolific and continuous until his death in 1929. It even enjoyed some success, as demonstrated by the four editions in less than twenty years of what can be seen as his main work: *Das Weltproblem vom positivistischen Standpunkt aus* (The Problem of the World from the Standpoint of Positivism, 1906, 1912, 1921, 1924).

Activism

Petzoldt's activism was not limited to writing books and articles. Over the years, he worked to set up a movement to campaign for an anti-metaphysical philosophy working closely with science. His first initiative was in 1912, with the foundation of the *Gesellschaft für positivistische Philosophie* (Society for Positivist Philosophy) in Berlin.[10] Its constitution was expressed in the following open letter, which was published in a range of philosophical and scientific journals.

> To bring forth a comprehensive Weltanschauung, based on the factual material that has been accumulated by the separate sciences, is an ever more urgent need; this is true first of all for science itself, but also for our era as such, which will only thereby have earned what we now own. But this claim can be achieved only through the common labors of many. Therefore we call upon all philosophically interested researchers no matter in which scientific fields they may be active and upon all philosophers in the narrow sense of the term whose expectation is to reach by themselves valid knowledge only through the penetrating study of the facts of experience, to join a Society for Positivistic Philosophy. The Society shall have the purpose of establishing lively connections among all the sciences, of developing everywhere the unifying ideas and thus press forward toward a contradiction-free unitary conception.[11]

This declaration had thirty-one signatories, including eminent figures such as Albert Einstein and Sigmund Freud; the mathematicians Georg Helm, David Hilbert and Felix Klein; the positivist historian Karl Lamprecht; the evolutionary biologist Wilhelm Roux; the botanist Henry Potonié; the philosophers Wilhelm Schuppe, Theodor Ziehen and Wilhelm Jerusalem; the sociologist Ferdinand Tönnies; the experimental physiologist Max Verworn; as well as Georg Elias Müller and Ernst Mach. Including the names of renowned scientists was not designed solely to stimulate dialogue between philosophy and the natural sciences but was also a move in the politics of academia. Evidencing their support to the higher echelons of academia, to those responsible for assigning professorships, was in fact designed to counter the resistance that positivist philosophers came up against in German universities from incumbents with more traditional philosophical leanings.[12]

The prominence of these signatories demonstrates the effort that Petzoldt had devoted to launching his society. Within a year it had 155 members, had hosted five meetings between November 1912 and March 1913, and had its own official publication, the recently launched *Zeitschrift für positivistische Philosophie* (Journal of Positivist Philosophy).[13] The journal was only published for two years, however. The outbreak of the First World War, which took Petzoldt to the front, and the economic crisis that engulfed Germany at the end of the conflict made running the journal increasingly difficult. In an attempt to salvage what he could from what remained of the society, in 1922 Petzoldt tried to turn it into a positivist subgroup of the less fragile *Kant-Gesellschaft* (Kant Society), with the hope of gradually growing its membership and increasing its influence to shift the focus of the *Kant-Gesellschaft* to his own agenda. The plan did not work as he had hoped, however. The positivist line always remained a largely irrelevant minority within the *Kant-Gesellschaft*, probably to some extent due to the difficulty of attracting scientists into a relatively traditional philosophical society. Petzoldt therefore decided to set up a new organization to advance dialogue between philosophers and scientists. A new declaration appeared in 1927, the manifesto for an *Internationale Gesellschaft für empirische Philosophie* (International Society for Empirical Philosophy), initially consisting only of the Berlin chapter:

> Philosophical interests and creative philosophy have once again bloomed mightily in Germany after the war. But the field is dominated by narrowly logical tendencies, restricted to the pure analysis of concepts, and apriorist theories of knowledge, mystical-religious currents, romantic historical constructions. By contrast, there is little evidence of empirical philosophy cautiously evaluating the results of the individual sciences. Yet there is a great deal here awaiting evaluation, for example, the new results of atomic research and the theory of relativity, the science of heredity, brain research, *Gestalt* and developmental psychology, psychoanalysis and psychopathology. For this reason the undersigned decided to found a local chapter of the International Society for Empirical Philosophy, of which anyone may become a member who cares about the development of philosophy based upon scientific experience.[14]

A year later, the new society had more than a hundred members.[15] The greatest obstacle in terms of logistics and above all financing was setting up a journal that could serve as its platform. Petzoldt therefore decided to collaborate with Hans Vaihinger. Vaihinger had successfully founded the *Kant-Studien* in 1896 as well as the above-mentioned *Kant-Gesellschaft* in 1904, but was having problems with his new journal, the *Annalen der Philosophie* (1919). As specified in its subtitle, 'mit besonderer Rücksicht auf die Probleme der Als-Ob-Betrachtung' (with particular reference to the problem of the concept of the *as if*), the journal focused on the philosophy of the *Als-Ob* (as if) developed by Vaihinger himself.[16] This objective was so specific that it restricted the number of potential readers and contributors. A relaunch of the journal was therefore attempted in 1924, broadening its scope and changing its title to *Annalen der Philosophie und philosophische Kritik* (Annals of Philosophy and Philosophical Critique). Eventually, in 1927, the journal was handed over to the *Gesellschaft für empirische Philosophie*, conceding that it would become the platform for publications by its members. Petzoldt became a member of the board, as along with Vaihinger and his student Raymund Schmidt.[17] This change of direction became immediately evident. The 1927 issue, alongside papers on strictly philosophical themes, also contained papers that discussed scientific matters such as atomic theory, relativity, the foundations of mathematics and the concept of space.

Petzoldt and the Berliner Gruppe

Even though Petzoldt died just two years later, his work in Berlin – and in particular setting up the *Gesellschaft für empirische Philosophie* and the collaboration with the *Annalen* – was the foundation stone of what became known as the *Berliner Gruppe*, that is, the proponents of logical positivism active in Berlin, whose principal exponents were Hans Reichenbach (1891–1953), Walter Dubislav (1895–1937) and Kurt Grelling (1886–1942).[18] Even though only Dubislav appears in the list of members of the society, both he and Reichenbach organized conferences hosted by the society over the three years of Petzoldt's presidency.[19] Reichenbach also published a paper in the first issue of the *Annalen* when Petzoldt was its co-director.[20] On Petzoldt's death, Reichenbach and Dubislav took charge of the association, becoming its director and its administrator, respectively. In 1931 they changed the name of the society to the *Gesellschaft für wissenschaftliche Philosophie* (Scientific Philosophy Society), to better reflect its new orientation. Even though one of its fundamental objectives was still engagement with science, the focus had in fact shifted from disciplines in the general field of medicine – such as psychology, biology and physiology – to those less associated with concrete empiricism, such as theoretical physics, mathematics, logic, probability calculus and geometry. The Berlin Group thus used the *Gesellschaft für empirische Philosophie* in much the same way as the Vienna Circle used the *Ernst Mach Verein* (Ernst Mach Association), as a forum in which the ideas of the group's members could be presented and discussed.

Petzoldt's legacy to the Berlin Group was not only his society but also the collaboration with the *Annalen*. On his death, the journal was taken over by Reichenbach and Rudolf Carnap, who turned it into the official publication of the *Gesellschaft für empirische Philosophie* and the *Ernst Mach Verein*. Once again, the management decided to mark the difference with a change of name. The new title was *Erkenntnis*.[21]

Situating Joseph Petzoldt in the history of philosophy

Petzoldt and the empiriocriticism of Mach and Avenarius

Having outlined this concise biography of Petzoldt, we can now address the question of his role in the philosophical milieu of the late nineteenth and early twentieth centuries and go deeper into the question of why he is worthy of attention.

First of all, Joseph Petzoldt can be considered the driving force behind the idea that there is a unitary philosophical line based on the ideas of Mach and Avenarius, whether this is termed 'empiriocriticism',[22] 'critical positivism',[23] 'neutral monism',[24] 'realistic empiricism'[25] or anything else. In fact, although these two thinkers have many points in common, they elaborated their concepts independently of each other and engaged in only sporadic correspondence. Petzoldt, on the other hand, developed close personal relationships with both Avenarius and Mach, whom he regarded as his mentors. He saw them as pioneers in a new direction in the history of thinking whose imminent consolidation was a historical necessity, notwithstanding the lingering resistance of traditional philosophy. He therefore set himself the objective of fighting for the triumph of these new ideas, on the one hand through his scientific work, which provided a theoretical elaboration of the thinking of Mach and Avenarius, and on the other taking on the role of the public face of this philosophical approach in the academic debate that shook Germany at the start of the twentieth century. As a result, if even today Mach and Avenarius are assigned the same label and cited as the exponents of a specific line of thinking, this is largely due to Petzoldt's efforts to align and promote the ideas of his two mentors. A historical and philosophical understanding of empiriocriticism cannot therefore ignore Petzoldt's role in the definition of this framework.

The main aim of this volume is therefore to reconstruct Petzoldt's intellectual evolution, showing how and why he arrived at the philosophy of Mach and Avenarius. This is important not only for an understanding of Petzoldt's thinking per se but also to provide a more general frame for the philosophical leanings of Mach and Avenarius in the prevailing context of the late nineteenth and early twentieth centuries. As we will show, Petzoldt's complex relationship with Kant and the neo-Kantianism of the day reveals an important aspect of these philosophical trends, which at times seem to be so close as to almost overlap each other, and at other times so far apart as to appear incompatible.

Petzoldt's interpretation of the thinking of Mach and Avenarius also discourages any automatic inclination to interpret their philosophical approach in a subjectivist or phenomenalistic manner. There is no doubt that Petzoldt's interpretation of the thinking of Mach and Avenarius is just one of many possible interpretations. Nevertheless, it cannot be denied that his close links with these thinkers allowed him to present himself as a privileged interpreter of their approach. This effect persists to this day. Moreover, even where his own stamp on the concepts of Mach and Avenarius is evident, it sheds light on elements and trends in their works that might otherwise have been undervalued or dismissed.

Petzoldt and the relativity debate

Petzoldt did not limit himself to systematizing the philosophy of his mentors. His aim was also to insert their concepts into the changing landscape of the first decades of the twentieth century.

The second half of the nineteenth century was characterized by a reaction against the romantic philosophy of nature (*Naturphilosophie*), which had exalted the living organism to the status of a paradigm for understanding reality, postulating the existence of a finalistic 'vital force' and the general aliveness of the world. In contrast, new directions in science held that all of nature, including living beings, was subject to the explanatory tools of physics. Medical research had therefore turned to principles in biology and physiology that were based on the conservation of energy, on experimental methods and on a quantitative mathematical approach. However, drawing living beings into the domain of physics had highlighted even more forcefully the problem of the mind.[26] If organisms could be treated as complex mechanisms subject to the laws of chemistry and physics, how should the mind be addressed? Was it possible to turn even psychology into a science, using experimental methods and a quantitative mathematical approach? Or was it necessary to distinguish between physical and mental spheres, and therefore between the aims and methods of the natural sciences (*Naturwissenschaften*) and those of the sciences of the mind (*Geisteswissenschaften*)?

These were the questions that permeated academia in the years in which Mach and Avenarius were studying and embarking on careers in philosophy. The elevation of physics to the status of a model for the natural sciences had extended its scope to the study of living organisms. The focus was therefore on physiology and psychology, and the disciplines that spanned these such as the physiology of the sense organs and a psychology that was based on experimental physiology. Even though they had different backgrounds, it was no coincidence that Mach and Avenarius had addressed these topics. Mach had trained as a physicist, but his research into optics and acoustics had stimulated his interest in the physiology of the senses and thus to more general physiological issues such as the relationship between the body and the mind, and the empirical significance of scientific concepts. Avenarius, on the other hand, had received a more traditional philosophical education, but during his studies he was drawn to physiological psychology, seeing its potential as a method of resolving problems in the theory of knowledge.

This intellectual climate also influenced Petzoldt. His early works focused on the problem of scientific knowledge of the organic and psychological domain. However, the landscape had changed dramatically in the first decades of the twentieth century and was destined to be shaken up further. Although psychology was already established as an independent science, the ingenuous enthusiasm it had generated in the nineteenth century had by now been tempered, and the hope that it could come to the aid of philosophy by applying the tools of science to the long-standing question of the workings of the mind was somewhat diminished. The reasons for this relative decline of psychology can be found in the rise of arguments against 'psychologism' promoted by neo-Kantian philosophers, phenomenologists and logical positivists, as well as in the fragmentation of the world of psychology into a plethora of schools and directions that disagreed with each other about what the methods and aims of this science should be. In addition, while psychology started withdrawing into specialized research, theoretical physics was once again taking centre stage thanks to Einstein's formulation of the theory of special relativity in 1905 and that of general relativity in 1916.[27]

If scientific psychology was *the* topic for philosophers working on the theory of knowledge in the second half of the nineteenth century, a topic that required one to take a position, Einsteinian relativity became *the* topic of the first half of the twentieth century. This required a thinker like Petzoldt to take relativity into account, given his goal of demonstrating that the empiricism of his mentors Mach and Avenarius was the philosophy of the present and the future, the only philosophy relevant to the scientific advances of the times. This was not a particularly difficult step for Petzoldt, who became one of the first philosophers to discuss Einstein's theories. Indeed, a special link between empiriocriticism and the theory of relativity could be established on the basis of Mach's influence on Einstein. In Petzoldt's view, the theory of relativity was less an extraneous element to be integrated into empiriocriticism than something of a scientific confirmation of it.

A further objective of this volume is therefore to set out how Petzoldt used the philosophical framework that he had elaborated on the basis of the thinking of Mach and Avenarius to respond to the changes in the world of science and philosophy that the Einsteinian revolution had brought about. Petzoldt's work on adapting empiriocriticism to the theory of relativity, or rather, to take credit for the theory of relativity as the fruit of the philosophical approach that he and his mentors had devised and promulgated, is also the key to understanding the role he played in the world of philosophy in the early twentieth century. Against the complex and vibrant backdrop of phenomenology, neo-Kantianism and logical positivism, Petzoldt was a lesser figure with respect to the great names of the day, despite being recognized as a significant contributor to debates with those of other persuasions. He was in fact seen, in those times, as the most orthodox representative of the thinking of Mach, who was held in great respect. Moreover, he was one of the first philosophers to comment on Einsteinian relativity, which made him a key figure in the debate.[28] For these reasons, the history of the philosophy of the early twentieth century, and in particular the history of the philosophical debate on relativity in those years, would be incomplete without a chapter on Petzoldt and on his relationships with other philosophers of the day. The third and final objective of this volume is therefore to clarify Petzoldt's place in the broader context of early twentieth-century Germany.

A word of clarification

Even if Petzoldt has not been completely ignored in the history of the twentieth century, in particular with respect to his role in the debate on the theory of relativity,[29] there has so far been no work that addresses his thinking directly, with the aim of providing a unitary portrayal of it. As producing a philosophical profile of Petzoldt has little to build on directly, this volume cannot cover all aspects of his thinking, nor does it aim to, in particular when one considers his vast output and his extensive relationships with key figures in the philosophical and scientific circles of the day. We will therefore concentrate on the issues that are most closely related to the theory of knowledge, leaving to one side most of the ethical, political and pedagogical issues he also addressed over the course of his intellectual development.

The lack of studies on Petzoldt is in sharp contrast to the large and enduring research that has been undertaken over the years on his mentor Ernst Mach. Although the relationship between Petzoldt and Mach is central to this volume, we must avoid turning it into a work about Mach, which might have been the outcome if we had succumbed to the temptation of following every thread that led back to Mach. Accounting for the vast bibliography and extensive debate on the ideas of Mach, a veritable galaxy of work, would be an impossible task and one that is unnecessary given that we are focusing on a different thinker. We will therefore rely on Mach's stature, assuming to some extent that the reader will be aware of the key aspects of his thinking. Our decision not to examine Mach in greater depth is even more necessary given that we would not be able to do the same for Petzoldt's other mentor, Avenarius, given the lack of literature on his work. In fact, his secondary bibliography is even smaller than that relating to Petzoldt, and the general lack of awareness of the key aspects of his thinking requires us to devote more space to him whenever we need to shed light on his relationship with Petzoldt. The same is true of other writers whom we touch upon in this volume – such as Gustav Theodor Fechner (1801–87), Heymann Steinthal (1823–99) and Wilhelm Schuppe (1836–1913) – whose ideas we need to present in greater detail, given that we cannot assume significant awareness of them on the part of the reader.

In conclusion, we have opted to focus only on sources that address aspects relevant to learning about and understanding Petzoldt. This is just the first step towards the further objective of providing a more complete historiographical account of the philosophical line of which Mach was certainly the most eminent proponent, but which also involved other thinkers worthy of consideration, such as Petzoldt and Avenarius, as well as exponents of the closely aligned and partially overlapping immanentist school, such as Schuppe, Theodor Ziehen (1862–1950), Hans Cornelius (1863–1947) and Richard von Schubert Soldern (1852–1924). We hope that future research will focus on these thinkers and on the relationships between them.

2

Stability and Eindeutigkeit

Towards a unified science

To piece together Petzoldt's philosophical development, it is first necessary to identify the questions that had inspired him in order to understand what drew him to the ideas of Mach and Avenarius as possible answers to these questions. This requires examining the broader social context of the Germany of the second half of the nineteenth century. Bearing in mind that any generalization may lead to an oversimplification of the broad range of events and perspectives that define any period, we consider it reasonable to identify a number of macro-trends in the evolution of nineteenth-century philosophy that stemmed from debate stimulated by scientific advances initiated in earlier centuries.

If the scientific revolution had defined a vast programme of research into the physical world, based on mechanistic materialism, the nineteenth century was marked by extraordinary advances in the study of living organisms, due, among other things, to the perfection of the microscope and to discoveries in chemistry. These advances in the study of inanimate matter and of living organisms could not but raise the question of the relationship between these two spheres of reality and in consequence between these two domains of knowledge. Overall, there were two options. The first was materialism, which saw sentient living organisms as no more than matter in motion, and therefore to be understood using the principles of physics, and above all the principle of causality. The second was vitalism, according to which life and consciousness could not be explained simply as matter in motion but were dependent on some kind of finalistic vital principle.

Proponents of both positions can be found in every period. However, vitalism gained ground in Germany in the late eighteenth and early nineteenth centuries, embodied in the Romantic *Naturphilosophie*, whose view of nature was based on the concept of *Lebenskraft* (vital force).[1] In contrast, materialism resurged in the 1840s and 1850s thanks to the work of Jakob Moleschott (1822–93), Ludwig Büchner (1824–99), Carl Vogt (1817–95) and Heinrich Czolbe (1819–73). The triumph of materialism was illusory, however. It was not only what became known as *Materialismusstreit* (the materialism debate) that brought to light opposition to this world view based on philosophical, religious and moral themes; it was also the scientific developments themselves that provided arguments against materialism.[2] Advances in organic chemistry and electrophysiology, together with the Darwinian revolution, appeared to demonstrate the possibility of understanding living organisms within the framework of mechanistic materialism.[3] However, initial scientific research into the relationship between the physical world and the mind appeared to give new life to idealistic conceptualizations. Research into the sense organs and the nervous system suggested that there is an inescapable difference between reality and how it is perceived and organized in the human mind. This undermined faith in the possibility of knowing the true essence of the material world, of going beyond the limits of subjectivity.[4]

A third macro-trend was thus established in the second half of the nineteenth century. Increasing numbers of thinkers turned to the teachings of Kant in order to resolve and rationalize the long-standing conflict between materialism and antimaterialism by reducing it to methodological issues related to human cognition.[5] The typical position of this early neo-Kantianism was described by Klaus C. Köhnke as 'seeking a philosophical justification for an attitude of *'neither-nor'*, or one of '*this as well as that*', leading to a 'dualism [...] clearly identifiable in the form of scepticism'.[6]

In other words, given the conflict between the advocates and the decriers of the possibility of extending the materialistic-mechanistic approaches of physical science to other fields of knowledge, the new movement turned to the Kantian concept of the limits of our faculties, thereby acknowledging the legitimacy of different approaches, each functioning in its own domain. This fully validated the materialistic-mechanistic assumption on how to understand nature, but at the same time emphasized its limits, thus vindicating the need for it to coexist with other assumptions, such as freedom and finality, when other areas were being addressed, including living beings, consciousness and the mind. Following Kant's line, the possibility of the coexistence of determinism with freedom, of causality with finality, of matter with the mind, was founded on the fact that human consciousness was incapable of capturing reality itself, being constrained by its cognitive functions. This view suggested that these antinomies were not grounded in reality but derived from human reasoning. This neo-Kantian view was therefore sceptic as it claimed the impossibility of truly knowing reality, and dualistic because it combined materialism and antimaterialism, the mechanical understanding of nature and the recognition of the specific characteristics of the mind.

This was the context in which Petzoldt embarked on his studies. We cited, in the Introduction, the passage where he describes the position he adopted in his youth as 'neo-Kantian and agnostic'. We can now clarify that 'agnostic' here should be read as the 'scepticism' noted above with respect to the possibility of perceiving things directly. This framing of his education allows us to identify the philosophical questions that set him on his path. The great question of the day, one that was debated fervently, was in fact that of the unity of science. Is there one fundamental overarching kind of knowledge? Is it physics? Is it therefore necessary to extend the causal materialistic-mechanistic position to all fields addressing reality? Or are there different fields of knowledge, each with its own distinct principles and methodologies? And what are these different fields? The physical world? The mind? Living organisms? The realm of cultural and spiritual creations?

The young Petzoldt saw the neo-Kantianism that was in vogue around the middle of the nineteenth century as a preliminary reply to these questions. However, its very fundamental features – the dualism and scepticism – soon left him dissatisfied. The common thread running through Petzoldt's philosophical development was, in fact, the search for a science that was positive and unitary, encompassing the contributions made by each individual science into a comprehensive vision, thereby providing a coherent and true understanding of reality. Given this aim, Petzoldt soon abandoned the neo-Kantian view that there is always an unknowable residual element that can never be grasped and that the roots of a series of unavoidable contradictions and antinomies lie in human cognition. Hence, Petzoldt embarked on finding a philosophy that could unify the sciences, basing them on new foundations. This philosophy was that of Richard Avenarius and Ernst Mach.

It was clear that the solution proposed by Avenarius and Mach could not be a retrograde step in which a return to naive realism replaced neo-Kantian scepticism and the reaffirmation of materialist or idealistic monism replaced dualism. In Petzoldt's view, what these two thinkers offered was more of an attempt to address old problems in new ways, thus opening the way to an alternative and definitive philosophical system: empiriocriticism.

The tendency towards stability

Fechner and the principle of the tendency towards stability

Having identified the starting point of Petzoldt's philosophical journey, we can follow its route by examining the opening page of his first scientific work: 'Zu Richard Avenarius' Prinzip des kleinsten Kraftmasses und zum Begriff der Philosophie' (On Richard Avenarius's Principle of Least Effort and the Concept of Philosophy), published in 1887 in the journal edited by Avenarius himself. The paper is an interpretation of the Avenarius monograph *Philosophie als Denken der Welt gemäss dem Prinzip des kleinsten Kraftmasses. Prolegomena zu einer Kritik der reinen Erfahrung* (Philosophy as Thinking of the World in Accordance with the Principle of the Least Amount of Energy, Prolegomena to a Critique of Pure Experience, 1876). Petzoldt introduces the work thus:

> Here, [Avenarius] applies, for the first time, a *fundamental proposition* [*Grundsatz*] to the mechanism of representational life; this proposition is endowed with *utmost validity* and is worthy of utmost consideration. After earlier pertinent intimations by Zöllner, it was Fechner – with his 'principle of the tendency towards stability' – who *generalized* and very effectively went into greater depth on the doctrine of Darwin, thus discovering, in a certain sense, the *essence of all evolution*.[7]

These brief words reveal the focus of Petzoldt's interest in the years when, as we have seen, the relationship between the different sciences and the question of whether they could be unified were at the heart of philosophical debate. We see that Petzoldt's focus is not on Avenarius's principle of least amount of energy as such but on the fact that it represents the first endeavour to apply a broader 'fundamental proposition' to physical events, namely, Fechner's 'principle of the tendency towards stability'. Avenarius's work thus gains stature as a further step in the process of identifying an overarching general principle governing *all* kinds of evolution, irrespective of which field they apply to. Darwin is thus seen as the one who explained evolutionary processes in the domain of living organisms, and Fechner as the first to formulate a fundamental generalization, integrating the processes of organic and inorganic evolution into a unified law: the principle of the tendency towards stability.

In 1873, in *Einige Ideen zur Schöpfungs- und Entwickelungsgeschichte der Organismen* (Some Ideas on the History of Creation and of the Development of Organisms), Fechner entered the debate on Darwinian evolution, presenting his proposal as a resolution of the sterile conflict between materialism and vitalism so as to construct a view of the world that could encompass both organic and inorganic processes.[8] The first step was to redefine the difference between these two domains, reformulating it as a relative difference rather than an absolute one. Fechner defined inorganic systems as those in which particles *did not* reciprocally modify their position, and organic systems as those in which their position was continuously modified as a result of the internal forces of the particles.[9] He cited salt crystals as an example of the former and – significantly – the solar system as an illustration of the relationships between the elements of organic systems:[10] not exactly what we would usually define as an organism.

Fechner's second step is key to understanding how the motion of the solar system could be considered organic. This was the formulation of the principle of the tendency towards stability, which, as noted also by Petzoldt, drew on concepts that had been formulated by Fechner's astrophysicist friend Johann Karl Friedrich Zöllner.[11] Fechner first provided a number of definitions, specifying that a system is in a state of 'absolute stability' (at rest) when there is permanent immobility of the parts of the whole, in a state of 'full stability' when 'there is mobility, but [the parts] reacquire not only the same relative positions at regular intervals, but also the same velocities, directions, and variations in velocity and direction', and in a state of 'approximate stability' when the parts of the system 'never return exactly, but only approximately, to their earlier relative positions'.[12] For Fechner, this last case is what characterized organic systems, and the solar system.[13]

Fechner's introduction of the concept of stability allowed him to define more precisely the sense in which an organic system is one in which the different parts undergo changes relative to other parts as a result of the internal forces of the system itself. In particular, 'the life of organisms is based completely on the *periodicity* of their functions, that is, on *stable* relationships', even though 'it is not always the same particles that return, but only equivalent particles', as in the case of the metabolism.[14]

Having defined his concept of stability, Fechner went on to formulate the principle based on the concept:

> In any system of parts of matter left undisturbed or exposed to invariant external conditions – including therefore the system of the whole material world, insofar it is regarded as a closed system – if we exclude infinitesimal motion, we discover a continuous progression from less stable states to more stable states until a final state of full or approximate stability is reached.[15]

According to Fechner, every system, including the universe itself, moves towards a state of stability. This led him to defining his own cosmogeny, where the universe was originally in a state of total instability and motion, which he termed 'cosmorganic',[16] from which relatively more stable forms progressively emerged (organisms), as well as fully stable forms (inorganic systems). This obviated the need to explain how organisms had developed from inanimate matter, as organic forms precede the inorganic forms that are the end state of the bodies of the universe. The classical notion that organisms had grown out of inorganic matter was, in Fechner's view, absurd, as it would require more stable systems to have become less stable ones.

Apart from this fascinating and unconventional vision of how the cosmos had developed, our interest (and that of Petzoldt) lies in how the principle of stability enabled Fechner to overcome the traditional conflict between materialism and vitalism, between the organic and the inorganic, and between causality and finalism, thereby leading to a unitary view of the world. It was not only the difference between organic and inorganic systems that was defined in relative terms as systems with a lesser or greater degree of stability between their parts but the tendency towards increasing stability meant that the former tended to change gradually into the latter. This replaced the hard line between the two with gradual and continuous change.

As well as blurring the line between the organic and the inorganic, Fechner's principle of stability also unified them under a single all-embracing law, similarly to what had been achieved a short time earlier by the definition of the principle of the conservation of energy. Fechner saw the two principles as complementary: the principle of the conservation of energy addressed 'quantitative relationships', whereas the principle of stability was its 'qualitative complement', and both were the basis for 'deduction with respect to the realization of consequences'.[17] The same laws governed the formation of the galaxies, the motion of the planets, the evolution of living beings, the life of individuals and the motion of atoms, albeit in different ways than in mechanistic materialism and in vitalism. Whereas the former sacrificed the concept of finality on the altar of causality, the second saw teleology as the cornerstone of an understanding of the universe. Indeed, Fechner states in the Preface to his book that one of his aims was an 'integration of the principle of causality with that of teleology for everything that happens'.[18] In particular:

> as the tendency towards stability operates according to the principle of causality, through the effects of forces that behave according to laws,[19] we find here the much sought-after unification of these two principles of physics, in that the difference between the two is simply that the principle of causality considers the foundations whereas the principle of teleology considers the purpose of one and the same lawful sequence.[20]

The critical issue of the relationship between causality and teleology was much debated in those times. This was, of course, a long-standing issue, and was at the heart of works by the likes of Leibniz and Kant, and the Romantic thinkers of the *Naturphilosophie* movement. However, new elements entered the old debate in the second half of the nineteenth century with the recent definition of the second law of thermodynamics and the spread of Darwin's theory of evolution. Rudolf Clausius had established that heat never passes from a colder body to a warmer body spontaneously, without added energy, which showed that physical events are not reversible but always have a specific direction.[21] Darwin's concept of natural selection had provided an explanation of how purposive structures can emerge in a world governed by the causal principle: something that originated without a purpose but happens to be purposive, by virtue of this very fact, is preserved over time.

The second law of thermodynamics and the principle of natural selection could thus be used to elaborate a new approach to the problem of teleology, an approach that could avoid the earlier metaphysical implications. The new concept of purpose no longer needed to presuppose an intelligence, whether human or divine, that sets goals and selects what is required to attain them, and no longer postulated final causes that overlapped with causal explanations. The new concept of purpose only implied that the processes at work in the universe had a direction and/or that purposive structures could accidentally emerge over time and be preserved. This broke the link between teleology and metaphysics or theology, re-establishing its position within the natural sciences. This was, at least, the intention, given that the degree of success, or coherence, that this position enjoyed varied from one thinker to another.[22]

It should be noted that, even if Fechner did not reference works on the definition of the second law of thermodynamics, there is no doubt that his *Ideen zur Schöpfungs- und Entwickelungsgeschichte der Organismen* conform to this paradigm. It is therefore necessary to go into further detail on how he elaborated the process of reconciling causality with teleology.

In the first place, so as not to be misled by his references to Darwinian evolution, it is worth emphasizing that Fechner did not see reconciling causality with finality as reducing the latter to the former, as in Darwin's theory. In fact, the success that Darwin enjoyed in Germany did not go hand in hand with the acceptance of his core concept, that is, the mechanism of natural selection as a causal explanation of how purposive adaptive traits emerge. German materialists certainly saw natural selection as an extraordinarily powerful argument for a mechanistic view of the world,[23] as demonstrated by the words of Büchner, who held that 'Darwinian theory' implied 'the definitive banishment of the harmful concept of finality from organic science and, in consequence, from science in general'.[24] At the same time, however, many other scholars rejected natural selection as a causal explanation of the transmutation of species, seeing it instead as no more than a subsequent step that sanctioned the survival or the extinction of emergent forms. The emergence of new forms was to be explained by orthogenetic theories, postulating the existence of laws of internal development with an explicit or implicit teleological nature.[25]

Fechner clearly belonged to this second group. In the introduction to his work on evolution, he asserted that his aim was to provide an 'in-depth analysis' of Darwin's theories by proposing 'a general principle that could unify all the laws governing organic development' so as to accommodate the 'need expressed by many opposed to Darwin and Haeckel for a unitary approach to organic development'.[26] Moreover, Fechner stated explicitly in his work that 'the principle of the struggle for existence only emerges as an adjustment of or a complement to a different higher principle',[27] namely, the 'principle of relative differentiation', which contrasts with the Darwinian concept of 'casual differentiation' in that descendants would not display fortuitous variation, but complementary variation that would lead to a greater overall degree of stability.[28]

In contrast to German materialists, we can see that Fechner's interest in Darwin's theory did not stem from the possibility of providing a causal explanation for the emergence of new forms of living beings but rather from a desire to place once again at the heart of philosophical debate the concept of 'evolution', understood as *Entwicklung*, that is, a progressive development based on natural laws that were at the same time causal and teleological, and able to unify the entire cosmos into a unitary process of change. Moreover, Fechner's roots were in Romantic *Naturphilosophie*, acquired from his study of the works of Friedrich Schelling and Lorenz Oken, who – as noted in an autobiographical note – had enabled him to 'find a general and unitary perspective for a conceptualization of the world'.[29]

Herbart and Steinthal's psychic mechanics

Fechner's deliberations on the tendency towards stability stated in passing that 'even the domain of the mind appears to be governed by this principle […], which can be thought of as related to the increasing stability of the material processes that underlie the life of the mind'.[30] Despite this observation, Petzoldt's view was that the only true and proper extension of the principle of stability to psychology was in the work of Avenarius, with the formulation of the principle of least amount of energy.

However, the insertion of Avenarius's principle into the line sketched by Fechner was the fruit of Petzoldt's interpretation. Petzoldt observes that 'the introduction of Avenarius's fundamental proposition' is 'somewhat unexpected': 'after a brief explanation of the guiding principle' Avenarius turned to other issues, even though 'the relatively new fundamental proposition' of least amount of energy 'needs some clarification about its actual core, the relationship between means and ends, and its *generality*'.[31] In particular, according to Petzoldt, 'in the text and in the annotations Avenarius refers to other fields that could be framed in the light of the principle in question, but without illustrating the reasons that might be the basis for this extension of its validity', whereas going deeper into this topic might have allowed him 'to arrive himself at the ideas developed by Fechner'.[32]

Indeed, despite its title, the principle of least amount of energy was not even the main subject of Avenarius's work but rather part of a broader approach designed to demonstrate that – on the basis of the principle in question – it is necessary to eliminate the remaining metaphysical elements inherent in the concepts of causality and substance, so that our world view can increasingly be reduced to only what is in fact 'pure experience', that is, experience that has not been contaminated by elements that are not truly empirical. Moreover, the idea of a principle of least amount of energy was itself far from original, as Avenarius had adopted it from contemporary Herbartian psychology, in particular as developed by Heymann Steinthal (1823–99).

One of the first attempts to use the same principles to unify knowledge of the physical world with that of the mind was in fact undertaken by Johann Friedrich

Herbart (1776–1841).³³ His foundational work on psychology was based on the idea that turning the study of the mind into a science required adopting the methods that were used in the study of the physical world. It required starting with experiential data, but then going beyond these to elaborate a metaphysics that could lead to a mathematical elaboration of consciousness. Therefore, just as physics started from experience, but then elaborated metaphysical concepts of force and matter, leading to the science of 'mechanics', expressed in mathematical terms, so Herbartian psychology adopted the metaphysical concepts of soul and representation in order to build a 'mechanistic science of the mind', which – just like its physics counterpart – consisted of two sub-branches: 'statics' and 'dynamics'.

In particular, Herbart held that the purpose of metaphysics was to resolve the apparent contradictions manifested in experience, such as that between the unity and identity of objects on the one hand, and the multiplicity and mutability of their properties on the other hand. His metaphysics therefore postulated the existence of 'reals', which in themselves had no properties and remained permanently unchanged, but which, in their interactions with other reals, responded with acts of 'self-preservation', which appear as their properties.

As everything remains identical to itself in Herbart's metaphysics, and every change is exclusively due to an interaction with something else that stimulates a self-preservation response, the same is true of the mind. In psychology, representations are the self-preservation acts of the 'real' that is the mind. The self-preservation of the mind and, in consequence, of its representations is the basis of the dynamics of the mind. Every representation tends to remain identical to itself in consciousness, but, when other representations appear, either it merges with them to form a 'representational mass' or it is inhibited by them, hence slipping temporarily below the threshold of consciousness. This leads to the process of 'apperception', when new mental contents are grasped by merging with the old representational masses, which remain latent until they are brought into consciousness as a result of their fusion with the new mental contents.

As we can infer from the concept of 'psychic mechanics', Herbart's ideas implied a clear refutation of any teleological approach. In his 1806 *Hauptpunkte der Metaphysik* (Main Points of Metaphysics), Herbart had already attacked Schelling forcefully for having reintroduced teleology with his organic conception of nature. Hence, Schelling had disavowed Kant, who had negated the concept of purpose in his *Critique of Pure Reason*, only to allow it under certain conditions in his *Critique of Judgment*.³⁴ In his theory of psychic mechanics, Herbart harked back to the causal explanations of physics and even more to chemistry, which had made great leaps forward in the late eighteenth and early nineteenth centuries. The dynamics of representations, which, when they collide, can either repel each other or merge into 'representational masses', was clearly formulated on the model of chemical reactions.

The second half of the nineteenth century therefore saw two different approaches to psychology. Herbart had based the explanation of mental processes on a mechanics of representations, modelled on the metaphysics inherent in physics and chemistry. The other approach came out of the physiology laboratories of Johannes Müller and his students, of which the foremost was Hermann von Helmholtz. This approach was later systematized by Fechner in his *Elemente der Psychophysik* (Elements of Psychophysics, 1860). Fechner held that making the study of mental activity scientific required starting with an analysis of its physiological conditions. Since these can be investigated experimentally, measured and expressed mathematically, if we establish a connection between them and their psychical correlates, we can describe mental processes in function of their physiological correlates, that is, quantitatively, mathematically and therefore scientifically.

Given the growing favour enjoyed by the psychophysical approach, the adherents of Herbart's school of thought found themselves needing to update their mentor's approach to psychology, stripping it of its metaphysical foundations and bringing it more in line with the novel physiological trends. The work of Steinthal is relevant here. Despite adopting Herbart's 'psychic mechanics',[35] in his *Abriss der Sprachwissenschaft* (Outline of the Science of Language, 1871), Steinthal put aside metaphysical assumptions, stating that his aim was to address solely 'facts of experience', putting aside the question of the 'real principle' of mental activity.[36] From the outset, this work saw consciousness in the light of its biological importance, as an instrument at the service of the life of the organism for nutrition, reproduction and defence against external threats.[37] At the same time, however, Steinthal did not fully embrace Fechner's psychophysical approach, placing himself in a certain sense halfway between Herbart and Fechner.[38] Whereas Fechner saw a way forward for the study of mental processes in the possibility of analysing their physiological correlates, Steinthal asserted that his theory remained unchanged in the case of the discovery that 'cerebral fibres play a role', and even in the case that 'all the laws that govern the development of the representations that we consider to be reactions of the mind are only laws of sympathetic participation or the conductivity of cerebral fibres'.[39] Relying neither on Herbart's metaphysics nor on Fechner's psychophysics, Steinthal assigned an essentially metaphorical sense to his 'psychic mechanics': given 'the impossibility of discussing mental activity without using analogies from the field of the behaviour of matter', it was important 'not to forget that we use images when we speak and think'.[40]

Now that psychophysics was no longer based on metaphysics, but on an analogy with the physical sciences, Steinthal also reinterpreted Herbart's theory of apperception. In Herbart's view, apperceptive processes stem from the tendency of reals to self-preservation, based in turn on their metaphysical self-identity. For Steinthal, on the other hand, apperception is based on the analogy of consciousness as a 'force' that can be transmitted from one mental representation to another. Psychic mechanics thus becomes a 'mechanics of the consciousness' that investigates 'the conditions under which a representation receives the energy of consciousness'.[41]

Unlike Herbart, Steinthal did not see consciousness as an intrinsic property of representations that falters only when some representations are inhibited by other representations, but as a state of 'extraordinary excitation that is transmitted to them'.[42] This makes the energy of consciousness scarce and highly unstable. There is only a limited amount of it, which passes continuously from one representation to another. In particular, it always passes to more normal representations, following a 'law of psychological inertia'.[43] Apperception processes, in which we use general and familiar representations to understand new mental content, are therefore based on the 'narrowness of consciousness' and on psychic inertia, both of which favour familiar mental contents.[44]

By assigning a purely analogical sense to psychic mechanics, Steinthal managed to steer clear of the problem of the causal and/or teleological nature of psychological laws. Thus he affirmed that:

> Psychology is a fully empirical science, and its goals cannot go beyond establishing the conditions that determine, through experience, a specific outcome [...], any additional step in the direction of causality or teleology belongs to metaphysics and the philosophy of religion.[45]

Even in the section on 'intention', Steinthal specified that this concept is not to be understood in a teleological sense, as the issue only concerned 'certain relationships between representations of motion and other representations and feelings', relationships that are no different from those that hold between 'a spark and gunpowder'.[46]

The influence of Steinthal and Fechner on Avenarius

Avenarius's first work continued the process of updating Herbartian psychology in the light of psychophysics as initiated by Steinthal. It is no coincidence that, as a student, Avenarius had taken courses by both Fechner and Steinthal.[47]

Steinthal had uprooted Herbartian psychology from its metaphysical bedrock, so Avenarius only had to take Steinthal's concept of 'force' and reformulate it as physical or physiological force to take the next step towards Fechner's psychophysics. He did this in one of the opening pages of *Prolegomena*, declaring that 'force is here understood above all in its physiological sense'.[48]

Placing the organism at the heart of his theory, Avenarius found himself needing to address the topic of the purposes of living beings. His work opens with these words:

> Regardless of how one may understand the mind and its relationship to the body, it is in any case necessary to attribute to the mind the *purposiveness* [*Zweckmässigkeit*] that we have long since ceased to fear to recognize empirically in the body. We may therefore wonder whether mental purposiveness should be seen as based on the purposiveness of its conditions in the living organism, or should even be considered identical to it, or whether it is particular in nature and based on relationships that are intrinsic to the mind; in any case, there can be no doubt that the functions of the mind must in general be purposive, given their fundamental importance to the survival of the individual.[49]

Thus, by making the theory of apperception of Herbart and Steinthal his own, Avenarius completed the process of distorting it. Apparently, he preserved its 'mechanical' approach to the mind, to be understood in a causal sense, but added a finalistic element. In fact, the principle of least amount of energy referred to in the title of the work on the one hand references Steinthal's law of psychic inertia and on the other transforms it into a teleological principle of organic evolution. Avenarius himself explicates this twofold aspect when he states that the principle of least amount of energy is both a 'principle of *persistence* [*Beharrung*]', according to which 'the changes that the mind communicates to its representations when a new percept is received is the smallest change possible', and also 'a principle of *evolution* [*Entwicklung*]', according to which 'from a multitude of possible apperceptions, the mind gives precedence to the one that can deliver the same outcome with least effort or *an improved outcome* with the same amount of effort'.[50] Avenarius even recognized the possibility of the mind delivering 'a temporarily *greater* force' in order to obtain a 'lasting advantage'.[51] As any wasted effort leads to a sense of displeasure and – to a lesser extent – every conservation of effort to a sense of pleasure, the mind tries to use effort in the most efficient way possible, thus reducing negative feelings and maximizing positive ones. This occurs in the mind through the increased elaboration and deployment of unitary, simple and familiar concepts. This would explain the emergence of mental constructs not only in the individual but also across human culture in general.

Although Avenarius was not explicit on this point, his work combined and merged a series of themes of the day. Debate on the works of Darwin had led to the idea that evolution was the result of the dynamics of the survival of organisms. However, the relationship between survival and evolution was not that of Darwin's theory of natural selection, in which the different rates of survival preserve in the line of descent any advantageous evolutionary trait that emerged by chance. Avenarius's propositions were closer to those of Lamarck, where the constant effort required for survival, or, more precisely, for the conservation of physiological energy, was the driver of the evolution of the organism through the emergence of more efficient structures and/or behaviours. This is a key difference, because in Darwin's theory variations do not arise *on purpose* but are preserved in case they *happen to* be purposive, in contrast to Lamarck's theory of evolution, which is based on purposeful variations.[52]

What Avenarius drew from Fechner, on the other hand, was the notion of a physiological force underlying mental activity, to which he added the concept of energy saving to explain the evolution of mental contents. Avenarius's view of the primary role of pleasure and displeasure was also Fechnerian in nature. Both saw pleasure and displeasure as playing a central role, as they were at the intersection of the physiological mechanisms of the organism, mental activity and the purposive behaviour of living beings. These feelings constitute the fundamental mental contents that are the *direct* result of the physiological stimulation of the sense organs. Moreover, they lie at the heart of everyday life, as the entire existence of an organism aims at maximizing pleasure and minimizing displeasure.[53]

Finally, Avenarius took from Steinthal's Herbartian psychology the idea that self-preservation was expressed in mental processes as the constant seeking out of familiar mental contents. Avenarius adopted Steinthal's theory of apperception in its entirety here, to the extent that the first footnote in the *Prolegomena* refers the reader directly to the *Abriss der Sprachwissenschaft* for further details on the function of apperceptive processes.[54] The difference is that while Herbart based his theory of apperception on metaphysical concepts, and Steinthal on a psychic mechanics that was built on an analogy with the physical sciences, Avenarius grafted it definitively onto Fechnerian physiology.

A further confirmation of the importance of Steinthal in Avenarius's *Prolegomena* can be seen in what Petzoldt writes in a brief autobiographical note on his first encounter with Avenarius:

> Without being aware of the significant role that Steinthal's *Abriss der Sprachwissenschaft* had played in Avenarius's development, and in particular in the writing of the work in question, I read Steinthal's work from start to finish [in the 1883/84 winter semester], and it impressed me greatly. I thus arrived well-prepared for my in-depth study of Avenarius's *Prolegomena*, which I embarked upon in the summer of 1886.[55]

These words clarify why we have focused on Herbart and Steinthal, given the significance of the influence that this line of thinking had on Avenarius and Petzoldt.

Petzoldt's principle of minimum disturbance

After this extended preamble, we can turn to Petzoldt's 'Zu Richard Avenarius' Prinzip des kleinsten Kraftmasses und zum Begriff der Philosophie', now that we are equipped with the tools to appreciate the relevant issues and the explicit and implicit references.

As we have seen, Petzoldt is in search of a universally comprehensive principle governing the inanimate world, living organisms and mental activity. He therefore harks back, in part, to Fechner's work on evolution, where he finds an approach that addresses both the organic world and the inorganic world, and in part to the work of Avenarius, where he finds a theory of the mind. Petzoldt links the work of these two thinkers on the assumption that Fechner's principle of the tendency towards stability and Avenarius's principle of least amount of energy are actually one and the same principle.[56]

In particular, Petzoldt starts with the hypothesis that a system is in a state of relative internal stability, in Fechner's sense, that is, in a state in which its parts move relative to each other with a regular periodicity. If an external event leads to a change, this can be considered a 'disturbance' that destabilizes the system. However, after a certain period of time, the system will return to stability, having in some sense absorbed the disturbance that had led to the change and thus reached a new equilibrium. The system will thus have 'adapted' itself, 'eliminating the disturbance'.[57] For example, if a foreign body enters a gravitational system consisting of two or more planets from elsewhere in space, this will lead to greater or lesser perturbation of the otherwise periodic motion of the planets. Sooner or later, however, the new body will become a stable part of the system, either by being pulled in by one of the planets and merging with it or by settling into its own regular orbit within the system.

The novel element that Petzoldt added to the work of Fechner is the use of the term *Störung* (disturbance) to indicate the resulting disequilibrium. This allows the creation of a bridge between Fechner's principle of stability and the Herbartian theory of apperception. Indeed, although Steinthal and, to a greater degree, Avenarius had worked on the relationship between Herbartian psychology and Fechnerian psychophysics, they had not addressed Fechner's principle of stability.[58] However, the introduction of the idea that stability can be reached by overcoming a disturbance allows Fechner's theory to overlap with Herbart's view that self-preservation requires the elaboration of new mental contents on the basis of more general and more familiar mental contents. Both cases involve systems that react to external change, returning to their original state of stability after incorporating the extraneous element. In his work, Petzoldt thus unifies the principles of Avenarius and Fechner as the 'principle of minimum disturbance'.[59]

Having introduced the concept of disturbance, thus unifying Fechner's theory with Herbartian psychology, Petzoldt turned to another fundamental concept, that of evolution:

> We refer to this whole process, from its onset to the establishing of the stationary state, as *evolution* [*Entwicklung*]. This consists of the progressive elimination of *disturbance* or *competition*. In other words: *An evolution is the result of competing tendencies, and the result of an evolution is a stationary state.*[60]

Evolutionary processes could be described as behaving according to the following schema: S, S + d, S'. That is, a stationary state, a disturbance and a new stationary state that is more 'evolved' than the preceding one as it has elaborated and encompassed the disturbance.

It is important to highlight the terminological shift introduced by Petzoldt, from Fechner's 'stability' to the 'stationary state' noted above. This concept is more closely associated with contemporary research on the second law of thermodynamics, which played an important role in philosophical debate on teleology. In physics, a 'stationary state' is defined as one in which mechanical events are fully reversible, such as, for example, the ideal case of a pendulum swinging under frictionless conditions. There is no dissipation of kinetic energy in such cases, nor any increase in entropy, and there is therefore no direction in which the system evolves. This means that a system that is in a stationary state and experiences no external changes or any decay towards other states will persist in that state indefinitely. This pure concept of stationary state is never found in nature, of course, where all spontaneous transformations result in an increase in entropy. This is why the concept of a 'stationary state of non-equilibrium' is also used, where a system's *internal* stationary state is guaranteed thanks to its interaction with the environment, from which the system acquires energy. In such cases, the system maintains constant entropy by releasing it to the environment. This concept is clearly fundamental in biology, where organisms are seen as systems in a stationary state of non-equilibrium that maintain that state through the exchange of energy and matter with the environment.

Petzoldt uses these physics concepts as the foundation for a mechanistic teleology of a sort, much as Fechner had done somewhat less explicitly.[61] The thermodynamic concept of 'evolution' as a system passing through states of ever-increasing entropy, that is, as the general direction of physical processes, is reinterpreted as a system's tendency to move towards a stationary state.[62] To do this, Petzoldt intersects the physics concepts of 'evolution' and 'stationariness' with Darwin's conceptual mechanism of natural selection. As a non-stationary system is a system in evolution, and a stationary system does not evolve, every non-stationary system will evolve until it reaches a stationary state, at which point – by definition – it will stop evolving. We will therefore see, over time, the emergence of an increasing number of stationary systems that tend to be 'preserved' and 'selected', in contrast to non-stationary systems, which, by definition, will continue to evolve. In other words, to borrow the 'little logical truism' formulated by chemist and popularizer Addy Pross: 'Unchanging things *don't change*, and changing things *do change* – until they change into things that don't.'[63]

Therefore, when Petzoldt writes that 'purposive phenomena can only be understood scientifically as the results of *evolution*,'[64] the term cannot be understood solely in the sense of Darwinian evolution but in this hybrid sense that also draws on physics and the contemporary debate on the second law of thermodynamics. This also clarifies how Petzoldt's concept of stability combines stability as the purpose and condition for biological self-preservation with stability as the outcome of purely mechanical processes: that is, as stationariness.[65]

The key point is that Petzoldt applies the concept of 'purpose' to the outcomes of mechanical Darwinian evolution.

> Relatively stationary states are what we primarily consider to be *purposes*, whereas we see the forces that serve to establish and maintain them as *means*. We consider a system the more purposive the less it generates disturbances in itself and in its relationship to external forces, that is, the more the forces in it exclusively serve its survival.[66]

According to Petzoldt, this point had eluded Avenarius, even though he had correctly highlighted the relationship between self-preservation and evolution, at least in the case of mental processes. Avenarius had not recognized that the essence of the principle of least amount of energy is the process of eliminating disturbance and of achieving ever-greater stability. He had therefore not grasped the universal significance of his principle, which went far beyond a simple law of psychology.

> Avenarius notes that the mind reacts with displeasure in the presence of an expenditure of energy that does not serve any purpose, of a waste of energy. An expenditure of energy that has no purpose is one that does not aim to achieve a stationary state in which the mind can rest. [...] The total use of energy to achieve a stationary state is its use in conformity to ends. *The whole of nature is necessarily governed by a tendency of this kind*, a 'principle of the tendency towards stability,' a principle of stationary states, of minimal disturbance, of maximal harmony. This principle represents the true sense of what Avenarius calls the 'principle of the least amount of energy'. *This is not a principle that is fundamental only to the life of the mind*, in Avenarius's sense, *but is a fundamental principle for everything that happens in nature*, insofar as 'evolution' and lasting states occur in nature.[67]

Having failed to grasp the generality of this principle, Avenarius had missed the key point, namely the possibility of bringing together 'natural purposes [*Naturzwecke*] in the physical and the mental world' thanks to a common principle (i.e. the principle of least disturbance).⁶⁸

The generality of the principle of least disturbance goes beyond the possibility of unifying the teleology of the physical world with that of the mental world. It is so general that it also covers the principle of least effort (or least constraint) formulated by Gauss in 1829:⁶⁹ that is, a law from the field of 'pure mechanics'.⁷⁰ Petzoldt sees the difference between the law of least effort and the principle of least disturbance as simply the fact that the Gaussian principle 'is ultimately just a description of the processes of motion resulting from the forces in play, without any involvement of "*naturphilosophisch* reflections"', whereas 'the principle of least disturbance or of minimum effort sees motion that is not subject to disturbance [...] as a *purpose*'.⁷¹ In other words:

> the [Gaussian] principle of least effort applies indiscriminately to motion as a whole, whereas our fundamental principle treats stationary states as purposes and active forces as means [...], so that the focus is always on the 'achieved' [*bezweckte*] fundamental result; a view that clearly lies outside the scope of pure mechanics.⁷²

Like Fechner and Avenarius, Petzoldt thus includes the concept of purpose in his conceptualization of nature. In fact, he even appears to reference explicitly the teleology of *Naturphilosophie*. In contrast to Avenarius, however, he does not see purposiveness as consisting of the actual ability of the mind and of organisms to act in such a way as to pursue self-preservation, but more as a possible way of interpreting the processes of the world. In the *Prolegomena*, Avenarius sees the mind as able to choose the most appropriate of a range of options, such as, for example, the option that produces the best outcome using the same amount of effort, or the one that obtains the same outcome using less effort, or even the one that requires a temporary increase of effort to deliver a future advantage. Petzoldt criticizes Avenarius explicitly on this point, stating that this view 'has no concrete meaning' as no choices of this kind can ever exist. There are no forces that can be spent or spared, as 'all available forces perform at every moment everything that they are able to perform'.⁷³ Indeed, 'from the point of view of pure mechanics, *and also of pure psychophysics*, there is no such thing as a greater or a lesser performance'.⁷⁴ Therefore, when we talk of purposiveness, it is necessary to clarify that it cannot be seen as the effective capacity to channel energy in one direction rather than another, selecting the one that best serves the purpose. Consequently, purposiveness cannot involve the capacity to put aside energy in order to 'save' it for future use. For Petzoldt, 'it is only *subjective teleological judgment* that differentiates between purposive and unpurposive, greater or lesser [performances], according to the degree to which they lead to stationary states'.⁷⁵

In a strict sense, no 'saving' or 'expenditure' of energy is possible, because, from the point of view of physics, energy cannot overperform or underperform. This means that the Gaussian principle of least effort does not imply – or, indeed, cannot imply – that nature proceeds on the basis of an ideal of parsimony. As Mach had already emphasized in *Die Mechanik in ihrer Entwicklung* (The Science of Mechanics), cited by Petzoldt himself, 'there is no need to seek a *mystical* or *metaphysical* reason for Gauss's principle' as 'all the work that can be performed in an element of time actually is performed'.[76] Therefore, all we can do is identify what we subjectively see as purposes and evaluate as more economical and efficient the process in which all the energy in play contributes to the attainment of the purpose, and as less economical or efficient the process that expends energy that either hinders the attainment of the purpose or does not contribute to its attainment. Avenarius appears to be aware of this twofold step, as, in the *Prolegomena*, he puts the purpose (the self-preservation of the individual) first and the principle of least amount of energy (the use of force in conformity to the purpose) second. However, he erred in supposing that energy could actually be directed towards a given objective, because purposiveness is only a way of subjectively evaluating the forces in play and not a way in which an individual can objectively determine the course of the world.

Petzoldt's position is again reminiscent of the neo-Kantianism of the day. He reconciles mechanism and finalism by regarding them as two possible ways of seeing events, one objective and the other subjective. The objective way, based on the natural sciences and in particular on physics, limits itself to describing the mechanical relationships between the forces in play (as per the Gaussian principle). The subjective, teleological and *naturphilosophisch* way evaluates the action performed by those forces in terms of its conformity to the purpose (as per the principle of minimum disturbance).

Summary, and an unanswered question

To summarize, in his first philosophical work Petzoldt sets himself the objective of identifying a maximally general law that would unify Avenarius's principle of minimum effort, Fechner's principle of the tendency towards stability, and – to an extent – the mechanical principle of least effort formulated by Gauss. The result is a synthesis of different fields of knowledge, as well as of different scholarly traditions. The principle of minimum disturbance can in fact be applied to the mind (as described by the Herbartian theories of apperception formulated by Steinthal and Avenarius), to organic and inorganic phenomena (as investigated by Fechner in his *Ideen zur Schöpfungs- und Entwickelungsgeschichte der Organismen*) and to purely mechanical processes (as captured in the Gaussian principle).

The interweaving of all these topics takes place against the backdrop of the specifically German reception of Darwinian evolution, which rejected the rigid mechanism of the natural selection of fortuitous variations in favour of the concept of '*Entwicklung*' as the unfolding of a teleological plan. Petzoldt himself is part of this movement and must therefore justify his openness to a finalistic conception. He therefore falls back on the neo-Kantian compromise between mechanism and finalism, where mechanism is the objective understanding of the world typical of the natural sciences and finalism is the subjective perspective that considers phenomena in terms of their purposes.

We have seen that these references were already present explicitly, to a greater or a lesser degree, in Avenarius's *Prolegomena*. However, it was Petzoldt's work that brought them to the fore. The core around which this thick fog of references condenses is the concept of disturbance (*Störung*). It is noteworthy that the concept of disturbance also plays an increasingly significant role in Avenarius's later works.

In Avenarius's *Kritik der reinen Erfahrung* (Critique of Pure Experience, 1888–90), cerebral evolution, and therefore mental evolution, is not determined by the tendency to save energy, as in the 1876 *Prolegomena* but by 'vital series' that consist of three stages. In the first, the 'value of vital self-preservation' of the brain is at its maximum, in that there is equilibrium between the two factors that determine the survival of the system, namely work (elaboration of stimuli) and nourishment (metabolic activity). In the intermediate stage, a condition emerges that induces disequilibrium and therefore the potential 'destruction' of the system.[77] Eventually, in the final stage, the system succeeds in re-establishing equilibrium, but an equilibrium of greater value than the original state, as it is no longer vulnerable to the conditions that had induced it to change. Thus, although Avenarius still sees self-preservation as the driver of evolution, self-preservation no longer consists of the conservation of energy but of elaborating any condition of disequilibrium that disturbs the system.

This shift is not missed by Petzoldt. In 'Maxima, Minima und Ökonomie' (Maxima, Minima and Economy, 1890), published after the *Kritik der reinen Erfahrung*, he notes that Avenarius had set aside the principle of least amount of energy in favour of an 'emphasis on the concept of *self-preservation*', thus arriving at 'a completely new position' which, 'at least from our position, adds further depth to Avenarius's work and presents the facts in an even more appropriate manner'.[78] Avenarius's new theory is in fact very similar to Petzoldt's principle of least disturbance. However, the origin of this similarity is unclear. It is certainly not a case of the teacher having been influenced by the student he had influenced earlier, as the first signs of Avenarius's new position can be found in notes written in the early 1880s, that is, before any of Petzoldt's other writings. In draft versions of the unpublished *Allgemeine Theorie der Erkenntniss* (General Theory of Knowledge), redacted around 1882, Avenarius already addresses psychophysical activity in terms of 'disruptions' (*Störungen*). Moreover, these pages contain a formulation by Avenarius of the following principle: 'the sum of a series of disruptions is as close as possible to zero, that is, to the minimum possible value.'[79]

The extraordinary similarity between this statement and Petzoldt's principle of minimum disturbance suggests that Petzoldt had drawn on the work of Avenarius. However, even if we accept this hypothesis, it is unclear how this transfer of ideas could have come about, given that Avenarius did not publish anything between the *Prolegomena* and the *Kritik der reinen Erfahrung*. Moreover, Petzoldt's autobiographical writings reveal that he wrote and published 'Zu Richard Avenarius' Prinzip des kleinsten Kraftmasses' before having ever met Avenarius in person. Finally, the extant correspondence between the two indicates that they did not address this topic even in the letters they exchanged before they first met. It is therefore plausible to exclude the possibility of Petzoldt having learnt of Avenarius's new ideas through the written word or the spoken word.

This question thus remains unanswered. Either Petzoldt learnt of Avenarius's new ideas in some way of which no trace remains or he was able to foresee the direction in which Avenarius's thinking was going from his reading of the *Prolegomena*. Another possibility is that he reached the same conclusions independently of Avenarius, perhaps facilitated by both thinkers having access to the same body of literature, such as the works of Steinthal and Fechner.

Eindeutigkeit and stability

Three years after working on the principle of least effort, Petzoldt published a substantial new article, once again in the journal edited by Avenarius. In 'Maxima, Minima und Ökonomie' (1890) Petzoldt returns to the subject matter of the earlier work, but structures it with greater precision, introducing some new fundamental elements. The work sees the first appearance of one of his most significant contributions, the concept of *Eindeutigkeit* (univocalness).

In contrast to his earlier work, here Petzoldt appears to want to differentiate rather than to unify. The work opens with a clear separation between two fields of investigation so as to allow the analysis of the concepts of maximum and minimum in each of the two domains, and to preclude undesirable misunderstandings. The first field concerns 'the processes of motion in general', and covers concepts such as Euler's principle of least action, Gauss's principle of least effort and Hamilton's variational principle.[80] The second field is 'the narrower one of purposive processes', which include Darwin's research into the origin of species, Zöllner's minimum principles, Fechner's principle of the tendency towards stability, Avenarius's principle of the least amount of energy and Mach's doctrine of the economy of thought.[81]

From maxima and minima to Eindeutigkeit

For a full understanding of Petzoldt's line of argumentation, it is first necessary to take a few steps back to briefly outline the history of the problems of maxima and minima.[82] Some of these issues had been part of physics and geometry since ancient times. The Greek mathematician Zenodorus addressed isoperimetric figures of greatest area, that is, figures that had the greatest area but the same perimeter. Euclid's *Catoptrics* had shown that the path taken by reflected light was the shortest possible. The idea that nature was parsimonious and acted in the most efficient manner, minimizing waste, was already widespread in the Middle Ages, even though it was not based on a rigorous knowledge of physics but on philosophical, teleological and aesthetic considerations.

In the late seventeenth and early eighteenth centuries, there was a resurgence of interest in the issues of maxima and minima now that they could be addressed using the recently developed tools of mathematical analysis. In the *Principia* (1687), Newton set himself the task of identifying the shape of a body that offered the least resistance to motion. Daniel Bernoulli formulated the brachistochrone problem in 1696, that is, the problem of determining the trajectory of a body travelling from one point to another in the shortest time. The same years also saw the spread of the problem of geodesics, that is, of the shortest path between two points on a curved surface. Addressing precisely these issues, first Euler and then Joseph-Louis Lagrange began elaborating mathematical tools that would in due course become the variational calculus, used to find the extrema (maxima and minima) of a function.

In parallel, these tools found application in physics, thanks to the discovery of natural phenomena that evidenced certain important quantities as minima, such as velocity, trajectory or duration. To provide a general account of these cases, Pierre-Louis Moreau de Maupertius formulated the principle of least action in 1744, which states that the quantity of action required for all changes in nature is as small as possible, where quantity of action is defined as the integral of the product of mass, speed and distance travelled. Taking this further, Lagrange reformulated Newton's laws of motion as minima problems in *Mécanique analytique* (17889), using the variational calculus he had himself developed. While Lagrangian mechanics was still based on the principle of least action, William R. Hamilton generalized the application of the variational calculus, using not only the minimal extrema but also maximal ones. Hamilton therefore renamed Maupertius' principle as the principle of stationary action, thus referring not only to minima but to all stationary points (i.e. to the points of a function where the derivative is zero, which are its maxima and minima), and the turning points (where the derivative of the function changes from negative to positive or vice versa). Hamilton's new approach thus contributed to the undermining of the metaphysical interpretation of the principles of minima, as he himself noted:

> But although the law of least action has thus attained a rank among the highest theorems of physics, yet its pretensions to a cosmological necessity, on the ground of economy in the universe, are now generally rejected. And the rejection appears just, for this, among other reasons, that the quantity pretended to be economized is in fact often lavishly expended.[83]

Petzoldt, too, addressing the physics principles of minima and maxima, emphasizes that these research areas, at least in the first instance, had stoked 'teleological and metaphysical speculation', giving new life to 'anthropomorphic thinking according to which nature is a parsimonious worker that achieves its ends using minimal means'.[84] However, like Hamilton, Petzoldt also observes that 'the watchful eye of more modern research' had begun to 'separate the wheat from the chaff', recognizing that the fact that 'mechanical processes can be described using analytical expressions in which the properties of maxima and minima appear' was irrelevant to 'suppositions of a principle of parsimony at work in nature'.[85]

As we have seen, 'Zu Richard Avenarius' Prinzip des kleinsten Kraftmasses' already contained a critique of the metaphysical interpretation of Gauss's principle of least effort. However, Petzoldt now extends his analysis to all variational principles and at the same time sets out to define what their correct interpretation should be. The error, in fact, is not that of pondering the philosophical value of such principles but starting from a misunderstanding of their mathematical significance, and thus incorrectly seeing them as proof of the parsimonious nature of the universe. It is therefore first necessary to ask what the fundamental property is of the functions contained in the variational principles.

The key feature of variational principles is that they do not express physical laws as simple equations but as solutions of variational problems. A certain function is provided, and the values of this function correspond to the range of possible cases, while the real case – the case we actually come across – corresponds to the values that minimize or maximize the function: that is, reduce its variation to zero.

When scholars first began to develop variational principles, their focus was on the correspondence between the actual case and the minimum of the function. Why, of all the possible values, it was in fact *the minimum* that occurred in nature? The reason for this clearly had to be sought in the internal order of the cosmos, not to say in the wisdom of the creator. According to Petzoldt, however, attention should not be focused on the minimum or maximum of the function but on the fact that the minimum and maximum bring to a halt the variation of the function. This means that 'if an interpretation is needed, it is not about the minimum – as Euler still held – or the maximum, or both, but, more generally, the disappearance of variation in the case of real motion'.[86] In other words, we use the maximum and minimum points of a function when we apply the mathematical tools of the variational calculus not because they have some special significance as maxima and minima but because they 'occupy a privileged position in that they are *singular, unique*'.[87]

The relationship between fact and the mathematical tool that describes it is therefore not that between a minimized function and the reality of nature proceeding parsimoniously, selecting from all of the available options the one that involves the least action, the least effort, and so on. The fundamental property of the minimum or maximum of a function is its *uniqueness*, and it is this uniqueness that serves to describe natural phenomena. Hence the appropriate question is why the uniqueness of the maxima and minima of functions helps us to describe reality.

Petzoldt's reply is that the uniqueness of maxima and minima, which brings to a halt the variation of functions, corresponds to the uniqueness of reality, since the real case is the only one that actually occurred from among all of the possible alternatives. In other words:

> In all of the principles and phenomena noted, we observe the univocalness [*Eindeutigkeit*] of natural processes. These are always specific singular cases in the infinite number of imaginable cases, and therefore they can be described analytically by the nullification [*nullwerden*] of a differential or variational formula and can in general be understood in terms of maximal or minimal properties. Maxima and minima are included into univocalness; it is univocalness that is the original property, common to everything that happens in nature, and it must therefore be considered the crux of the matter.[88]

This is where *Eindeutigkeit* (univocalness) first appears, a concept that Petzoldt will elaborate further in subsequent works. In one of the few works that analyse the emergence of this concept in the late nineteenth and early twentieth centuries, Don Howard notes that this term can be interpreted in different ways. The first sense of *Eindeutigkeit* is essentially what is termed 'categoricity' in model theory, that is, where a theory (in our case variational principles) describes a unitary model (a unique representation of reality). In other words, a theory is not categorical, univocal, *eindeutig*, if it can define *more than one* non-isomorphic representation of reality. In the second sense, *Eindeutigkeit* entails the assumption that reality itself is univocal, and thus that any theories that describe nature must also be univocal. The third sense of *Eindeutigkeit* concerns the question of how to choose from an ensemble of competing theories. In this case, a theory is *eindeutig* if there exist conditions (in reality, in experience, in usefulness, in transcendental necessity, etc.) that allow the selection of that theory from among all the different possible theories as the only valid description of a given set of phenomena. To distinguish these three senses, Howard refers to them as model-theoretic *Eindeutigkeit*, metaphysical *Eindeutigkeit* and epistemological *Eindeutigkeit*, respectively.[89]

Applying this categorization to Petzoldt's work, it is clear that we have the second sense of *Eindeutigkeit* here, the metaphysical sense. Petzoldt certainly sees *Eindeutigkeit* as a condition for selecting between theories, as when he writes that 'from an analytical point of view it must always be possible to find expressions which – when variation is nullified – return the differential equation of the motion'.[90] However, for Petzoldt, the requirement that the principles we use to understand reality are univocal derives directly from the fact that reality itself is univocal. Indeed, it seems almost impossible to distinguish between the univocalness of our scientific theories and the univocalness of the natural processes they describe. They are two sides of the same coin, as a univocal reality guarantees the possibility of univocal scientific propositions, and univocal scientific propositions presuppose a univocal reality. As Petzoldt writes:

> If there is truly something worthy of *awe* in nature, it is not the appearance of oft-invoked minima, but rather the *Eindeutigkeit* of all its processes. That is what makes science possible; if it was not present, it would be impossible to speak of laws of nature.[91]

Given the close link between the metaphysical and epistemological senses of *Eindeutigkeit*, it is no wonder that the concept ends up being identified with the principle of sufficient reason elaborated by Leibniz, which also has a hybrid metaphysical/epistemological nature. We also read that 'the fact that all processes are determined univocally goes hand in hand with the principle of sufficient reason', to the extent that 'the principle of least action and the other similar principles can be considered *analytic expressions of the principle of sufficient reason* within their own domains'.[92] Indeed, 'saying "all processes must have a sufficient reason" is the same as saying "all processes are determined univocally", that is: every process includes elements or means that enable what has actually been realized to be described uniquely'.[93]

In any case, in 'Maxima, Minima und Ökonomie' Petzoldt does not yet dwell on all the epistemological and metaphysical implications of the principle of *Eindeutigkeit*, as he does in later works. The key point, for the time being, is to clarify that the principles of maxima and minima have nothing to do with the *parsimony* of nature, given that parsimony does not exist in nature. What does exist is only the *intelligibility* of nature, which corresponds to the fact that the processes of nature occur in a univocally determined manner and can therefore be described univocally.

From economy to stability

As noted above, the second area that Petzoldt addresses in his examination of the concepts of maxima and minima is that of purposive processes. In particular, he divides this area into three parts: evolutionary processes, physical phenomena and considerations of an ethical and aesthetic nature. The first of these brings Petzoldt up against Zöllner and Fechner, and the second against Mach and Avenarius. For the purposes of this volume, we can disregard the third part.

Petzoldt points out that Zöllner's focus was on defining a 'principle of least displeasure'.[94] Zöllner linked mechanical processes to sensations, holding that every transformation of potential energy into kinetic energy produces sensations of pleasure and every transformation of kinetic energy into potential energy produces sensations of displeasure. At the same time, he associated pleasure and displeasure with purposive and unpurposive processes, respectively. This led Zöllner to declare that there was a general tendency in nature to minimize sensations of displeasure.

Petzoldt criticizes the 'mixing of the physical and the mental'[95] that characterizes Zöllner's work, whose main contribution was only to open the way for Fechner and his principle of stability. Although Petzoldt saw the principle of stability as laden with illusory hypotheses and somewhat metaphysical notions, he recognized that Fechner had succeeded in keeping separate his rigorous reflection and his incursions into the field of pure speculation.[96] Whereas 'Zu Richard Avenarius' Prinzip des kleinsten Kraftmasses addressed how Fechner's principle of the tendency towards stability had unified several fields into a single law, Petzoldt now emphasizes a different aspect of Fechner's work, the fact that he had focused on 'the *endpoint* of a series of evolutionary steps, the resulting *stability* of the system, the *lasting state* attained, its ensured *self-preservation*'.[97] In contrast to those who had concentrated on the process of evolution, the struggle for existence, Fechner 'simply notes the *fact* that evolution leads to stationary states, and it is this that is the key point and the huge contribution of the principle of the tendency towards stability'.[98]

For Petzoldt, the importance of the change of perspective proposed by Fechner is that it allows for a new way of formulating the problem of teleology. Up to that point, the focus had been on processes directed towards ends, with the key question being how it might be possible to proceed intentionally *towards* something. However, this is an anthropomorphic view of finality, where a *future* objective influences the present and the course of events. But the teleological problem cannot be about where processes 'want' to go but where they *actually* go. In other words, in order to understand the role of finality in nature one needs to analyse what the *objective* (i.e. necessary, lawful) endpoint of natural processes is. Once the endpoint has been identified, each step in the process that contributed to the endpoint can be considered as 'conforming to the purpose'.[99]

The point of departure of the anthropomorphic concept of finality is the start of the process, with the future end seen as a goal that can be chosen at will, as can the means to achieve it. This requires an intelligence and volition capable of imagining the future goal and making such choices. When, on the other hand, the focus is on the end, on what *we know to be* the endpoint of natural processes, namely stability, then the process that leads to the end has nothing to do with a subjective choice, or with the open-endedness of the future, but is as one with causality, with necessity, with the laws of nature. Fechner's contribution thus consists of having identified finality as stability, thereby defining objectively the concept of 'conforming to purposes' as 'conforming to the attainment of a stationary state'. In consequence, 'the principle of the tendency towards stability is in line with the teleological principle and is at the same time the alignment of the teleological principle with the law of causation'.[100]

Having identified finality with the tendency towards stability, Petzoldt then defines evolution as 'progress along the path from the initial conditions to the attainment of stability by the system'.[101] Given that the reconciliation of teleology with causality depends on the concept of stability as the objective and necessary end of the processes of nature, Petzoldt does not need to resort to Darwin's principle of natural selection to explain evolution in causal terms. Therefore, here too the 'struggle for existence' is reduced to 'a purely negative or regulatory factor' alongside the '*positive factors*', that is, the laws governing internal development.[102]

Petzoldt does not distance himself only from Darwin's view of evolution but also from that of Mach and Avenarius. We have already seen how Avenarius, in the *Prolegomena*, proposed his principle of least amount of energy as an evolutionary principle that explained psychophysical development in terms of the ceaseless attempt by organisms to optimize the forces at their disposal. We have also noted Petzoldt's criticism of this position in his highlighting of the absurdity of the concept of 'saving energy' in a universe where a force can provide neither more nor less than is in its power to do. Petzoldt now returns to the same arguments, and broadens his criticism to include Mach. The concepts of least amount of energy and economy of Avenarius and Mach are in fact equally problematic, as

> in the absence of a metric for the magnitude of the performance, neither the task of obtaining the maximum effect with the forces available nor that of attaining the envisaged goal with the least expenditure of force have any meaning. And the subjective concept of purpose cannot provide any such metric; we can only hope to find it in the corresponding objective concept, that is, in the stability of the system.[103]

As the concepts of economy and least effort have any meaning only if one evaluates forces in relation to their contribution to the attainment of a stationary state, '*the concept of persistence or of the preservation of the system is more important than that of minimum expenditure of energy or maximum economy*; the properties of maximum and minimum are only secondary, whereas the principle of the tendency towards stability goes deeper into the essence of phenomena'.[104] Petzoldt is even blunter elsewhere, stating explicitly that 'the concept of *economy* [...] *must absolutely not be used* for the *objective description* of things or processes'.[105]

This criticism of Avenarius and Mach also extends to their position on mental processes, and in particular on cognition. Petzoldt writes:

> We cannot accept that the specificity of the construction of concepts is founded on the limited forces of thinking, or that the roots of understanding must be sought

in the principle of least amount of energy. As soon as this occurs the principle ceases to be immanent and strays into transcendence. [...] The construction and application of concepts is a *reactive activity* [*Reaktionstätigkeit*]. There is something wrong with speaking of the overabundance of phenomena in relation to the 'limited powers' of thinking, and of the effort of thinking to 'to mirror in itself the rich life of the world'.[106] It is only when we consider the latter to be the task of thinking that that we become convinced of the fact that the means available are deficient. But thinking does not want to 'mirror' the world at all, it must not and it has no need to. Its *purpose* is to form a stable relationship with things and processes. But its mere existence [*Vorhandensein*] is, like all existence, *purposeless*, purely factual.[107]

In other words, when one regards the tendency to develop concepts that are increasingly unitary, general and empirical as stemming from the limitations of our cognitive faculties, which would thus be forced to favour a minimum use of energy and a maximum outcome, the flawed notion of finality re-emerges. Strictly speaking, thinking cannot set itself goals, not even that of understanding the world. The only goal that exists is the universal tendency of every system to increasing stability, so even human psychophysical activity must be interpreted under this notion of finality. However, if purposiveness is nothing more than the inevitable tendency towards stability, which imposes a direction on all natural processes, then the notion of thinking having limited means with respect to an over-ambitious goal becomes meaningless. It is simply that our psychophysical system is constantly engaged in processes that favour states that are more lasting and more stable. Therefore, rather than speaking of least amount of energy or economy, it should be recognized that 'even here it is the principle of stationary states that best describes the facts'.[108]

As we noted at the end of our analysis of 'Zu Richard Avenarius' Prinzip des kleinsten Kraftmasses und zum Begriff der Philosophie', Petzoldt recognizes that Avenarius later corrected the main errors he had made in his work on the principle of least amount of energy. In fact, in his *Kritik der reinen Erfahrung*, Avenarius focused on the self-preservation of the brain system as the basis of all mental processes, and thus on a concept that is very close to that of stability. Petzoldt also notes that even Mach 'appears to increasingly embrace the idea of stability'.[109]

Petzoldt summarizes his proposal in the conclusion to his work, namely, to remove all concepts that might be associated, to a greater or a lesser extent, with the old anthropomorphic notion of finality, and replace them with concepts that are better suited to the description of the true state of things. Therefore, in the field of physical phenomena, the concepts of maxima and minima should be replaced by univocalness, according to which the actual case is not 'chosen' from all possible cases by the parsimony of nature but is the only case that has been realized, and in this sense is determined univocally. In evolution, the concepts of least amount of energy, of maximal outcome, or more generally of economy, should be replaced by the principle of the tendency towards stability, which states that here, too, there is never a subjective 'choice' of ends or means but only the objective *fact* of the universal direction of the processes of the world towards lasting states. In his closing words, Petzoldt summarizes thus: 'It is not maximum, minimum or parsimony, but *Eindeutigkeit* and stability, that highlight those aspects of reality that must be at the centre of our attention.'[110]

Causality and *Eindeutigkeit*

From causes to functional relations

In 1895, Petzoldt published 'Das Gesetz der Eindeutigkeit' (The Law of Univocalness) in the journal edited by Avenarius. The article extends the concept of *Eindeutigkeit* that he had introduced five years earlier in 'Maxima, Minima und Ökonomie'. In the intervening years, Petzoldt had also published two other works: 'Über den Begriff der Entwicklung und einige Anwendungen derselben' (On the Concept of Evolution and Some of its Applications, 1894) and 'Einiges zur Grundlegung der Sittenlehre' (On the Foundations of Ethics, 1893–4). As its title indicates, the first of these two works returns to the theme of universal evolution towards stationary states, essentially recapitulating earlier arguments. The second article is of greater interest. Despite its title, it does not focus solely on ethical issues. It starts as a review of *Das Sittengesetz* (The Moral Law, 1887) by Franz Staudinger, a philosopher from the Marburg school of neo-Kantianism, but over the course of its 130 pages Petzoldt's reflections broaden in a range of directions, including an analysis of Avenarius's *Kritik der reinen Erfahrung*, a critique of Wilhelm Wundt's position on physical and mental causality,[111] and a first draft of a broader philosophical system based on three fundamental laws: the conservation of energy, psychophysical parallelism and the tendency towards stability.[112]

Leaving to one side the topic of the relationship between physical and mental phenomena, to which we will turn below, we will at this point focus on causality. After using the notion of univocalness to remove the metaphysical interpretation of the physics concepts of maxima and minima as the expression of the perfect parsimony of the universe, Petzoldt now extends the meaning and the field of application of this concept, turning it into the cornerstone of a broader reformulation of causality. Petzoldt wants to further the positivist movement in its battle against the inappropriate use of anthropomorphism in our understanding of the world. In the principles of maxima and minima, anthropomorphism meant that nature was organized according to the ideal of *parsimony*. In teleology, it meant that the course of events includes a *choice* of ends and means. In causality, it meant that the cause *acts* on the effect.

This criticism of the concept of cause is obviously not the first of its kind. Without going back as far as Hume, in the first half of the nineteenth century August Comte had rejected the notion, seeing in it the residue of the fetishistic thinking of the dawn of humanity.[113] Around the middle of the century, more strictly philosophical debate on the topic had become intertwined with debate on the law of the conservation of energy. One of the earliest formulations of this law was produced by Robert Mayer, who saw it as equivalent to the principle of causality in that both expressed the fact that no effect occurs without a cause and that, conversely, no cause occurs without an effect; or, in the words of the famous Latin motto: '*Causa aequat effectum*'.[114] If Mayer was still bound to the metaphysical notion of force as an indestructible entity, others with a more warily critical approach derived the concept from the simple regular connection between quantitative aspects of phenomena. In this sense, the principle of the conservation of energy merely affirms the existence of mathematical equations that can express the conversion constants between different forms of energy. Hermann von Helmholtz was the other German scientist who formulated this principle. In the pamphlet entitled *On the Conservation of Force* he clarified that 'we define force as the law recognized as objective power',[115] adding that 'the principle of causality is, in fact, nothing other than the presupposition of the lawfulness of all natural phenomena'.[116] Emil Du Bois-Reymond, in the Preface to his *Untersuchungen über thierische Elektricität* (Investigations of Animal Electricity) also targeted the use of the concept of force as a causal explanation of phenomena, stating that force 'does not exist' in reality and 'is simply the brainchild of the irresistible tendency to personify that is within us'.[117] Therefore, 'for us, the word *force* can have no meaning other than that which has been so valuable in analytical mechanics'; that is, 'force is the *measure* of motion, not its cause'.[118]

To summarize, the concept of causality experiences a twofold crisis in the nineteenth century. On the one hand, it is subjected to a philosophical critique, in the tradition of Hume and founded on empiricist and anti-metaphysical positions. On the other, it undergoes rethinking even by scientists, thanks to the formulation of the law of the conservation of energy. In particular, the principle of causality is first reformulated as the principle of the indestructibility of force, and then simply as the assumption of the existence of constant quantitative relationships between phenomena, relationships that can be expressed as mathematical equations. In the light of this process, the reformulation of the relationship between cause and effect in terms of functional relations between events, launched by Fechner and followed up by Mach, seems a completely natural development.[119]

In the incipit of 'Das Gesetz der Eindeutigkeit', Petzoldt does not lay out the entirety of this evolution but refers only to his most immediate predecessors, Gustav Robert Kirchhoff and Ernst Mach, who had taken 'the final important step' in the long process that had led to the rejection of causal *explanations*, and to recognize that all that science can do is *describe* relationships between phenomena.[120] In particular, Mach had 'brought to completion' this 'inevitable evolution', thus 'definitively freeing himself from substantiality and causality'.[121]

In this respect, therefore, Petzoldt returns to Mach's doctrine that in the final analysis causal relationships are nothing more than functional relations of the form *if A then B*.[122] This means that 'The law of causality is identical with the supposition that between the natural phenomena $\alpha, \beta, \pi, \delta, \ldots \omega$ certain *equations* subsist' and that the goal of physics 'is to represent every phenomenon as a *function* of other phenomena'.[123]

In his outline of Mach's views, however, Petzoldt soon introduces his own terminology, such as the concept of *Eindeutigkeit*:

> If there exists an equation between two variable such that for every specific value of the first there is a unique value for the second, then we say that the second value is determined by the first, or that it is a *univocal* function of the first value, or that it depends on the first value.[124]

In line with the thinking of Mach, once the relationships between cause and effect have been stripped of their superfluous metaphysical elements, they are simply relationships between variables. However, because the relationships between the variables are univocal, the principle of causality is simply the principle of *univocalness*, that is, 'the hypothesis of the *universal and complete determination* [...] of all processes'.[125] Petzoldt goes into further detail:

> for *any* process it must *always* be possible to find a means of determination [*Bestimmungsmittel*] such that *only* this process is determined [*festgelegt*], so that, *for any other process that one might wish to consider as determined via the same means, it is possible to find at least one other that could be determined in the same way.*[126]

Here Petzoldt returns to the concept expressed earlier in 'Maxima, Minima und Ökonomie', according to which we come to understand a process when we are able to describe it univocally, that is, when the tools we use to describe it (the 'means of determination', such as, points, lines, trajectories, velocities, temperatures etc.[127]) address the *uniqueness* of the actual case with respect to all other imaginable cases. In other words, our means of determination operate correctly when they return a single model for the actual case, and several equivalent models for which no selection criterion can be established for cases that are merely possible. We can also say that they provide a description of actual reality that is univocal (*ein-deutig*) and a description that is ambiguous (equivocal or, better, plurivocal, *viel-deutig, mehr-deutig*) for possible cases.

Petzoldt illustrates his concept with some examples, such as that of a falling object subjected to the force of gravity:

> We usually determine the trajectory of a falling object using the acceleration that the mass of the earth 'communicates' to it without considering the implicit presupposition that we also assume, and to which the magnitude of *g* is completely irrelevant. This presupposition consists of the fact that we see the direction of the acceleration [...] as coinciding with a line that runs from the object to the earth's centre of gravity. The presupposition expresses the principle of *Eindeutigkeit* in that, in principle, every other direction of acceleration is conceivable, but only the one that actually occurs is determined univocally: for every other direction that we might wish to assign to the motion, an infinite number of others could be found that – with respect to means of determination *g* – would be 'equally justified' with respect to it. The actual trajectory is unique, singular or – as Ostwald puts it, having also addressed this principle – is 'the exceptional case from among the possible cases.'[128]

As we have seen, univocalness is therefore an epistemological principle, in the sense that it places a requirement on our cognitive apparatus (the means of determination must be able to describe the actual case univocally and unambiguously), but it is also a metaphysical principle, because the fundamental presupposition is that it is natural processes themselves that have a kind of uniqueness, because the actual case stands out against all the others.

Univocalness and determinism

The principle of univocalness is not, however, limited to providing an alternative formulation of Mach's notion of causality as it also covers other presuppositions beyond that of the existence of functional relations between phenomena. Petzoldt's position differs here from that of Mach and other scholars who had based causal relationships on quantitative regular connections, as his aim is to maintain a 'strong' notion of the dependencies that tie nature together.

Some maintained that it was possible to reformulate causality as the principle of the conservation of energy, maintaining a certain degree of indeterminacy in natural phenomena so as to guarantee some space for human will. Fechner wrote, in his *Elements of Psychophysics*:

> The principle or law of the conservation of energy tells us nothing about the direction or way in which the reciprocal transformations between living force and potential force occur; nor, in this respect, what the state of a system must be at given point in time. This depends on the specific conditions and relationships of each system, which cannot be determined by any general principle, but can only be addressed through experience. Apart from telling us about how the transformation between actual force and potential force occurs in a system left to its internal effects, the principle of the conservation of energy merely states that the transformation can only take place in a manner that maintains the overall constant sum of the two, which ensures *freedom* with respect to the infinite number of possible ways in which such transformation might occur.[129]

Thus, if we say that natural phenomena are 'determined', this only means that they are subject to constant quantitative relationships, not that their course is fixed by the laws of nature. This allows Fechner to maintain that they can be fully defined scientifically, while rejecting 'strong' mechanical determinism. All that is required for a scientific understanding of phenomena is that they are measurable, that they can be described as quantitative equations. It is unnecessary to assume that they are fully predetermined by their causes.[130]

A position similar to that of Petzoldt is held by Wundt, whom Petzoldt criticizes at length in 'Einiges zur Grundlegung der Sittenlehre'. In the section on 'Freedom and determination of the will' in the third edition of the *Grundzüge der physiologischen Psychologie* (Principles of Physiological Psychology), Wundt writes:

> [The causality of the material world] allows for different levels, with essentially different senses of the principle of equivalence [the conservation of energy].[131] *The only thing that is guaranteed is the quantitative relationship*, which ensures that for every amount of energy that disappears from the cause an equivalent amount of energy must appear in the effect. But it is only in the simplest forms of mechanical interaction that [...] the causal process corresponds to an equation that is fully *univocal*. In any more complex system, only the sum value of each side of the equation is fixed. How this is composed of the individual values can

only be determined after the actual process itself, as there are incalculable relations with other systems which could affect the development of the system at any point. Every causal equation is therefore *plurivocal*, in the sense that the quantitative sum is determined, but how this sum is subdivided into different values remains to be determined on the basis of the actual effect. It would only be univocal if we could take into consideration all of the relevant relations, but given that the number of such relations is infinite, the equation can never be univocal.[132]

Petzoldt's summary of Wundt's position (which also applies to Fechner's position) is that 'it is *only for the magnitudes* of the types of energy that a law exists, *not for the forms* in which the types of energy spread. It is at this point that "free volition" or the "creative agency" of any voluntary action comes in'.[133] Wundt thus 'on one hand acknowledges that energy is constant in the natural sciences', and on the other 'allows for "free" volition at the expense of the law of the univocalness of all phenomena, of which the law of [the conservation of] energy is just a special case'.[134]

Petzoldt rejects this attempt to reconcile the principle of the conservation of energy with volition because it would lead to the negation of the necessity of nature, its univocal determination. Moreover, an 'indeterminist' basis for freedom misunderstands the very concept of freedom. For Petzoldt, the notion of freedom is not in contrast to that of causality, but to that of 'constraint' (*Zwang*).[135] In this, he harks back to the Hobbesian line of thinking that resolves the apparent contrast between freedom and necessity by postulating a negative concept of freedom: 'freedom *from* …' rather than 'freedom *to* …'.

It is clear that the concept of univocal determination advanced by Petzoldt expresses more than just the existence of functional relations between phenomena, and thus constant quantitative relationships between them. On an epistemological level, our cognitive apparatus must be able to identify the actual case as unique with respect to the many imaginable cases because on the level of the real it is the conditions of a phenomenon themselves that make that phenomenon the only actual one from among all the possible ones. For Fechner and Wundt, given the initial conditions of a system, *several* outcomes are possible, and what matters is that a quantitative equivalence holds between antecedent and consequent, but for Petzoldt, given the initial conditions of the system, there can be *one and only one* outcome. However correct it may be to strip the concept of causality of its extraneous metaphysical and anthropomorphic elements, it is important not to throw the baby out with the bathwater. In this case, the baby is the universal determination of natural phenomena, and the concept of univocalness gives an account of this aspect of reality, which the simple concept of functional relations does not recognize, thus opening the door to indeterminism.

Univocalness and unidirectionality

Fechner and Wundt are not, however, the only thinkers for whom the conceptualization of causality in terms of functional relations is accompanied by a certain degree of indeterminism. Petzoldt notes that even Mach concedes that there are 'two situations that indicate indeterminacy in nature'.[136]

We have already noted that for Mach 'The law of causality is identical with the supposition that between the natural phenomena α, β, π, δ, … ω certain *equations* subsist'. He also specifies:

> if the number of the equations were greater than or equal to the number of α, β, π, δ, … ω, all the α, β, π, δ, … ω would thereby be *overdetermined* or at least *completely determined*. The fact of the varying of nature therefore proves that the number of equations is less than that of α, β, π, δ, … ω. But with this a certain *indefiniteness* in nature remains.[137]

Whereas Fechner and Wundt see indeterminacy as stemming from the fact that the principle of the conservation of energy can at best indicate the constant *quantitative* relationships in the evolution of phenomena, for Mach it is the result of the fact that only a certain number of equations between the properties of phenomena can be identified, so something is always omitted and thus remains indeterminate. However, if some aspects of phenomena are always beyond the scope of functional relations, and therefore remain undetermined, this is even truer of the world as a system, understood as the ensemble of all phenomena. Mach therefore holds that it is impossible to foresee how events will unfold in general. This leads him to criticize the assumption that a consequence of the second law of thermodynamics is that the universe is proceeding towards a state of entropic death.

> For instance, such a theorem is that defended by W. Thomson and Clausius, according to which after an infinitely long time the universe, by the fundamental theorems of thermodynamics, must die the death of heat, that is to say, according to which all mechanical motion vanishes and finally passes over into heat. Now such a theorem enunciated about the whole universe seems to me to be illusory throughout. As soon as a certain number of phenomena is given, the others are co-determined, but the law of causality does not say at what the universe, the totality of phenomena, is aiming, if we may so express it, and this cannot be determined by any investigation; it is no scientific question. This lies in the nature of things. The universe is like a machine in which the motion of certain parts is determined by that of others, only *nothing is determined* about the motion of the whole machine.[138]

Obviously Petzoldt does not share Mach's view, not solely because this indeterminism conflicts with the principle of the *Eindeutigkeit* of all natural phenomena but also because it negates the existence of the universal tendency towards stability that, as we have seen, Petzoldt identifies with the entropy principle.

According to Petzoldt, the fundamental problem with Mach's position is that, yet again, his concept of functional relations is insufficiently informative. Mach holds that causal relationships can be reduced to propositions of the form 'if A then B', and therefore that 'the law of causality is sufficiently characterized by saying that it is the presupposition of the mutual dependence of phenomena'.[139] Petzoldt's contention is precisely that causality cannot be 'sufficiently characterized' in this manner. The problem is that the 'dependence' described by Mach 'has nothing to do with temporal order, it is rigorously immediate, simultaneous, purely logical, and therefore unable to provide a precise expression of the standard representation according to which "cause" precedes "effect".[140]

The fact that Mach does not take temporal order into account is clearly not accidental but a consequence of his rejection of the notions of absolute space and time. Mach's view is that spatial and temporal determination can be expressed exhaustively in terms of relations between phenomena. For example, 'To say that this or that "is a function of time" means that it depends on the position of the swinging pendulum, on the position of the rotating earth, and so on'.[141] The indeterminate nature of the evolution of the universe is also explained by this absence of absolute coordinates, because when we speak of 'the universe in itself [...] we have nothing over to which we could refer the universe as to a clock'.[142] In fact, Mach himself specifies that his concept of indeterminacy differs from that of Fechner in that it 'results immediately from the law of causality by the elimination of space and time'.[143]

Even though he did not support the concept of absolute space and time, Petzoldt does not share Mach's position, a position that leads to the elimination of any temporal determination from functional relations. Mach only accounts for 'simultaneous dependence', that is, a dependence where functional relations, the equations that are formulated, refer to quantitative aspects of a single point of time in a process. However, simultaneous dependence 'does not exhaust the determinateness of nature'.[144] Petzoldt's aim is therefore to show how it might be possible to assume 'sequential dependency [*succedane*]'[145] without needing to refer to the metaphysical concept of absolute time.

It is important to note that when one applies the concept of functional relations to processes that involve temporal dependency, as in the case of 'sequential dependency', it does not mean that the value of variable x at time t_0 is a condition of parameter y having a certain value at time t_1. Functional relations between quantitative aspects always exist *at the same time* in a process. The temporal element comes from the fact that the values of the two variables at each time in the sequence $t_0, t_1, t_2, ...$ change according to the relationships expressed in the function. In other words, in sequential dependency the function does not express a dependency relation of the type *if x_0 then y_1*, but one of the type *if x_0 then y_0, if x_1 then y_1, if x_2 then y_2* and so on, where the relation between x and y remains constant. That is, 'the two sides of an equation never refer to different moments in a process, or to different processes, but always only to the same moment, rigorously simultaneously [*gleichzeitig*]'.[146] Petzoldt does not see this as a particularly novel understanding of the issue. Indeed, he emphasizes that '*no physicist* would think of linking the two sides of an equation to different times in a process', as the temporal aspect derives from the fact that 'the *same* equation holds for *every* time'.[147]

Having established that even in the case of 'sequential dependency' functional relations must *not* be interpreted as if the value of the independent variable was the condition for the value of the dependent variable *at the subsequent time*, we can now turn to an analysis of how Petzoldt captures this type of dependency. We will examine how he characterizes sequential dependency with respect to simultaneous dependency without running the risk of returning to the metaphysical notion of absolute time.

The first property of causal relationships that take place over time is irreversibility. How can we maintain irreversibility if we discard the notion of absolute time running inexorably and never turning back on itself? For Petzoldt, the irreversibility of the course of nature remains intact even when, following Mach, temporal coordinates are reduced to mere relationships between phenomena. Indeed, it is never possible to reverse (*umkehren*) a phenomenon once it has taken place; it can at best be repeated (*wiederholen*). However, any such repetition is never exact, and never maintains the same reciprocal relationships with all of the other processes underway in nature. Even when a phenomenon appears to repeat, as in the case of a swinging pendulum, this is only the result of abstracting away from the relationship between the individual phenomenon and all of the others.[148]

The unrepeatability of phenomena also allows for the possibility of establishing a difference between the values of a function such that 'one is the function of another, but not vice versa'.[149] Indeed, the independent variable appears only in the context of certain specific concomitant phenomena, and the dependent variable only in the context of certain other phenomena. This preserves the asymmetrical relationship between the antecedent and the consequent of functional relations.

The second property of the relationships between cause and effect that take place over time is continuity. How can continuity be interpreted rigorously if we have discarded the continuity afforded by absolute time? Petzoldt's answer appeals to the mathematical continuity of the quantitative values of the function: 'before a means of determination with a determined value reaches a different value of finite magnitude, it must have passed through all of the intermediate values'.[150] In Petzoldt's view, 'if nature permitted variations by leaps, it would no longer be possible to speak of univocalness, even if *simultaneous* dependence could be fully guaranteed'.[151] Indeed, if the quantitative parameter of a phenomenon could acquire a completely different value *at a stroke*, then any other value would be equally justified, and we would therefore have plurivocality. The principle of the continuity of natural phenomena is thus yet another aspect of the principle of univocalness.

The concepts of irreversibility and continuity are, however, not enough to complete the rigorous reformulation of the temporal aspects of cause and effect. A third aspect needs to be taken into account: unidirectionality (*Einsinnigkeit*). The traditional view of time is that it flows not only in a continuous and irreversible manner but also in a single direction. Once the concept of absolute time has been rejected, unidirectionality is captured by the second law of thermodynamics, according to which 'all differences in temperature, all electrical differences or differences of level, left to themselves, become smaller and not greater'.[152] It is no longer absolute time that gives direction to natural processes, but the entropy principle governing the direction of spontaneous change.

Petzoldt clearly saw the *uni*directionality (*Einsinnigkeit*) of natural processes as an aspect of the principle of *uni*vocalness (*Eindeutigkeit*). This means that the principle of univocalness is merged with Petzoldt's other fundamental principle, that of the tendency towards stability. Petzoldt writes: 'the determinateness or univocalness of processes is the condition [...] for all evolution, for all tendency towards stability, whether physical or mental.'[153] All of the issues addressed by Petzoldt in his early works thus end up converging on univocalness: the law of the conservation of energy, the second law of thermodynamics, the interpretation of the physics principles of maxima and minima, the rigorous reformulation of the principle of causality and the replacement of teleology with the principle of the tendency towards stability.

To summarize, then, we see that Petzoldt is at the very heart of the historical reappraisal of the concept of causality that led to its reformulation in terms of functional relations between phenomena. However, he differs from other proponents of this view, such as Fechner, Wundt and Mach, in that he sees the concept of functional relations as insufficiently expressive in itself, and therefore unable to capture the positive legacy of the notions of cause and effect. In particular, Petzoldt holds that the concept of functional relations only captures simultaneous relations and opens the way for indeterminism. His alternative proposal is therefore to reformulate the principle of causality in terms of the notion of univocalness, which encompasses several aspects: the univocal functional relation between variables, the determinateness of all natural processes, the continuity of change and the irreversibility and unidirectionality of the course of events.

The principle of Eindeutigkeit: A priori or a posteriori?

Given the importance that Petzoldt attributes to the principle of univocalness, it is necessary to examine the foundations on which it is built and its epistemological status. Should it be considered an a priori principle, like a demand of reason, in Kantian sense? Or is it an a posteriori generalization induced from experience? Moreover, what is the relationship between such a maximally general principle and the specific reality that it is applied to? Does it allow us to use deduction to foresee the course of events? Or does it leave the task of determining what happens to experience?

With respect to the first issue, Petzoldt states that the principle of univocalness is both an a priori demand (in the sense of a postulate established by human beings as a condition for organizing experience) and a truth that can be demonstrated as such a posteriori, on the basis of experience. For living organisms, this demand is based on evolution, on adaptation seen as the ability to enter into *stable* relationships with the environment. If the events that occur in the environment were not determined univocally, if *Eindeutigkeit* did not hold in nature, then the evolution of organisms, and thus their very existence, would not be possible. This means that univocalness is a requirement imposed by organisms, but also that the actual existence of organisms demonstrates that univocalness exists in nature. As Petzoldt puts it, 'we must approach nature with a premise without whose confirmation we could not live, whether as minds or bodies'.[154] This premise is so powerful that 'we cannot consider that even a single process might be exempt from this premise without immediately descending into mental unease, without finding ourselves in the greatest danger of an, albeit partial, involution [*Untergang*]'.[155] Indeed, if we were to acknowledge 'the indeterminacy of a process, then we would have to doubt the understandability of nature completely and give up on any investigation, eventually descending into madness'.[156] To paraphrase Kantian terminology, we could say that the principle of univocalness is not so much a condition of the possibility of *experience* as one of the possibility of *existence*: we need it, as living organisms evolving relative to the environment, let alone as rational beings.

If, therefore, 'the univocal determination of all processes is a research principle [*Forschungsprincip*]', it also 'derives its compelling power from the fact that the actual existence of conscious beings of higher order cannot be envisaged without univocalness'.[157] The fact that we exist, that we have evolved thanks to a stable environment and continue to survive in it, demonstrates that univocalness is not just a proposition but a fact. Petzoldt writes: 'the univocal determination of all processes starts as a postulate, but it is a postulate that is "founded" on extremely sound grounds'.[158] In one sense, the very circularity of the argument ensures the grounding of the postulate: our psycho-physical apparatus evolved thanks to the univocalness of nature, thus incorporating into itself the need for univocalness, and by presenting nature with its demand for univocalness it always finds the demand fulfilled.

Petzoldt expresses himself even more clearly on this point in his next work, the two-volume *Einführung in die Philosophie der reinen Erfahrung* (Introduction to the Philosophy of Pure Experience, 1900–4). In the section on univocalness, he writes 'its power does not stem from a sum of individual experiences, but from the fact that we require nature to confirm its truth', so that it becomes 'a principle we use to address reality, a postulate that is valid relatively *a priori*, independently of specific experiences'.[159] However, this a priori character should not be seen as 'a return to the worst kind of metaphysics', as univocalness is confirmed by experience, even if not by 'individual experiences obtained through experiments' but by 'somewhat general experience', such as 'the fact of our own existence and that of the world; the fact that the cosmos exists rather than chaos; the fact that we exist as thinking and acting beings; and the fact that evolution exists'.[160] Thus, since univocalness is based on a 'fact', the incontrovertible fact 'that we ourselves are relatively stable mental systems', it is just as valid as 'the conviction that two plus two equals four'.[161]

Having established that *Eindeutigkeit* is a well-founded postulate, we can now consider its application to specific events. It would be natural to think that the assumption of the full and univocal determination of natural processes implies that specific cases can be deduced from general laws, or at least implies that knowing the relevant conditions might allow the course of events to be known in advance. In other words, Laplacian determinism should hold, according to which 'An intellect which at a certain moment would know all forces that set nature in motion, and all positions of all items of which nature is composed […] for such an intellect nothing would be uncertain and the future just like the past would be present before its eyes'.[162] However, Petzoldt rejects this view, stating that 'nothing can be established "*a priori*" with respect to reality: *Eindeutigkeit* can only provide, at most, a partial and general view of specific events, and nothing that is complete or specific'.[163] He adds that '"*a priori*" everything is possible, even that a body at rest begins to move "by itself". No conception is ever necessary "*a priori*"'.[164]

These statements are rather surprising in the light of Petzoldt's position on the *Eindeutigkeit* of nature and our knowledge of it. It is clear that Petzoldt's ambiguity here derives from his attempt to postulate the principle of the intelligibility of reality without descending back into rationalism or Kantian apriorism. That this attempt mires him down in a series of contradictions becomes even clearer in his discussion of the necessity of everything that happens in *Einführung in die Philosophie der reinen Erfahrung*. Here Petzoldt observes that the concept of the 'necessity of nature' is weighed down by a 'confused and mystical' element, the 'representation of *constraint*' where 'effect *must* follow cause'.¹⁶⁵ Just as cause is interpreted anthropomorphically, endowed with agency, so the concept of necessity implies an anthropomorphic view of effect, seen as 'a slave in the service of a master'.¹⁶⁶ The idea that 'an effect [...] cannot occur freely but *must* occur, occurs *by necessity*' thus represents a residual element of 'animism' and 'fetishism'.¹⁶⁷

Petzoldt's unsurprising answer to how the necessity of natural events should be seen is once again univocalness:

> The univocal determination of all events is the fact that is at the heart of the usual confused representations of the *necessity of nature*. The necessity of nature is a fact and nothing other than a fact. [...] There is no further problem hidden behind this pure factuality. This is an *ultimate* fact, not reducible to other *antecedent* facts. And ultimate facts have no cause. There is only one way to consider them: by ascertaining them, by acknowledging them.¹⁶⁸

The necessity of nature is thus identified as the 'fact' of the univocal determination of nature. However, the consequence of this is simply that Petzoldt's ambiguity on the necessity of nature is shifted to the concept of univocalness. These pages reveal in the clearest way possible that this notion actually has three different senses that are not easily reconcilable.

In the first sense, univocalness is the determinateness or determinability of events, seen as their intelligibility. This can be interpreted in a weaker purely epistemological sense, as the possibility of developing a cognitive apparatus, such as a scientific model, able to provide an explanation of the events that occur. Alternatively, it can be interpreted in a stronger metaphysical sense, that is, as the existence in reality of conditions that explain the occurrence of events. It is this first sense that emerges when Petzoldt writes, returning to the definition provided in earlier works, with some minor changes:

> For every process, means of determination can be found that show it to be univocally determined, in such a way that, for any variation of this process that can be considered determined using the same means, at least one other can be found that is determined in the same way, and is therefore equivalent.¹⁶⁹

Let us consider the example that Petzoldt himself often uses: that of the parallelogram of forces. What does it mean to say that, when two forces with directions AB and AC act on the same object, the object moves along a trajectory that corresponds to the diagonal of the parallelogram constructed from AB and AC? With respect to this first sense of *Eindeutigkeit*, it means – given the resultant motion of the object, which we can observe empirically – that we can find a model (the use of linear vectors as means of determination) that can describe this motion as unique among all the possible motions, that is, a model that can describe the actual case univocally. If, however, we ask *why* the object moves along the diagonal rather than along any of the other possible trajectories, the reply is that the question is meaningless. The object moves along the diagonal because it moves along the diagonal. It is a fact, and as such can only be observed empirically. In this case we can say that 'necessity is assimilated to pure factuality'.[170]

> There is no need or logical constraint for the mind to assume that the resultant motion should follow the diagonal, and therefore the law of the parallelogram of forces can never be *demonstrated* but must instead be acknowledged as a fact of experience. We could in principle have envisaged that the object might follow a different trajectory, one whose direction depended in a range of other ways on the magnitude and direction of the forces.[171]

Had Petzoldt limited himself to this sense of *Eindeutigkeit* there would have been no contradictions. The problem is that he adds two further senses. In the second sense, the principle of *Eindeutigkeit* goes beyond stating that everything that happens in nature does so for a reason that can be determined univocally, because it now claims to state what this reason is *in general*. According to this second meaning, the principle of univocalness affirms that the real cases always have some extraordinary and exceptional properties that make them unique. Petzoldt himself states, more or less explicitly, that this sense is different from the preceding one when he writes:

> *However*, equivalent possibilities from which to choose do not always exist in nature, as in a large number of cases – metaphorically speaking – natural events are not only determined after a choice is made, but also *before* it: here nature is in effect left with no choices; its determinateness means that from among all possible cases it must choose the one that actually occurs. It is precisely these cases that are of particular importance to understanding what we call the determinateness of nature. They are all those cases that involve a spatial event, that is, all processes of mechanical motion, as well as optical and electrical processes that involve the propagation of light or electrical excitation through a medium.[172]

At the start of the text cited, Petzoldt appears to refer to a very specific set of natural processes, while towards the end it becomes harder to understand which events might not be covered by the definition. In any case, the key point is that this sense of univocalness states that the actual case does not occur by necessity simply because it is the actual case, but because there is something exceptional about it that makes it the one that is chosen from among all the other possibilities. While the possible cases are 'equivalent', in that they have 'the same *right* to occur', the actual case has something exceptional that gives it a *greater* right to occur.[173]

Returning to the issue of the trajectory of an object subjected to an external force, Petzoldt specifies that in this case 'the question "why does nature not choose any of the infinite number of other directions?" is not an illogical one'.[174] In fact, nature chooses the case that is unique in some way, the case that has an 'extraordinary position relative to all the imaginable cases'.[175] Thus, while in the first sense of *Eindeutigkeit* the trajectory is only determined a posteriori, as it actually occurs, in this case it is also possible to determine a priori what the trajectory must be, at least in general, because we *already* know that nature will always choose the exceptional case from among all the possible cases.

To this second sense of *Eindeutigkeit* is added a third: the constancy of natural events. This sense specifies that nature must continue to behave as it has always behaved. As Petzoldt writes, 'Thinking is in principle indifferent to which of the infinite number of possible cases is chosen each time by nature from among those that are available for any type of event: the only thing is that nature must stick to what it has once decided to do.'[176] This view holds that the necessity of nature is not the same as its mere actuality or its being the unique and exceptional case, but the very necessity of the regularity of what happens. Nature is therefore a priori determinable, in the sense that we can expect it to continue to behave as it has always behaved.

Returning to the example of the object subjected to external forces, the first sense of univocalness states that nothing can be said about its trajectory before the force is applied. The second case states that we know in general that the trajectory will be the one that is unique among all the possible trajectories. The third case states that the trajectory, in general and in particular, will be the same trajectory as the object has always taken.

As we can see, there are three different senses of univocalness, leading to three different interpretations of the necessity of nature and three different answers to the question of whether the course of nature can be known a priori, knowing in advance which of the different possible outcomes will actually occur.[177]

Univocalness and mental contents

The meaning of physical and psychical

Having analysed Petzoldt's concept of univocalness, we now turn to a more detailed analysis of its application to mental activity. To do this, it is first necessary to establish how Petzoldt defines the terms 'physical' and 'psychical'. The relationship between the two has a pre-eminent role in his works since at least his *Sittenlehre* of 1893, in which – as we have noted – three fundamental laws are presented: that of the conservation of energy, that of the tendency towards stability and that of psychophysical parallelism.[178] However, it is not until 1904, in the second volume of *Einführung in die Philosophie der reinen Erfahrung*, that we find Petzoldt's explanation of the meaning of the terms 'physical' and 'psychical'. As late as in 1902, in his long article 'Die Notwendigkeit und Allgemeinheit des psychophysischen Parallelismus' (The Necessity and Generality of Psychophysical Parallelism), he refrains from any definition of the terms:

> As investigation must proceed independently of any fundamental conceptualization of the world, it will suffice to distinguish between corporal and mental processes, between physical and psychical processes, between natural processes and spiritual ones. Indeed, it is not necessary to know the *true nature* of these two

kinds of processes or *their essence*, or what sets them apart, or in contraposition. Indeed, it is irrelevant to the treatment of our subject, which only concerns the *connection* between the two areas, just as the relationship between the centimetre and the gram can be clarified without knowing anything about the origins or the definition of our system of measures. Even the humblest electrician understands the relationship between the volt, the ohm, the ampere and the watt; but what does he know of their definition? Researching the *relationships* between things is an endeavour that is totally different from investigating the essence of such things, and completely independent of it.[179]

Petzoldt refrains from providing definitions in order to avoid descending into essentialism, which contraposes the physical and the mental as two dualistically distinct elements. This anti-metaphysical requirement survives into the second volume of the *Einführung*, where Petzoldt finally decides to address the issue of what these two terms mean. Indeed, he specifies, from the outset, that 'the opposition between the physical and the psychical is not to be understood in an *absolute* sense, as would be the case if the totality of everything that is found could be separated *into two parts*'.[180] However, things do not go better with the position that distinguishes between the physical and the psychical 'as different *sides*' of the same reality, given that they 'might recall too closely Spinoza's absolute attributes'.[181] The solution is to differentiate the physical and the psychical 'as different ways of shedding light on the same thing [*verschiedene Beleuchtungen derselben*], different ways to regard [*Auffassungsweisen*] the same content'.[182]

Petzoldt thus goes back to the teachings of Mach and Avenarius, and of course Fechner, whose influence we have already noted. In *Elements of Psychophysics* (1860) Fechner had stated that 'the relationship between the material world and the mental world' is just one of the many cases 'where what is in fact one thing appears as two when observed from two different points of view', like the surface of a circle, which appears concave from the inside and convex from the outside, or the solar system depending on whether it is observed from the earth or the sun, and thus being either the universe of Ptolemy or that of Copernicus.[183] Along the same lines, Fechner reaches the conclusion that 'what you see from an internal point of view as your mind, because, for you, you are this mind, from an external point of view appears to be the corporeal substratum of this body'.[184] Mach, in his *Beiträge zur Analyse der Empfindungen* (Contributions to the Analysis of Sensations, 1886), partially followed Fechner's line in denying the existence of a distinction between the physical and the mental, holding that 'we do not find the gap between bodies and sensations described above, between what is without and what is within, between the material world and the spiritual world', as reality consists of fundamentally homogeneous 'elements' that form 'a single coherent mass only'.[185] This perspective means that 'the antithesis of ego and world, sensation (phenomenon) and thing, then vanishes, and we have simply to deal with the connexion of the elements'.[186] In this sense, for Mach the terms 'physical' and 'psychical' do not refer to objects or experiences of different types but to two different 'directions of investigation' (*Untersuchungsrichtungen*). That is, given that elements are connected by dependency relationships, they can be seen as physical or psychical, depending on which relations we focus on. In particular, if we consider elements ABC in terms of their dependence on the set of elements KLM that make up the human body, they appear to be mental elements, whereas if we consider *these same elements* ABC in terms of their dependence on other sets of elements, they take on the appearance of physical elements. For example, 'a color is a physical object so long as we consider its dependence upon its luminous source, upon other colors, upon heat, upon space, and so forth. Regarding, however, its dependence upon the retina (the elements KLM), it becomes a psychological object, a sensation'.[187]

Avenarius, too, had tried to develop a philosophical system that could overcome the dualistic opposition between the physical and the mental. Avenarius saw this opposition as the result of a fundamental error which he called 'introjection'. This is the idea that experience happens *inside* people. As a result, the original unit of experience gets separated into two opposed realities with unclear reciprocal relationships: the internal world and the external world, subject and object, psychical and physical. Avenarius therefore rejected the notions of the soul, of consciousness and of mental events, as they only have a meaning that derives from that distinction, that is, only from a framework that has already been falsified by introjection.[188] In his view, if we go back to the original experience, it is true that we can distinguish between a series of experiences that we call 'I' and a series of experiences that we call 'the environment', but 'both complexes of elements, in terms of how they are given to us, are definitely on one and the same plane', they 'are not given differently and separately'.[189] For Avenarius, the positive non-metaphysical concept of 'psychical', as 'the subject of psychological consideration', should not be grounded on the distinction between the 'I' and 'the environment', as – in a certain sense – the environment can itself be regarded as something psychical. As he wrote in his last work:

> Even the 'tree in front of us,' the 'movement of the leaves' and, more generally, the 'physical world in motion' can become the subject of psychology. That is, *insofar as* we can think of them as in some kind of *connection* with the speaking individual and in some way as (logically) *dependent* on the determinations of this individual, such that, for example, the 'tree,' being somewhat *conditioned* (in the logical sense), can be *fully determined* on the basis of its conditions only by *taking into account this 'dependency on the individual'*.[190]

Thus, like Mach, Avenarius holds that there is nothing that is essentially mental, but it is rather how we consider things that makes them appear in this light, as we observe them from the point of view of their dependence on the individual and, more specifically, on the individual's nervous system and brain. However, unlike Mach, Avenarius does not see this kind of 'psychological' approach as being in opposition to the approach that is appropriate for the physical sciences. Instead, he distinguishes between the 'relative perspective' of psychology and the 'absolute perspective', about which he writes:

> I merely observe an object in the environment [...] and I consider it as something absolute that I simply 'take' as it 'gives' itself and which I 'describe' as I 'find' it. In

this case 'I' am not taken into consideration and the object in the environment is not in a relationship with me, the one that observes and describes.[191]

This is what he writes about the relative perspective:

I find that I myself or another man do not lack significance with respect to what is 'given' as an item in the environment; observing such item, I now reflect at the same time on the properties of the 'I' that is 'engaged' in this act of observing and I no longer observe the item in the environment 'in itself and for itself' but 'in itself and for me', as something that is relative, which I must 'take' as party to a relationship, in which the individual, the so-called 'I', is the other party, and which I can no longer describe as something that 'gives' itself independently of this relationship.[192]

To summarize, Avenarius rejects the traditional opposition between the physical and the psychical, the internal world and the external world, where these terms are understood as referring to distinct parts of reality or as empirical fields that present themselves in different ways in our experience. In place of this opposition, he distinguishes between (1) myself and the environment, as different empirical content but nevertheless homogeneous in how they are given; (2) absolute and relative perspectives, where the former is limited to observing empirical contents as they appear, whereas the latter, the psychological perspective, observes them relative to the observer.

The next part of Petzoldt's deliberations reveals the influence of both Avenarius and Mach. Petzoldt specifies:

We can observe *things* and *processes* (that is, primarily the complexes of elements that are perceived or found, and the changes they undergo) in two ways. Once as items that are simply found and therefore present themselves to our *intuition* in exactly that way, actually one next to another and one after another. In this case we must define them as the contents of consciousness, phenomena of consciousness, processes of consciousness, mental contents, mental experiences, etc. We may then, in turn, observe *exactly the same complexes of elements* in terms of the interconnections they display, in their reciprocal functional dependencies, in their univocal determination. In this second case, they appear to us as physical objects: things and processes. We distinguish these two *perspectives* as the *psychological* perspective and the *physical* perspective; but these extend immediately to *everything* that exists.[193]

As we can see, the first perspective corresponds exactly to what Avenarius defined as the 'absolute perspective'. However, according to Petzoldt, it is this first perspective that characterizes the psychological approach, in contrast to the view that Avenarius held that it was the relative perspective.[194] This means that, for Petzoldt, psychology does not concern itself with what is given in terms of its dependence on the individual, or on any other empirical content, but is limited to addressing what gives itself simply in its being-given. On the other hand, like Mach, Petzoldt contrasts the 'psychological' perspective with the 'physical' perspective, assuming that the latter focuses on the functional relations that hold between different forms of experiential content. However, Petzoldt distances himself from Mach in the extent to which he also assigns to the physical perspective the task of addressing dependence on the physical individual. He specifies that his position implies 'a shifting of the actual boundary between psychology and physiology' such that all research that investigates 'the connection between mind and body' becomes a branch of 'physiology' – that is, ultimately 'in the realm of physics in general'.[195]

We can therefore draw some conclusions about Petzoldt's position on the difference between the physical and the psychical. There is no doubt that he takes the same line as his mentors Mach and Avenarius when he defines the physical and the psychical as two different ways of regarding the *same* content, and not as two distinct substances or two kinds of experiences that are given in fundamentally different ways. However, he distances himself from both Mach and Avenarius in that he does *not* identify the psychical with the perspective that regards content in terms of its dependence on the individual, but with the perspective that simply regards that content as it is given. Therefore, in contrast to his mentors, he considers the study of the dependence of mental contents on cerebral activity to belong to physics, which is the science that studies reality in its univocal determination.

The role of Eindeutigkeit and regularity in the psychical domain

We have seen that the topic of psychophysical parallelism runs through practically all of Petzoldt's work from the 1893 'Sittenlehre' onwards. The issue is examined in increasing depth over the years, but two cornerstones remain unchanged, namely the conviction that (1) mental contents *are not* determined univocally *by each other*, and (2) mental contents are fully conditioned, and therefore univocally determined, by their physiological correlates. These two propositions appear to have the force of postulates in Petzoldt's approach. In fact – especially in his early works – he introduces them without even attempting to validate them, seeing them as almost self-evident. It is only in the *Einführung in die Philosophie der reinen Erfahrung* that Petzoldt proceeds to justify the two assumptions in a systematic way.

Regarding the absence of reciprocal dependency between mental contents, Petzoldt appeals to everyday experience to show that they do not exhibit the characteristics of univocally determined processes: uniqueness (*Einzigartigkeit*), continuity (*Stetigkeit*) and unidirectionality (*Einsinnigkeit*).[196]

The problem with the uniqueness requirement is the lack of *mental* means of determination (e.g. sensations) displaying functional relations such that, given the value of one of the two variables, the other has *one* and *only one* value, as is the case of physical means of determination (weight, velocity, mass, temperature etc.). The lack of *mental* means of determination also stems from the problem that psychical contents are not measurable. Since 'magnitude has no place in the psychical domain', mental events cannot be expressed in mathematical form, and therefore, unlike in physics, it is also impossible to describe the relationships between them in terms of functional relations.[197] Petzoldt therefore rejects associationism and its claim that simple sensations can be used to explain the nature of mental constructs. He also rejects the attempts of the psychophysics of the day to measure sensations by their intensity. He sides instead with those who emphasize that there are only qualitative differences when it comes to the psychical domain.[198]

As for the continuity of mental constructs, Petzoldt swiftly dismisses the issue. As he puts it, 'even the most superficial self-observation' shows that 'what happens in the mind consists of abrupt events [*Plötzlichkeiten*]', hence with a complete lack of continuity.[199]

The issue of unidirectionality is more complex. Petzoldt concedes that 'many mental processes usually follow one another in a fixed direction', as in the case of 'the individual events in a story, or the individual notes of a melody'.[200] However, Petzoldt resolves this 'series of associations' by invoking the concept of 'exercise' (*Übung*).[201] It is worth going into further detail on this point, as it is crucial to Petzoldt's position on the relationship between the physical and the psychical. Petzoldt holds that psychology is victim here to the common error of 'confusing the *regular* [*regelmässig*] connection between specific mental contents, which appears after a certain period of exercise [*Übungszeit*], with the *univocal dependency* that holds between these contents; that is, an *evolutionary outcome* is confused with a *law of nature*'.[202]

As we have seen, from his earliest works onwards, Petzoldt follows Fechner's line in holding that living organisms exhibit the tendency towards stability as the ability to develop activities that are periodic and recurrent. He therefore enthusiastically embraces the theory of psychophysical connections proposed by Avenarius in the *Kritik der reinen Erfahrung*, largely based on the concept of exercise (*Übung*). We have already noted that Avenarius saw the functioning of the brain as based on the constant re-equilibration of two factors: work, that is, the elaboration of stimuli; and nourishment, that is, metabolic activity.[203] The activity of the brain can therefore be seen as consisting of 'vital series' that consist of three steps: an initial one when the two factors are in equilibrium, an intermediate one when a condition emerges that leads to a disequilibrium that threatens the survival of the system and a third one when the system succeeds in re-establishing equilibrium. The key point here is that when the brain identifies familiar stimuli it is able to redeploy the vital series it had previously used to address them. Thus, every time a vital series is *exercised*, every time that it is repeated, it becomes more usual (*geübt*) and more efficient. In particular, with each repetition, it tends to dispense with any steps that are superfluous to addressing the stimulus and re-establishing equilibrium. Even more importantly, it becomes more prompt, and thus able to be deployed more quickly. This means that the brain will *tend* to react to a given stimulus by using the most usual vital series. In other words, of the infinite number of possible responses by the brain, the usual one is the one most likely to be realized. However, this *regularity* comes from the system practising the series, not from a *law* of nature. Nevertheless, however 'practiced' or 'usual' a vital series might be, thereby appearing *regularly* in response to a given stimulus, it is always possible for the brain to be in a certain state at a certain point which leads to the series *not* being realized. Moreover, the regular tendency for a given vital series to be realized only manifests itself in brains that have *developed* in this way through practice. Regularity is therefore unable to say anything about which brain state will be realized or will tend to be realized when a given stimulus appears in a 'virgin' brain, so to speak.

To express this using Petzoldt's terminology, we can say that the regularity of brain activity is unable to determine a response by a system to a stimulus as *unique* with respect to all the possible responses, but only, at best, as *more likely* than any of the other possibilities. In this sense the regularity of cerebral processes is not the same thing as the *univocal determination* of cerebral processes.

In any case, given that the brain is a physical organ subject to the principle of the conservation of energy, any change in its state must be assumed to be univocally determined by other physical conditions. In other words, brain activity is *always* determined by conditions that can explain why that *particular* actual case had to be realized as *unique* from among all the possible cases. However, the brain is also able to *develop* regularities that establish which case tends to be realized *in general* (but *not always*).

For Petzoldt, then, *cerebral processes* exhibit both regularities and univocal determination. If, however, we examine the relationships that hold *between mental contents* we see that regularities exist, such as associative series, but univocal determination does not. The problem of psychology is precisely that:

> the ease with which *regular* connections in mental processes can be demonstrated raises the hope of also being able to use proximity or similarity to *explain* sequences of thoughts that have not yet become usual [*geübt*] but appear for the first time [...]. Vain hope! What we find here is always only stability, never laws![204]

That is, acquiring scientific knowledge requires more than merely demonstrating regularity or generic connections between events: 'it is obvious that mental events are in general dependent on preceding ones; what is disputed is their *univocal* dependence'.[205] For Petzoldt, 'only someone who mistakes a *rule* for a *law*, an *evolutionary outcome* for an *original and exceptionless connection*' could confuse the type of relationship that holds between physical events with the type that holds between mental events, as Wundt and anyone else who talks of 'mental causality' does, claiming that this can explain sequences of mental contents.[206] The relationships that hold between physical events are indeed ones of univocal determination, which is a precondition for any true scientific knowability as it is a requirement of the complete intelligibility of events, that is, of the existence of a sufficient explanation for why the actual case is the *only* case that is realized from among all possible cases. In contrast, the type of relationship that holds between mental events is at best that of regularities established over time. These regularities do allow the formulation of empirical generalizations, but do not lead to true and proper scientific knowledge of mental events. Therefore, if we were to limit ourselves to considering regular connections between mental contents, we would have to conclude that true scientific knowledge of mental processes does not exist.[207]

Even if Petzoldt's view is that no univocal determination exists *between mental contents*, but at best a certain degree of acquired regularity, this does not mean that they should not be considered as univocally determined. We know, in fact, that univocalness is an unavoidable postulate for human beings, and that it must therefore also be applied to the mind. As Petzoldt writes, 'it is impossible to assume that at a given moment one could perceive red but *just as well* [*ebensogut*] perceive green', or 'that instead of a thought that is present at a given moment one might *equally* have, at the same moment, a different thought'.[208] We must therefore assume that 'the mental state that exists at a given moment is the only possible one at that moment and must therefore be seen as univocally determined'.[209] However, as we are unable to find *mental* means of determination that allow us to see mental states as univocally determined, there is no alternative but to seek *physical* means of determination.

It is therefore the need to subject mental events to the principle of univocalness that constitutes 'the strongest reason to assume "psychophysical parallelism"',[210] according to which 'no sensation or representation, no feeling or thought, in general not even the slightest change in the human mind can occur without a simultaneous process in the brain, without which it could not occur'.[211] Needless to say, Petzoldt states that the converse is invalid. To suggest that 'every physical process is accompanied by a mental process' would be to descend back into '*metaphysical* parallelism'.[212]

Indeed, it is not only the case that not every *physical* process is accompanied by a mental process but not even every *cerebral* process is accompanied by a parallel mental process. According to Petzoldt, the relationship between changes in the brain and those that involve mental processes must be seen as follows:

> If we represent the phases of a series of processes in the brain with the letters:
> $\quad a\ b\ c\ d\ e\ f\ g\ h\ i\ j\ k\ l\ m$, and use the letters:
> $\quad \beta\quad \delta\ \varepsilon\quad \gamma\qquad\qquad \lambda$
> *to represent the corresponding mental acts, then the first series, in which each member can be thought of as dependent on the preceding one, proceeds in a continuous manner, whereas the second is discontinuous.*[213]

The sequence of physical changes in the brain thus meets the requirements of univocal determination and continuity. In contrast, the sequence of mental acts does not exhibit continuity, as mental acts only depend on *some* of the changes in this physical sequence. Even more importantly, as mental acts depend on parallel cerebral processes, there is no univocal determination between them. The condition for the realization of δ does not lie in its mental antecedent β but in its cerebral correlate d.

We have seen that alongside the univocal determination of physical processes and the dependence of mental events on the brain, Petzoldt also recognizes the existence of *regular* sequences of mental events, acquired over the course of evolution according to the principle of the tendency towards stability. Even if these regular sequences are found in mental processes, the conditions that determine them are once again to be found in the brain. As the brain evolves, developing ever more stable relationships with the environment which lead to periodic and recurrent physical processes, so the mental events that depend on the brain will also become regular, periodic and recurrent. The regularity of mental events is therefore the result of their dependence on the regularity developed over time in the brain. However, as regularity only indicates what happens *in general*, and not what happens *by necessity*, the specific case can always differ from the norm. In such cases, the mental constructs realized must be seen as univocally determined by cerebral activity, and this cerebral activity must be univocally determined by a series of physical conditions. The point of univocal determination is in fact that *even where the individual case does not adhere to the rule and deviates from what occurs in general, it must always be possible to find means of determination that can show why the exception* must *have occurred instead of the normal case*. Exceptions to the rule can never be exceptions in an absolute sense, as events that occur against the lawfulness of nature, but must only be seen as deviations from what happens in most cases, in the *usual* course of events that is the result of evolution towards stability.

This forms the basis on which Petzoldt sets out how the study of the connections between mental activity and the physical processes of the brain should in general proceed:

> The simple thought at the heart of psychophysical coordination is that the regularity of mental events must be seen as dependent on the regularity of cerebral processes, and that any deviation from this psychical regularity must be viewed as determined by a deviation in the regular progression of variations in the central nervous system. As the emergence of regular sequences of changes in the central nervous system can be understood in terms of physiological conditions that are primarily linked to the evolutionary history of this organ, it follows that parallel mental events can also be seen as univocally determined.[214]

Petzoldt's position is that addressing the relationship between mental and cerebral activity in terms of univocal determination allows for the elimination of potential materialistic misunderstandings of psychophysical parallelism. The concept of univocalness is based on the notion that relationships between events should not be understood in the old anthropomorphic sense of causality, where the cause *acts upon* or *produces* the effect, but in the sense of the simple functional relation: *if x changes then y changes*. Therefore, saying that mental contents are univocally conditioned by the brain does not mean that they originate in the brain, or that they are under the yoke of the brain, but only that when something changes in the brain, something also changes in the mind.

Petzoldt thus refutes Wundt's accusation of materialism against Avenarius and, more generally, against all those who supported psychophysical parallelism as the affirmation that all mental content is dependent on the processes of the nervous system. For Petzoldt, one can speak of materialism only if 'the equal gnoseological legitimacy of mental and physical events' is denied.[215] However, the doctrine of the univocal determination of mental events via cerebral correlates in no way implies that 'mental events are of inferior validity, not autonomous, produced by things that are material'.[216]

If one believes in the concept of functional relations between changes, neither variable precedes the other. Both are given together, and the relationship consists of how they change. Therefore, there is no gnoseological or ontological priority between the physical world and the world of the mind. Indeed, the parity of the physical and the mental becomes even clearer if one considers that these terms do not refer to two different domains, but simply to two perspectives on an entity that is fundamentally homogeneous. The dependence of mental content on the physiological activity of the nervous system is not even a specific form of dependence but is simply yet another case of the innumerable univocal relations that define reality. Therefore, the lawful connection between the brain and mental contents 'is no more extraordinary than the existence of laws *in general [überhaupt]*'.[217]

Rejecting the causal efficiency of psychical contents does not imply that 'the mind must be seen as the most superfluous thing in the world. If this was the inevitable consequence of psychophysical parallelism, it would be so absurd that this alone would be enough to reject the notion'.[218] Psychophysical parallelism only states that mind depends on the physical activity of the brain, but it has nothing to do with 'the metaphysical interpretation' of this dependence, according to which 'the world out there would be in some sense absolute, […] and the mind would be superfluous to the natural understanding of the world'.[219]

Once we accept that the physical and the mental are both part of a single reality, or rather two aspects of a single reality, it becomes evident that understanding this reality requires both. Psychophysical parallelism does not entail dismissing mental activity as a mere epiphenomenon. The mind retains all its dignity and coherence, on a par with the body. The only notion that is rejected is that mental processes can be explained through a univocal determination that exists *between* mental processes. Indeed, as we have seen, the absence of univocal determination in some sense becomes the very definition of the mind, given that Petzoldt defines the psychological perspective as that which considers empirical events 'as items that are simply found and therefore present themselves to our intuition in exactly that way, actually one next to another and one after another' rather than 'in their reciprocal functional dependence'.

Furthermore, even though Petzoldt holds that strictly 'scientific' understanding 'cannot be reached using rules, but only using laws that have no exceptions',[220] this does not mean that there is no value in investigating the regular connections that can be found between mental events. Beyond the scientific understanding of the univocal dependence of mental events on the brain, Petzoldt concedes that 'we understand a mental process when we are able to associate it with regularities observed in our mental events; or, at least, this usually provides a sufficient surrogate for understanding'.[221] This can be shown by considering that 'in our everyday lives, we do not notice the absence of univocal determination between mental events', as the regularities 'are sufficient to achieve the goals of our actions'.[222] Petzoldt therefore emphasizes that 'research into mental regularities of this kind is not the slightest bit less important than research into natural laws'.[223]

The conservation of energy and psychophysical parallelism

The preceding section demonstrates that Petzoldt ends up seeing psychophysical parallelism as one and the same thing as the postulate of the univocalness of the mind, formulated as the assumption that 'if we want to fully understand mental life then we must correlate [*zuordnen*] each of its phases to the univocal processes of the central nervous system'.[224]

Together with this more strictly theoretical requirement, Petzoldt also presents empirical reasons in support of psychophysical parallelism. Indeed, at the start of the *Einführung in die Philosophie der reinen Erfahrung* he concedes that 'usually' the hypothesis is accepted not 'because of some need to explicate mental events' but 'to agree on a series of facts and a general conceptualization of natural events'.[225] There are essentially two 'facts' here. The first is the lack 'of sensation or voluntary movement when the connection between certain parts of the brain and the peripheral sense organs is severed'.[226] The second is the observation that 'the development of the mind proceeds hand in hand with the development of the nervous system',[227] as a result of which organisms with a more highly developed brain are also those with superior mental faculties. The 'general conceptualization of natural events' that is added to these 'facts' in support of psychophysical parallelism is the principle of the conservation of energy. The discovery that 'the amount of energy in all the transformations of a process remains constant' leaves 'no free space' for 'mental properties' in material processes.[228] Mental events can therefore only be considered as an 'accompaniment [*Begleiter*] to given physiological processes'.[229]

As well as using the principle of the conservation of energy in support of the claim that mental events depend on the brain, Petzoldt also appeals to it to distance himself from any conceptualization that states more or less openly that mental processes can determine physical phenomena. In the previous section, we outlined his criticism of 'indeterministic' interpretations of the principle of the conservation of energy, such as those proposed by Fechner and, above all, by Wundt. Fechner and Wundt held that the conservation of energy was only relevant to quantitative relationships between phenomena and did not determine the specifics of what occurs. They believed that this allowed them to ensure that volition (and thus, definitively, a purely mental factor) could influence physical events, but without coming into conflict with the principle of the conservation of energy. This, however, was not the only strategy used to defend the causal role of volition in a physical world where energy remained constant. The other route appealed to the concept of potential energy.

This concept was first introduced by Leibniz in *A Brief Demonstration of a Notable Error of Descartes and Others Concerning a Natural Law* (1686). According to Leibniz, a body that is raised to a certain height, even if at rest, latently conserves a given amount of 'dead force' (potential energy), given that, if moved, for example by falling, the body is able to release a certain amount of 'living force' (actual energy or kinetic energy, as it is defined today). Leibniz had thus provided a first formulation of the principle of mechanical energy, which establishes that the sum of living force (actual/kinetic energy) and dead force (potential energy) is constant.

The concept of potential energy thus became particularly useful as a way of capturing the fact that an immobile body could nevertheless contain a certain amount of energy, which could be activated by a relatively small amount of force. An example is the potential energy of a mass of snow at the top of a mountain. A minor tremor could trigger its descent, resulting in the transformation of this enormous amount of potential energy, which had been lying there, dead, into the living and devastating energy of an avalanche. In the German of the day, this was defined as *Auslösung* ('activation', 'triggering', 'release'), that is, 'the process by which, thanks to a small amount of work, a large amount of potential energy is freed, and is transformed into mechanical work'.[230]

The principle of the conservation of energy was universally accepted. This meant that even those who held that mental factors such as volition could influence the course of events, such as voluntary movements of the body, had to accept the assumption that this principle governed the human organism. The concept of *Auslösung* as the triggering of potential energy was therefore called upon to prevent the mental determination of physical events from coming into conflict with the conservation of energy. The assumption was that the transformation of energy that occurs in the course of processes that begin with a physical stimulus leads to an accumulation of potential energy in the brain, principally in the cerebral cortex, which was considered to be the part that was least exposed to external stimuli.[231] When the potential energy of the brain is static, dead, latent, like the accumulation of snow on a mountain, there is room for purely mental activity. Following this mental activity, volition can then proceed to the *Auslösung* of the potential energy of the brain, turning it into the kinetic energy required to activate the movements of the organism. As Petzoldt summarizes the conception of his opponents:

> Every psycho-physiological process of this kind therefore has five phases: 1) a *purely physiological* centripetal conduction process from the periphery of the sense organs to the central nervous system; 2) a *psychophysical* process whose mental aspects consist of sensations and of something like groups of low-level representations; 3) a *purely mental* process in which – to use the terminology of a well-known school of psychology – abstract high-level representations, their apperceptive connections and associated feelings appear, that is, a process reserved for the most advanced mental functions, for the highest forms of thinking and feeling, and for volitional decisions; 4) a further *psychophysical* process whose mental aspect must be seen as the representation of the muscle movements to be realized or of the future sensation of movement; 5) and finally a further *purely physiological* centrifugal process, with the innervation of muscle groups.[232]

Under this view, the transformation of kinetic energy into potential energy takes place in the second phase. Throughout the third phase, the potential energy remains latent in the brain, leaving space for *purely* mental deliberative processes, that is, processes that are *not conditioned by physical cerebral activity*. Then, in the fourth phase, the volitional mental impulse triggers the potential energy in the brain (*Auslösung*), turning it into kinetic energy that is transmitted centrifugally to activate the muscles.

This model of the process of the volitional reaction of the psychophysical organism underpinned Wundt's approach to psychology, as alluded to by Petzoldt when he refers to 'a well-known school of psychology'.[233] Indeed, according to Wundt, the 'apperceptive' processes of response to stimuli take place in five phases:

> 1) Transmission from the sense organ to the brain, 2) entry into the field of consciousness [*Blickfeld*] or perception, 3) entry into focal point of consciousness [*Blickpunkt*] or apperception, 4) the time required for volition [*Willenszeit*] to

launch the corresponding movement in the central organ, and 5) the transmission of the resulting motorial excitation to the muscle.[234]

Petzoldt rejects this position, pointing out that 'it takes the law of energy into account only outwardly'.[235] Indeed, although it is true that even in this formulation 'every phase of the process contains the same amount of energy', the fact remains that it contradicts both the spirit and the substance of the principle of conservation.[236] The 'tacit assumption' of the principle amounts to 'the full determination of material processes'.[237] However, more importantly, the principle of the conservation of energy states that 'something that is at rest cannot leave that state *by itself*; the transformation of potential energy into actual energy requires the use of a specific quantity – perhaps minimal, but in any case finite – of energy, an impetus [*Anstoss*], a triggering event [*Auslösungsvorgangs*], which can only be material, physical'.[238] This approach therefore fails to validate the step from phase three to phase four, as the potential energy in the brain cannot be released by a mental trigger but only by a physical/physiological condition. In consequence, the concept of potential energy cannot be used to defend volitional action, as 'at no point in cerebral processes, however complex they are, can space be found for the insertion of mental events, in the sense of the schema outlined above'.[239]

It is worth noting that Petzoldt's criticism of this attempt to reconcile the law of the conservation of energy with the causal efficacy of volitional impulses is a prime example of his position on the relationship between philosophy and the individual sciences:

> One gets the impression that psychologists who adopt [the above schema] do not see the law of energy as an important facilitator of the understanding of nature, warmly embracing its consequences in the field of the physiology of nerves, but rather see their position as a peace treaty with the victorious opponent. [...] It is a bad sign for philosophy when the firmest achievements of natural science are only recognized reluctantly, instead of being used to forge new tools and weapons.[240]

In other words, rather than starting with philosophical assumptions only to begrudgingly 'come to terms' with the natural sciences, we should instead welcome what the natural sciences reveal without fearing the consequences of developing them further. As Petzoldt writes in another work: 'it is preferable not to address problems with *epistemological assumptions* but rather to derive *epistemological consequences* from their specific solutions'; our 'conceptualization of the world' will thus 'not be the foundation stone of the edifice, but the last brick to be laid'.[241]

In the case of research into conscious organisms such as human beings, embracing the results of the natural sciences means recognizing that behaviour can be fully explained as an ensemble of *physical* processes, on the basis of the closed causality of nature, which is a consequence of the law of the conservation of energy, and without resorting to assumptions about mental conditions.

The process that occurs when we *involuntarily* retract our hand when it encounters a sharp object can be fully explained in physical terms by the law of energy, without resorting to the sensation of pain. But the same thing happens when we avoid, by the skin of our teeth, a menacing mass of rooftop snow that may come down on us at any moment. The *perception* of the mass of snow, the corresponding *representation* of the danger of being struck by it, and the *wish* to escape this danger, all these mental acts in no way contribute to explicating the physical process. We can understand this physical process only through physical elements, and we would find ourselves in great difficulty if we were required to find space in them for psychical phenomena. We can only see them as accompanying events, as parallel phenomena.[242]

This is, in fact, the fundamental lesson that Petzoldt learnt from Avenarius, who, in the first volume of the *Kritik,* had set out to develop a purely physiological theory of human behaviour based solely on the activity of the brain. As Avenarius wrote in a note at the end of the work:

> The preceding propositions [...] set the reader the potentially *disconcerting* task of regarding changes in human beings that help them to survive in a less than ideal environment without resorting to the concept of 'consciousness' for once. [...] The backwards step requested of the reader here is exclusively methodological; it can be accepted in its entirety regardless of the systematic question of whether *in fact* 'consciousness' is to be assumed at the same time. Just as we have *learned* to *be able* to think of the 'marvellous construction' of vegetal and animal organisms, of how they are born and grow, [...] how they feed, how their wounds heal, how they recover from illness, how they adapt to the environment, etc. without the 'participation' of a 'spirit' in general or of 'the soul' in particular; it is therefore also possible to acquire the ability to *be able* to consider the so-called 'purposive' variations in system *C* [the brain], and variations of variations, without immediately resorting to a 'mind' as an explanation; all the more so that its 'changes in psychical states' would all in turn remain in need of an *explanation*.[243]

A consequence of the law of the conservation of energy and psychophysical parallelism is that it must be possible to see the human organism as a purely physical entity whose actions (whether practical or theoretical) consist entirely of processes that are chemical, electrical, thermodynamic, etc. We could therefore potentially conceive of ignoring all mental events, given that they cannot be assigned any causal role without negating the principle of the conservation of energy.

Therefore, in the Preface to the second edition of the *Kritik*, published after Avenarius had died, Petzoldt praises him for having based his work on two pillars of modern psychology: 'the elimination of any efficacy [*Tätigkeitsmomentes*] from the field of the mind and the hypothesis that there is no psychological process without a corresponding biological process'.[244] Avenarius had thus 'completely excluded any activity [*Aktivität*], any efficacy [*Tätigkeit*], any apperceptive or volitional factors from the life of the mind, subjecting even the highest forms of mental activity to the rigorous methods of the natural sciences, that of the simple ascertainment of facts, which had hitherto taken place only with respect to sensations'.[245] Avenarius's *Kritik* therefore represents 'the *first extended and in-depth endeavour to provide a univocal determination of mental events*'.²⁴⁶

It is important to note that Petzoldt's recognition of the importance of the work of Avenarius comes with a fundamental criticism. Avenarius had written the *Kritik* as two volumes, analysing the purely physiological activity of the brain in the first and the mental events that depend on this activity in the second. The motive for this structure was that cerebral processes are the *independent* variable, that is, the antecedent, whereas mental events are the *dependent* variable, that is, the consequent.

This separation came under attack by Wundt, who emphasized that Avenarius had not used an empirical research method but one that was metaphysical and speculative. In the great majority of cases, what is experienced is the 'dependent vital series' consisting of mental events. However, instead of limiting himself to this line, Avenarius 'hypothesizes' an 'independent vital series' consisting of changes in the brain, which was not actually experienced and turns out to be a mere duplication of the first series. Therefore, according to Wundt, it is not true that Avenarius starts with the brain in order to derive the mental events that are dependent on it, as the two-volume structure of the *Kritik* would have us believe. He starts, in fact, with mental events, to which he traces back presumed cerebral activity which was never experienced, being simply a 'hypothetical and transcendent' 'auxiliary representation' that was elaborated to provide an ad hoc explanation.²⁴⁷ Avenarius thus based what is known (experienced mental events) on what is unknown (brain activity, not experienced but constructed artificially), an approach that inverts the direction of the process of valid explanation.

Petzoldt agrees to some extent with Wundt's criticism, acknowledging that the claim in the *Kritik* of starting with the brain in order to arrive at mental events is in fact 'impossible', and that 'the approach that Avenarius had in practice adopted was inconsistent with that objective'.²⁴⁸ In fact, Avenarius had embarked on 'an investigation of the psychical state of affairs free of any presuppositions', which had led him to two 'discoveries of great importance': (1) 'that mental events occur in series [...] which however have nothing to do with associative series', and (2) 'that the traditional separation of mental events into sensations, representations and feelings, and potentially also volitional acts, is unsatisfactory'. It was therefore necessary to replace this approach with one that was both 'more complete and simpler', distinguishing only between 'elements' (sensations) and 'characters' (everything that characterizes sensations: feelings in the strict sense, such as 'pleasant' and 'unpleasant', but also in a broader sense, such as 'known', 'unknown', 'beautiful', 'repugnant' etc.).²⁴⁹ Avenarius had started from this *psychological* standpoint in an attempt to imagine how the functioning of the brain should be envisaged so as to explain these psychological facts. It is this route that led Avenarius to his 'biology of the central nervous system', based on the mechanism of vital series.²⁵⁰ However, this does not mean that Avenarius had used 'metaphysical assumptions', as Wundt maintained, but only that he had 'hypothetically constructed superempirical conditions', in the same way that 'any hypothetical integration' of what we experience is 'superempirical'.²⁵¹ Petzoldt also points out that Friedrich Albert Lange had already attributed 'the unfruitfulness of investigation of the brain' to 'the entire absence of any workable hypothesis, or even of any approximate idea, as to the nature of cerebral activity'.[252] The hypothetical nature of Avenarius's theory of how the brain works is not therefore a limitation but a thing of great value as it answers to the problem highlighted by Lange.

The evolution of the brain and of mental events

As Petzoldt does not agree with the separation of the physiological and the psychological analysis across the two volumes of the *Kritik*, he must necessarily adopt a different thematic structure in his introduction to the philosophy of Avenarius. The first volume of *Einführung in die Philosophie der reinen Erfahrung* therefore addresses 'The Determinateness of the Mind' (*Die Bestimmtheit der Seele)*, while the second is entitled 'Towards Perdurability' (*Auf dem Wege zum Dauernden*). Petzoldt's ultimate goal in this work is in fact to show that mental events are conditioned by the nervous system and its evolution (as the first volume demonstrates), and that they therefore proceed towards the achievement of lasting states (as illustrated in the second volume). Once Avenarius's hypothesis is accepted – that the biological activity of the brain consists of vital series designed to address stimuli that threaten the survival of the nervous system – all Petzoldt needs to do is assert that the brain is in a more lasting state at the end of a vital series.[253] This makes Avenarius's entire system an integral part of the principle of the tendency towards stability, which becomes the key to understanding theoretical and practical human behaviour.

Petzoldt rejects the idea that practical human behaviour is based on feelings of pleasure and displeasure. Since these are indeed mental events, and therefore at best an accompaniment to organic processes, they cannot determine them. The condition for the activity of a physical organism cannot be the desire to flee or to seek certain psychological states. What guides organic processes is rather the tendency to achieve stationary states. Feelings of pleasure and displeasure come into play only to the extent that lasting states are accompanied by feelings of pleasure and unstable ones by feelings of displeasure. In other words, it is the tendency towards stability that is perceived in the mind as a tendency to reduce displeasure and increase pleasure.[254] Petzoldt therefore states that 'everything that is wanted is wanted for the sake of stability', and 'stability is the ultimate "purpose" of every endeavour and instability the ultimate "motive" for every action'.[255]

As the brain is a physical organ and therefore unable to pursue psychological goals except indirectly, it is impossible to attribute to it the *intention* to maximize stability in order to safeguard the body in general. 'The life of the brain must be understood on its own terms'; the brain 'is not at the service of the rest of the organism', its activity is based only 'on the preservation of the central nervous systems themselves […] preserving these is an *end in itself*, with nourishment and bodily movement being merely a means to realize that end'.[256] In other words, the brain does not *aim* to safeguard the body; rather, as it is intimately connected to the body via the nervous system, any threat to the body becomes a threat to the brain itself. Therefore, *in order to ensure its own survival*, the brain ends up also ensuring the survival of the body, or rather, *by ensuring its own survival* it also ensures that of the body. This obviates the need to attribute to the brain a kind of 'higher' intelligence, a more sophisticated intentionality that would enable it to pursue 'external' goals, putting itself at the service of *other* organs, given that, like every other organic structure, the brain aims only to ensure *its own* survival. As the brain is the key protagonist here, Petzoldt goes as far as to declare that 'the principal task of nature is not the preservation of mankind as such, but the preservation of the higher-order nervous systems'.[257]

This situation also holds for theoretical work. According to Petzoldt, 'all research and striving for the "truth" are simply a tendency towards lasting states'.[258] The interesting issue, however, is how this tendency of the brain to stability is expressed. For Avenarius, the nervous system worked in a somewhat passive manner: the brain was constantly engaged in reacting to stimuli, and everything that had happened in the past determined the possible responses available to the system at any given point. For Petzoldt, on the other hand, the brain does not limit itself to 'defending itself' from external stimuli, elaborating responses to them once they are required, but is driven by a kind of impetus towards the environment, which makes it elaborate potential responses to potential external stimuli *in advance*. Then, from among all of the *potential* responses to *potential* stimuli, the environment *will select* the *actual* responses when it presents the organism with *actual* stimuli that will only activate certain neural circuits and not others. As Petzoldt puts it:

> Just as a rhizopode extends its protoplasmic projections in all directions in readiness for any nutritional particles that may present themselves, so the human brain uses its tireless sense organs to explore the darkness of the future in all directions, thereby enabling it to swiftly grasp whatever matches any of these neural *possibilities* and thus adapt even to environmental conditions that differ considerably from actual conditions. [...] However, as soon as actual reality has made its choice from among the available neural *possibilities* – that is, as soon as a theoretical question with several possible answers obtains a response determined by an experiment, the discovery of a source, etc. – physiology requires the onset of the regression of the unselected and unused neural alternatives. The selected neural formation, the one that matches the environment, is therefore durably deployed thanks to the stimulus encountered. As a consequence of this additional practice, it takes the lion's share of the available nutrients – exactly as in Roux's sense of the struggle between the parts.[259] This stimulus leads to the functional structuring of this part of the central nervous system, while the other partial systems that had been *possible* regress as a result of the lack of stimuli and nutrients.[260]

As we can see, the para-Darwinian view of evolution that we analysed in the preceding chapter is part of this process.[261] For Petzoldt, the brain has an internal innate tendency to develop, which leads it to produce a variety of possible neural responses to external stimuli. The struggle for existence (which Wilhelm Roux maintained also exists between the parts of an organism in the form of competition for nutrients) therefore comes into play at a later stage, selecting, preserving and reinforcing in a lasting way the reactions of the system that match actual environmental stimuli, while those that are not used regress through starvation. This process of neural variation and selection finds its psychological counterpart in the continuous use of the 'imagination'

(*Phantasie*), that is, in the 'creative games of thinking', where the *possibilities* are imagined, pending the world's confirmation of what is *real*.[262] Petzoldt calls imagination 'the noblest of our mental faculties, our innate guide alongside critical assessment, which carefully challenges thoughts and weighs them up against each other'.[263]

It is even more important to note, however, that this process contains the same conceptual mechanism that is found in the principle of univocalness. It is based on the relationship between a range of possibilities and the uniqueness of the actual case. This clarifies what Petzoldt meant when he stated that this principle is intrinsic to our organic existence, as we have evolved in conformity with a world in which univocalness exists.[264] Indeed, the development of the brain is itself based on the consolidation of the actual case from among the range of possible cases. In this sense, *Eindeutigkeit* turns out to be what quite literally shapes the brain, and therefore how we think. The brain is capable of elaborating an infinite number of possible variations, but the environment does not present it with an *infinite* number of different stimuli but only certain actual stimuli, thus leading to only some of the different cerebral responses being selected.

This also clarifies the centrality of experience in Petzoldt's system, which harks back completely to Avenarius in this respect. If the development of the brain depends on the environment, only knowledge that is consistent with the environment will be 'selected', that is, empirical knowledge. As Petzoldt writes, 'the only fount of knowledge and the only testbed for any theory is experience, that which has been found'.[265] It is the experience of actual reality that determines the development of our knowledge, because 'without actual reality, there would be nothing else that could choose which of the innumerable possibilities offered by thinking to preserve'.[266]

For this reason, as advocated by Avenarius, our view of the world cannot but tend towards pure experience, that is, towards a state in which all of the responses of the brain that are not determined by the environment, together with all of the associated non-empirical mental events, will be eliminated as a result of insufficient exercise, succumbing in the intraorganic struggle for nutrition. Hence, the theory of the biological functions of the brain proposed by Avenarius was also a 'philosophy of the history of philosophy' that described the development of our worldviews towards pure experience.[267] Indeed, in empiriocriticism, 'the history of theories of reality is at the same time the history of the corresponding parts of the brain'.[268]

Cerebral and psychological evolution therefore leads to pure experience. However, in the 1893 *Sittenlehre*, Petzoldt also emphasizes the importance of deriving maximally general concepts, as the 'tireless and violent drive of thinking towards stable states' is reflected in the 'tendency of every concept towards its most generalized form possible'.[269] In the 1900 *Einführung in die Philosophie der reinen Erfahrung*, Petzoldt does not argue against searching for a unitary and general understanding of reality, but at the same time he emphasizes the importance of 'differentiating concepts, which leads to increased sustainability, rigour and stability'.[270] According to Petzoldt:

> As evolution proceeded, so the richness of the view of the world had to conflict with rigid schematism. [...] As we observe nature in increasing detail and even more extensive analysis, so our eye becomes more sensitive to the nuances of aesthetic stimuli and the richness of their form and colour, to the slightest variation in tonality.[271]

The richness of the world does not, however, mean that we must be sceptical about our ability to grasp the reality around us in all its detail. The ultimate goal is not to grasp the entirety of reality but to enter into a stable relationship with it.[272] Consciousness is driven by biology, not by logic:[273]

> The task of describing facts as exhaustively as possible is inherently self-contradictory. When someone set himself this task, he imposes it on himself from the outside, without taking into account his own essence, thus creating an irreparable fracture between nature and the human mind [...]. Is that any way to talk of purposes? Is it the goal of the stomach to digest stones? Or of the lungs to create blood from carbon dioxide? Should not the *purpose* align with the normal function of an organ, and even coincide with it? When we require of thinking that it goes beyond what is *in its nature*, we enter into an irresolvable conflict with ourselves.[274]

As 'the human environment is finite' and 'in a finite system, no advance can be infinite', the development of the brain and of human knowledge must have a limit, and not just a limit to be reached asymptotically but a limit that is actually achievable.[275] Clearly, reaching this limit 'does not mean the end of thinking, but the end of evolution'.[276] For Petzoldt, the fundamental difference between mankind and other livings being such as plants and animals is that development has not yet completed; evolution is still *in progress*. More precisely, what is still in evolution is the human brain.[277]

Petzoldt's impassioned interventions in German society had the objective of vindicating this philosophy based on experience, which he saw as humanity's most advanced form of biological and psychological development. He was also very active in education, organizing special schools for outstanding students.[278] For Petzoldt, 'a genius is a man of evolution [*Entwicklungsmensch*]', one who furthers human development and has a following of 'prodigies' that allow novel ideas to spread.[279] Indeed, even if 'humanity's *Weltanschauung* is a *social* product, the result of the thoughts and feelings of the masses', to 'rise above a social phase', requires 'singular individuals.'[280] New ideas are expressions of variations (in a quasi-Darwinian sense) of exceptional individuals who are not constrained by prevailing norms. Therefore, as evolution usually occurs in species with particularly large populations,[281] where there is greater probability of the emergence of favourable individual variations, so, for Petzoldt, 'new *Weltanschauungen* only appear in large populations', where there is greater probability of geniuses being born.[282]

If human evolution is to be stimulated, it is necessary to cultivate geniuses and prodigies, identifying them and allowing them to express their full potential. That is, human evolution is not simply a theory – it is a task, and a task to which Petzoldt devoted his entire career as philosopher and teacher.

3

Subjectivism and relativistic positivism

The rejection of subjectivism

The elimination of the concept of substance

In the preceding chapter, we outlined the development of some of the areas that are key to Petzoldt's thinking, namely, those that centre on the principle of the tendency towards stability and of *Eindeutigkeit* as fundamental propositions of maximum generality. We touched on a number of issues related to these two principles, such as the non-anthropomorphistic re-elaboration of causality and teleology, and the assumption of psychophysical parallelism. We will now turn to Petzoldt's philosophical position on the relationship between the individual and the world, between subject and object. This will reveal the escalation of his polemic against what can broadly be called subjectivism, and in particular Kantian subjectivism. Some of these issues have already been alluded to in the preceding chapter, of course. Our aim in addressing these two aspects of Petzoldt's thinking separately is to provide a structure for the huge body of work and large number of topics he addressed, but this distinction is not a true reflection of the content of his work. When dealing with thinkers whose ambition was to propose an all-embracing system that covers absolutely everything, any attempt to dismantle and reassemble that system in order to examine it in greater depth cannot but be somewhat arbitrary. It would have been equally plausible to have addressed the topics examined in this chapter first, before then proceeding to the principles of stability and *Eindeutigkeit*. Given the possibility of multiple reconstructions of this kind, it is not surprising that we will now rewind the tape, returning to Petzoldt's first works in order to lay out the development of these other topics.

As we have seen, in his first work, 'Zu Richard Avenarius' Prinzip des kleinsten Kraftmasses und zum Begriff der Philosophie' (On Richard Avenarius's Principle of Least Effort and the Concept of Philosophy, 1887), Petzoldt addresses Avenarius's *Prolegomena*, which postulates – on the basis of the principle of least effort – that our conception of the world tends to progressively dispense with any content that is not empirical, thereby moving towards pure experience. In particular, Avenarius identifies causality and substance as two residual metaphysical notions doomed to disappear over the course of human development.

Petzoldt sides unreservedly with Avenarius on this point. Leaving to one side his criticisms of the notion of cause, which we addressed in the preceding chapter, we will now focus on substance. Petzoldt writes:

> Fetishism, naive realism and solipsism are no more than individual attempts to grasp the mysterious essence of substance, steps on an evolutionary path that must lead to an absolutely negative outcome. The subject matter of this research can never in fact become an object of experience. The external world and the internal world are undoubtedly *only complexes of sensations and representations*; we can never know what is hidden behind these – assuming indeed that there is something hidden behind them – and fortunately there is no need for us to know it. *It is obvious that that possibility of the transcendent existence of a substance can never be denied*; but we must concur with Avenarius that requiring philosophy to answer the question 'What is it, in reality, that we see and feel?' is a spurious problem. [...] *We will never do anything other than compare representations with perception of objects, that is, with other representations. Objects are our representations and our representations are our objects.* This is the circle within which we must move without rest: just as we cannot leave the third dimension in favour of the fourth, so *we cannot escape from the 'sphere of our subjectivity'*. Our experience, which Avenarius sees as a system of perceptions that integrate and correct each other mutually, and can be supplemented by deduction, *is no more than a representation* and is not 'how the external world or in general what exists actually is, but only how it is conceived'.[1]

For Petzoldt, eliminating the concept of substance means abandoning the idea of a reality that is independent of our consciousness and which our consciousness should try to grasp. Everything that exists does so only as our experience of it. Petzoldt's argumentation and his terminology do not appear to differ significantly from the philosophical position launched by Kant. Even though he sees the Kantian notion of the 'thing-in-itself' as a last refuge in which the idea of substance as the 'unknown and indeterminable cause of sensations and thoughts' can survive,[2] he appears to move even closer towards a perspective that is very similar to Kantian 'agnosticism'. That is, although we are unable to deny the existence of substance, the impossibility of accessing the substance of things through experience means that we must limit ourselves to the subjective phenomenological domain of representations.

Avenarius on idealism and realism

'Zu Richard Avenarius' Prinzip des kleinsten Kraftmasses und zum Begriff der Philosophie' was published in 1887 and is thus based entirely on Avenarius's 1876 *Prolegomena*. However, between 1888 and 1891 Avenarius published his more mature works, the *Kritik der reinen Erfahrung* (Critique of Pure Experience) and *Der menschliche Weltbegriff* (The Human Concept of the World), in which he adopted a different position from that of his early work. As these differences are reflected in the works that Petzoldt publishes from the 1890s onwards, we need to examine them in greater detail to understand how they were received by Petzoldt.

In the Preface to the *Weltbegriff*, Avenarius himself provided some indication of how his thinking had evolved when he writes that in his first work, the *Prolegomena*, he had adopted 'an "idealist" standpoint', before deciding to take 'a "different path" [...] that probably one is apt to label as "realist"'.[3] In particular, the change of direction in the *Kritik der reinen Erfahrung* was 'to abandon the classical position that assumes the "immediate givenness of consciousness", as it is the (anything but "certain") result of a *theory* that is all yet to be proved'.[4] Indeed, as we have seen, in the *Prolegomena* Avenarius had held that eliminating the concept of substance implies the need to relate human knowledge to the content that is immediately available to the consciousness, that is, sensations. This was an 'idealist' position, as all human knowledge was a matter that was purely internal to the subject, making it impossible to posit an external reality as the origin, measure or reference point of knowledge. However, this 'theoretical idealism' adopted in the *Prolegomena* turned out to be 'unfruitful' when applied 'in the field of psychology', where the 'facts of consciousness' cannot be addressed by starting from the immediate givenness of consciousness but must be regarded from the perspective of the 'the relationships between the environment and the human central nervous organ', thus requiring a 'realist' perspective.[5] In the *Kritik*, therefore, Avenarius had rejected the 'idealism' of his first work in favour of addressing 'experience and knowledge' from 'the perspective of experimental psychophysicists' or of 'the doctor specialising in physiology and psychiatry that does not concern himself with what schools of philosophy have to say' when he simply attributes certain mental content to certain cerebral processes.[6] Avenarius thus came to elaborate a theory of knowledge that was based not on the epistemological precedence of consciousness but on its dependence on cerebral activity.[7]

Having abandoned the 'idealism' of the immediate givenness of consciousness in favour of the 'realism' of the dependence of consciousness on the interaction between the brain and the environment, in the *Weltbegriff* Avenarius returns to the problem of the relationship between the two positions. In particular, this problem arises from the need to recognize that 'idealism [can be] considered an absolutely inevitable consequence precisely of the physiological conception of the relationship between our "sensations" and stimuli, that is, between our "consciousness" and the environment'.[8] It is no coincidence that the Kantian transcendental idealism that saw a revival in the nineteenth century, especially among scientists, was sparked by the discoveries made in the field of psychophysiology by Johannes Peter Müller and his students.[9] It is this background that Avenarius refers to when he writes:

> I believe that I can state, on the basis of personal observation, that there is an entire legion of advocates of philosophical idealism, *trained in the natural sciences*, that would be relieved to see their earlier 'realism' re-established, and would happily let this happen if they only knew how to escape "idealism" with a clear conscience – from a logical point of view. But for them is an unavoidable fact that, as soon as one reflects on things, one arrives at the schema of cause and effect, where things

are the cause and 'sensations' = 'perceptions' = 'phenomena of consciousness' are the effect; and it is an unavoidable fact that these are 'idealist' values and these 'idealist' values are 'what is immediately given,' and therefore 'the only thing that is given' from which one might be able to 'deduce' 'what exists outside consciousness', even though 'what is deduced' would once again only exist 'in our consciousness'.[10]

Avenarius thus conceded in the *Weltbegriff* that sticking coherently to a 'realist' path had led to him 'going round in circles', finding himself once again 'in the "arid landscape" of philosophical idealism'.[11] In an endeavour to escape this vicious circle, he decided to put aside the various gnoseological conceptions and go back to the 'natural concept of the world' that precedes all philosophy:

Philosophy attempts to use particular *theories* to teach me what existed at the start of my 'spiritual' evolution; but I myself can immediately say what existed at the start of my philosophising [...]: I found myself, together with all my thoughts and feelings, in an environment. This environment consisted of numerous components in innumerable dependency relationships one with another. Other men with innumerable assertions were also part of this environment; and what they said was also largely in a relationship of dependence with the environment. Moreover, these men spoke and behaved as I did: they answered my questions as I did theirs; they sought out or avoided various components of the environment, modified them or sought to keep them as they were; and they described in words what they did or did not do and explained their reasons and intentions in doing what they did or did not do. All this just as I did, for which reason I had no other thought than that other men are beings just as I am and I myself am a being just as they are.[12]

A natural concept of the world of this kind, one that precedes any philosophical interpretation, becomes the fundamental content on which philosophical positions are based, in the sense that every philosophical position is a 'variation' on this view of the world. Avenarius does not see this natural concept of the world as immutable dogma. However, the key point is to consider whether any change is actually necessary, and whether theories that propose variations to it are able to provide a more advanced concept of the world. According to Avenarius, the history of philosophy demonstrates that idealism and metaphysical realism are inherently unable to provide a lasting answer to the question of the nature of the world, as they always create new unresolvable problems that require further variations, without ever arriving at a *stable* solution. The goal of Avenarius's position is therefore to definitively reinstate the natural concept of the world, neutralizing the elements that lead to variants of it.

In particular, the source of possible variation lies in the fact that the natural concept of the world does not only include experiences but also a hypothetical element contained in the assumption that 'other men are beings just as I am and I myself am a being just as they are'.[13] Avenarius therefore devoted a large part of the *Weltbegriff* to establishing how this hypothesis should be interpreted to avoid falling into the trap of 'introjection'. The risk is to assume that other men are beings like me in the sense that their words and actions have meaning as they refer to what happens *inside* them. This would lead to the idea that there is something like a special dimension of interiority that is distinct from external reality, whereas the natural concept of the world knows nothing of any supposed distinction between the external world and the internal world.[14]

Subjectivism as a theory that alters the original datum

The rejection of the 'idealist' perspective expressed by Avenarius in the *Kritik* was immediately adopted by Petzoldt when he read the work. A further significant role was probably also played by his face-to-face conversations with Avenarius in the summer of 1888, which Petzoldt spent in Zurich as Avenarius's guest. In the words of Petzoldt, his time with Avenarius was what allowed him to free himself 'from the last vestiges of neo-Kantian idealism', leading him to definitively set out on the path of 'the great shift towards the perspective of pure experience'.[15] Further confirmation comes in the form of a letter Avenarius wrote to the young Petzoldt the following year, in which he ironically asks 'whether things are still going well, from a philosophical point of view, without the convenient "idealist" formula "things = representations"'.[16]

Petzoldt notes his mentor's revised position as early as his 1889 review of the recently published *Kritik der reinen Erfahrung*. He initially writes:

> It is impossible to know, and it will never be possible to know, what the 'world out there' is like independently of my own self, what it is like 'in itself'; indeed, we cannot even know whether there truly exists anything beyond our own existence, independently of 'our internal self'. It is an idle question, we must settle for the world we have. [...] We cannot know of any other kinds of existence other than existence as sensations. Simple sensations are the elements of the 'world', and we ourselves are the creators of this world, in that thinking immediately intervenes to shape an image of the world out of the chaotic and disorderly aggregation of sensations. An inevitable consequence of this conception is that there is no reality beyond our representations; but even we ourselves, for ourselves, are merely specific complexes of sensations and representations.[17]

We appear to be reading the same words as those used in the passage on the rejection of the concept of substance cited above. However, this position is now presented rhetorically, with the sole aim of refuting it. The problem of the conception that 'immediately resolves things and processes into sensations' is that 'it is adopted too swiftly' and 'without a truly cogent reason' for discarding the earlier point of view.[18] Therefore, as Avenarius claims, idealism is an unnecessary variation on the natural concept of the world.[19]

The first place where Petzoldt goes into greater detail in his critique of what is now also defined as a 'subjectivist position' is in 'Einiges zur Grundlegung der Sittenlehre' (On the Foundations of Ethics, 1893–4). Here Petzoldt points out that the position that 'starts with "consciousness"' as what is 'immediately given' is 'the outcome of a century of evolution that has become ever more established, and is now accepted by most philosophers'.[20] Petzoldt's objection to the idea that 'it is the subject that is primary, the only thing that actually exists, and objects are available only subjectively, as objects of the "consciousness"' is that 'the difference between subject and object is the outcome of long historical evolution', and is therefore 'a theory' and not a 'fact of experience'.[21] A separation of this kind between the subjective and the objective is completely absent from primary experience. Everything that exists empirically is on the same plane. There is no such thing as privileged content, and there is nothing that is true reality while something else is mere semblance. Subjective idealism therefore errs, because the data of the consciousness have no precedence over our knowledge of external objects: 'I am immediately certain of the objects and processes of my environment.'[22] However, materialistic realism also errs in its view that physiological processes connected to the consciousness are primary. What we must recognize instead is the unity of experience, that is, that 'the red of the blotting paper on the notebook does not exist in some way that differs from the cerebral process on which it "depends"', and 'their mutual dependence is merely a connection between things that are *of the same kind* [*Gleichartigem*]'.[23]

However, there does not appear to be a fundamental difference between saying that all of reality originates as the content of consciousness and saying that all of reality originates as empirical content. Indeed, since the times of Berkeley, radical empiricism was seen as contiguous with if not indistinguishable from subjective idealism. Mach himself was criticized by his contemporaries because his view of science as the efficient structuring of the 'elements' of the world was in fact simply a different form of subjectivism, with the reality of the world dissipating into mere complexes of sensations.[24]

Petzoldt shows that he is aware of the affinity between these two philosophical movements when he concedes that 'even idealism does not hold that there is a fundamental duality between "things" and "thoughts"',[25] or when he states that he 'sees positivism as the most recent variant of idealism'.[26] Indeed, he had himself held a subjectivist position before changing his view. The key point, therefore, is the blurred boundary between subjectivism and the position later adopted by Petzoldt in the footsteps of Mach and Avenarius. In particular, it is important to understand why this position is not identified as subjectivism despite the many similarities, and is in fact adopted precisely in order to go beyond it.

Avenarius and Petzoldt agree on what the fundamental problem of subjectivism is, namely, that it leads to the devaluation of reality, an idea that human beings cannot take on board. This means that subjectivism can never become a world view that is lasting, stable and stationary. The words Avenarius uses to congratulate Petzoldt on the birth of his son are significant in this respect:

> I trust that you are grateful to empiriocriticism for the fact that the wriggling, screaming, hungry yet so adorable little being who carries your name is not just a 'mere complex of sensations projected outwards', is not a mere 'semblance', but is a magnificent example of true and proper 'actual reality', an 'existence' and a 'becoming' who will become the pride and joy of his parents, and their good fortune![27]

Petzoldt expresses the same criticism – in a more canonical form, from a philosophical point of view – in the work that we are analysing, when he outlines the difference between his own position and that of subjective idealism thus: 'We distinguish between "mental" and "material" events only for methodological purposes, *as does idealism*; but in the final analysis we *do not* consider what exists as a "fact of consciousness"; for us, it is *not* the case that there is something "subjective" about primary experience.'[28]

Unlike subjectivism, which identifies empirical data with the content of consciousness, the empiriocriticism developed by Avenarius and adopted by Petzoldt refrains from characterizing experience one way or another. We must not try to explain or interpret what exists as subjective, objective or anything else, but must limit ourselves to a purely descriptive approach to the content that we experience. As Petzoldt writes, 'The primary task of science is to provide a full description and analysis of objects and processes. But this is also its ultimate task. There is no explanation of the world [*Welterklärung*] beyond the description of the world [*Weltbeschreibung*].'[29]

Further errors of subjectivism

We have so far seen how Petzoldt, influenced by Avenarius's change of position, criticizes subjectivist approaches that substitute a particular *theory* for primary experience. Instead of limiting itself to describing what actually exists, subjectivism holds that what we experience consists of the content of consciousness, of representations, or whatever one might call them. This proposes an unnecessary variation of the natural concept of the world. Moreover, this is an unsustainable variation, one that human beings can never accommodate, one that can never be an enduring stable assumption as it cannot avoid raising new problems.

However, Petzoldt's arguments against this position do not end with the 'Sittenlehre' of 1893–4. They also open and close the two-volume work *Einführung in die Philosophie der reinen Erfahrung* (Introduction to the Philosophy of Pure Experience, 1900–4). In fact, the Introduction to the first volume opens with an attack on the position that 'the world and mind are separated by an insurmountable barrier, in that the world I know is only my representation of it, behind which is found the actual world, which is unknowable'.[30] Elaborating his argument, however, Petzoldt introduces a new element. As he puts it, 'a doctrine of this kind' is in fact 'the latest offshoot of the very old notion [...] of the impotence of the human mind'.[31]

In some ways this criticism is a return to a theme we have already seen in 'Maxima, Minima und Ökonomie' (Maxima, Minima and Economy, 1890), albeit directed against Avenarius and Mach. In this paper Petzoldt argued against the economical conception of knowledge, according to which we need to use the meagre mental forces available to mankind parsimoniously. Petzoldt already noted then that it was 'something of an error to speak of the over-abundance of phenomena with respect to the "limited means" of the mind'.[32] The return of Petzoldt's criticism of the presumed 'impotence' of the human mind thus reveals one of the driving forces of his philosophical position: his unreservedly positivist desire to assert mankind's possibility of knowing the world. Even though he had adopted 'agnosticism' at the start of his intellectual journey, over the years he will come to condemn it ceaselessly for undermining trust in the objective validity of our knowledge.[33] As he writes in the *Einführung*:

> Is it not downright illogical to set the mind a task that *by its very nature* it can never fulfil? The world would have to be something that the mind can never understand! In other words, something unthinkable! [...] The task of the mind can only be set by its own nature, it cannot be imposed on it from the outside. [...] All of the questions that mankind can reasonably put to itself can also be answered, and questions that mankind is convinced of never being able to answer, that have an unimaginable answer, are questions that have been put badly *from a logical point of view*, and will inevitably vanish over the course of evolution. The power of the mind with respect to its task is limitless. [...] Nothing is incomprehensible to the mind, there are absolutely no "limits to our knowledge of nature".[34]

The last part of this quote is a reference to Du Bois-Reymond's *The Limits of Our Knowledge of Nature*, in which he formulated his famous '*ignoramus et ignorabimus*' ('We do not know, and will not know'). We can see this reference as a desire to lay to rest once and for all this 'agnostic' position and its proponents, which had initially set Petzoldt on his philosophical path.

Petzoldt hastens to clarify, of course, that denying that there are limits to human knowledge must not be interpreted as a desire to descend back into 'Hegelian intellectualism' and its affirmation of the omnipotence of the mind.[35] His aim is rather to declare his faith in 'a new philosophy that has developed gradually and almost unnoticed over the last century' – a philosophy that 'makes experience the substantial and solid cornerstone of its edifice' and relies 'on research that is based on the natural sciences and on psychology'.[36]

From this perspective, the philosophies of Mach and Avenarius should not be interpreted as variations on the theme of the 'impotence' of the human mind but as expressions of an unreservedly positivist position. Petzoldt thus refutes any strictly phenomenalistic and/or pragmatic interpretation of their ideas,[37] emphasizing instead their epistemological optimism. Mach's principle of economy does not require us to resign ourselves to seeing our knowledge as no more than practical tools with no actual grasp of reality but is a call to base knowledge on the actual datum of our biological constitution and its specific needs. Similarly, the emphasis that Avenarius places on 'pure experience' is not an invitation to fall back on the subjective dimension of experience but a reaffirmation of our primordial ability to grasp the real world. Petzoldt therefore sees Avenarius and Mach not as part of the long subjectivist tradition but as the architects of a new philosophical movement that reinstates the human mind's claim to be able to grasp reality.

In the first volume of the *Einführung*, Petzoldt argues against the agnosticism and epistemological pessimism implicit in subjective idealism, and in the second he introduces two further arguments against this philosophical position. An indication of the first of these arguments appears in an article published between the two main volumes, *Solipsismus auf praktischem Gebiet* (Practical Solipsism, 1901). Before turning to the 'practical solipsism' of the title, the article addresses 'theoretical solipsism', which is just another way of referring to idealism and subjectivism, that is, the notion that 'I am unable to observe anything other than my own representations'.[38] The error attributed to this position is here a 'logical' one: 'if *everything* is a representation of mine and nothing but a representation of mine [...] defining something as a representation becomes meaningless' in that, as every other concept, 'the concept of representation' also only has any meaning 'in opposition to and in relation to something that is not represented, something that in its essence is *not* a representation'.[39] To illustrate this further, Petzoldt cites the image that is projected onto the retina: as *all* the images are inverted *none* is inverted, or rather, to apply the concept of inversion is meaningless given that it can only have any meaning in contrast to something that is upright.[40]

This criticism of the logical error of defining everything as something returns in the closing sections of the *Einführung*, accompanied by the same example of the retina. However, here Petzoldt adds a step that is highly significant given the later development of his position from the empiriocriticism of Avenarius to what he himself will define as 'relativistic positivism'. Indeed, he writes, 'We are only ever dealing with *relative* concepts. In the final analysis, all our concepts are relative. Trying to apply them in an *absolute* manner must therefore lead to error'.[41] Given the relative nature of our concepts, 'there is no logic in asking what the world is like as a whole, as the concept that would indicate its essence lacks an opposite concept'.[42] This means that 'The world as a whole, the totality of what is found, what is originally and immediately given is neither inside nor outside, neither appearance nor thing, neither representation nor object, neither conscious nor subconscious, neither physical nor psychical, neither I nor not-I'.[43]

The second argument against subjectivism first put forward at the end of the *Einführung* relates to the now-familiar concept of *Eindeutigkeit*. Petzoldt holds that mental constructs are not univocally determined or determinable through other mental constructs. This means that *Eindeutigkeit* can only be preserved if 'alongside the coordinated emergence of things in the mind, we also recognize the actual existence of a reality that is independent of our perception of it, which we call physical'.[44] In other words, as mental constructs do not depend univocally *on one another*, we must assume that they depend univocally on the brain, and in particular on the brain as an organ that is in the physical world where, conversely, the principle of univocal determination *does hold*.

Petzoldt anticipates the possible objection that 'this physical connection is nothing other than a representation, an auxiliary representation created solely to see all the complexes of sensations as univocally determined'.[45] His reply is that, given that univocal determination does not exist between mental constructs, if we were really to assume that everything that exists consists only of representations, there is no way that order could emerge, that a world could emerge, from this chaos, from our disorderly stream of representations. It is possible only because the world itself is not in chaos but is univocally determined, and because our brain, on which these mental constructs depend univocally, is part of that univocal determination, and in particular has evolved as a consequence of it, having thus etched univocalness into itself. Petzoldt uses an example to elucidate his argument:

> Consider the onset of an expected lunar eclipse. For an unbiased individual, this is a simple matter: the sun, the moon and the earth exist independently of his

perceptions of them, and they change their relative positions in a certain specific way. After a given period of time, that is, after a quantifiable number of rotations and orbits, the moon must enter the shadow of the earth. The day and the hour are announced, so anyone can observe the event in a completely natural way. Numerous factors, of which the observer is unaware but which are no less concrete for that, are in play here so that the perception of the natural event can occur. For the subjective idealist, all these factors must be excluded completely. For him, the only actual data are his own few complexes of sensations, and everything else is just an interpretation serving to connect them. The sun, the moon and the earth are mere conceptual constructs, as is their motion relative to one another. The work of the astronomers, their instruments, tables, maps, publications, research results, the entire history of astronomy, etc. are all only representations whose purpose is to make the few and insignificant complexes of sensations *understandable*. Perhaps the idealist has not spent weeks thinking about the expected 'natural event' – therefore his past experience has not prepared him for it – and it is only 'by chance', via an unexpected 'announcement' that 'catches his attention' (which in turn requires an enormous representational apparatus if it is to be understood) that he actually perceives the 'eclipse', that is, he experiences a sequence of visual sensations that could be described as the decrease and subsequent increase in the size of a whitish 'disc' against a dark 'background'. Every connection is obtained only by constructing 'empirical concepts' from the chaotic sequence of sensations. But how is this possible, unless we want to say that subconscious representations are the means of determination of real events? Clearly, it is possible only if that chaos of actual impressions is in fact not chaos at all, but is summoned to the playhouse of our perception by some stage director, so that our representations are guided, ordered and linked, and merge with the complexes of sensations until a meaningful complex is created. The idealism of sensations thus inevitably exceeds the constraints imposed by its own premises.[46]

In other words, if subjectivism needs the 'conceptual structures' of the planets, the laws of gravity and so on to explain the occurrence of mental content, like the perception of an eclipse, it fails to demonstrate that everything can be explained through representations, that everything happens in the mind. On the contrary, the need for these 'conceptual structures' proves that even subjectivism must refer to something that is *not* in the mind (the planets, the laws of gravity etc.) in order to explain mental constructs univocally. If subjectivism were *truly* based on subjective data *alone*, it would not be able to explain any lawfulness or any univocality as emergent from these subjective data. As Petzoldt puts it, 'The *object* cannot be constructed from the *subject*'.[47]

The principle of *Eindeutigkeit* thus requires the subjectivist view of the world to be renounced, because, if we reject the existence of a real order beyond the sequence of representations (which depend on it), then we cannot explain why, from among the infinite number of possibilities, *one* specific mental representation appears at a given moment, followed by *another* specific mental representation, again from among the infinite number of possibilities and so on. The only way that subjectivism could provide a univocal explanation of the disorderly sequence of mental content would be by arbitrarily assuming other subconscious content to fill the gaps in the sequence of representations, so as to univocally explain those that actually exist in consciousness. However, subconscious representations, as they are not and cannot be experienced, cannot be a valid scientific explanation.

Therefore, rather than limiting the scope of our reflections to the subject and the subject's mental content, we should simply assume the existence of reality, of the world, not as what exists absolutely *beyond* this mental content, which would mean falling back into materialistic realism but as what exists *together* with it.

The evolution of relativistic positivism

The coining of the term 'relativistic positivism'

Having outlined Petzoldt's negative arguments against subjectivism, we can now turn to the positive part of his reasoning, which aim at the definition of a new philosophical system. This task is entrusted to the closing sections of the 1904 *Einführung* and the 1906 *Weltproblem*. In later years, Petzoldt continued to expound this position without effecting any substantial changes to it,[48] if we exclude his attempt to integrate it with Einstein's theory of relativity, which we will address in the next chapter. In defining his philosophy, Petzoldt now abandons the term 'empiriocriticism', which had grown out of the circle of Avenarius and his students,[49] choosing instead the term 'relativistic positivism'. We will see below how this expression describes Petzoldt's philosophical system. At this point, we will simply show how he gradually arrived at this label.

The term 'positivism' first appears in the second volume of the *Einführung* (1904), where Petzoldt refers to Mach and Avenarius's 'new direction of positivist thinking' in opposition to idealistic thinking. Petzoldt now also mentions Wilhelm Schuppe as an exponent, to whom we will return below.[50] Although it is only mentioned twice in the whole volume, positivism earns its own entry in the index to the work, a sign that Petzoldt attributes a greater weight to it than its scarce mention in the text might suggest. Indeed, when *Das Weltproblem vom positivistischem Standpunkt aus* (The Problem of the World from the Standpoint of Positivism, 1906) is published two years later, the term is promoted and appears in the title, en route to contributing to the definition of Petzoldt's philosophical position.

The decision to favour the term 'positivism' over Avenarius's 'empiriocriticism' could be explained by the measures Petzoldt took at the start of the century to gain his *Habilitation* from the Technical University of Berlin. As we have noted, the term 'positivism' does not appear at all in the first volume of the *Einführung* of 1900. Petzoldt first tells Mach of his 'personal plan' to gain this professional qualification from TU Berlin in May 1901, having resigned himself to the impossibility of securing a position as a lecturer at the more prestigious Friedrich-Wilhelm-Universität (today, the Humboldt-Universität), given the resistance he came up against.[51] In later letters in which he tells Mach about how his project is progressing, Petzoldt frequently refers to the effort he is putting into convincing the upper echelons of the TU Berlin not only of the value of offering courses in philosophy, which did not exist at that point, but in particular of the value of offering courses in positivist philosophy to students of the natural sciences.[52] In the same correspondence with Mach, Petzoldt also frequently expresses his impatience over the publication of the *Einführung*, which he needed in order to prepare the application for his *Habilitation*.[53] We can therefore hypothesize that Petzoldt wanted to portray himself not only as a member of Avenarius's school, which would probably have been the case if he had retained the term 'empiriocriticism', but more generally as an exponent of positivist philosophy, in order to make his *Habilitation* more palatable to the university authorities. This would explain the first occurrence of the term 'positivism' and its subsequent growing importance in Petzoldt's works.

As for the term 'relativism', it is interesting to note that Petzoldt already uses it in a letter to Mach in 1894, during one of their early exchanges. Thanking Mach for having sent him one of his works,[54] Petzoldt refers to the new philosophical direction launched by Mach and Avenarius, and – expressing his appreciation of Avenarius – he adds that he reached 'a perfect [*vollkommen*] relativism'.[55] The same opinion reappears in the final chapter of the second volume of the *Einführung*, a chapter devoted to the general importance of the work of Avenarius. Replying to Wundt's accusation that Avenarius's position was an unwelcome return to metaphysics, Petzoldt emphasizes that the key feature of metaphysics is the assumption of something 'absolute', a feature totally alien to Avenarius's philosophy, 'whose principal position was full relativism and who saw any metaphysical speculation as a futile waste of time, concurring fully with Mach's anti-metaphysical work'.[56]

Apart from these two sporadic but nevertheless significant instances, the term does not reappear in Petzoldt's works until the publication of the first edition of the *Weltproblem* in 1906. As we will see below in our more detailed examination of this work, Petzoldt regards relativism as the philosophical position that argues against metaphysics and its endless search for the substance of things. Indeed, relativism is added to the title in the second edition of the work, which becomes *Das Weltproblem vom Standpunkte des relativistischen Positivismus aus* (The Problem of the World from the Standpoint of Relativistic Positivism, 1911).

The decision to emphasize the 'relativistic' element of his conception was probably also driven by Petzoldt's goal of underlining the inherent connection between the philosophical position that he saw himself as the main advocate of and the theory of relativity proposed by Einstein in 1905. Einstein is cited explicitly in the second edition of the *Weltproblem* to claim that the 'relativization of time' proposed by Minkowski and Einstein proves that Mach's ideas were not an obstacle to physics research, as Max Planck had affirmed. Indeed, it was precisely the relativization of time and space first proposed by Mach that had opened the way for Einstein's proposals. The link between the Mach-Avenarius 'relativism' advocated by Petzoldt and Einsteinian relativity is also addressed in Petzoldt's next work, *Die Relativitätstheorie im erkenntnistheoretischen Zusammenhange des relativistischen Positivismus* (The Theory of Relativity in the Epistemological Context of Relativistic Positivism, 1912). This work marks the start of a long period, the last decades of his life, during which Petzoldt promotes Mach's philosophical position as the only position that was appropriate for the recent developments in physics, not only because it was the only position able to give an account of Einsteinian relativity but also, above all, because it was the only one to have supported it and even anticipated it.

To summarize, the terms 'positivism' and 'relativism' already appear, sporadically, in some of Petzoldt's works before the 1906 *Weltproblem*. However, it is in this work that the two become the subject of more detailed elaboration, with the term 'positivism' now being used to define Petzoldt's philosophical position. However, it is only from the second edition of the *Weltproblem*, in 1911, that Petzoldt settles definitively on the term 'relativistic positivism', a decision also influenced by the growing popularity of Einsteinian relativity.

Relativistic positivism in the history of philosophy

The origins of philosophical and scientific thinking

As we have noted, Petzoldt sees Mach and Avenarius as the exponents of a new philosophical direction that will undermine and prevail over older philosophical standpoints. The *germ* of this notion can be found in Petzoldt's earliest works. However, in the second volume of the *Einführung* and in the *Weltproblem*, Petzoldt extends his considerations to a reconstruction of the entire history of philosophy. We will not go into the details of Petzoldt's full narrative, but will focus on its most important points, given our objective of clarifying the relationship between Petzoldt's position and subjectivism.

As early as in the Preface to the *Weltproblem*, Petzoldt expresses the aim to 'clarify, from a psychological point of view, the history of reflections on the world as a history of reasonable [*sinnvoll*] errors'.[57] We have seen that Avenarius had identified the fundamental error of human thinking as introjection, which makes experience an internal process in counterposition to the supposed external nature of the world. Petzoldt concurs on this, adding, however, that there is not just one form of introjection, but two. There is the form noted by Avenarius, namely, 'the introduction […] of a soul into other human beings and animals', 'into their internal self', and more generally 'into every entity', but in addition there is 'the introduction of substance'.[58] This second form of introjection 'is unrelated to the first because it is not a consequence of dreaming or of ghosts, but of the need for stability in human thinking in the face of the multiplicity and mutability of things'.[59] The two notions, soul and substance, turn out to reinforce each other, in that substance, being 'the underlying and immutable element', needs the soul as the 'initiator of change', while the mutability of the soul requires a 'substantial fixed point'.[60]

As human beings are inherently in constant search of 'finding something that is old and familiar in anything that is new and unfamiliar',[61] they inevitably tend to relate the diversity of the world to a fixed and unchangeable substratum that serves as a bedrock for everything that is varied and variable. Therefore, 'there was no route to society and truth except through this error', and it is in this sense that the introjection of substance and of the soul should be seen as 'psychologically necessary',[62] and therefore 'reasonable' errors.

This twofold introjection characterizes human thinking from the outset. However, human thinking begins to shift thanks to the rise of scientific knowledge and its commitment to being founded on experience, to observe things without being influenced by preconceptions, that is, by the old ways of understanding things based on the notions of substance and soul. Along with science, however, comes philosophy, since there is no difference between the two disciplines except for the fact that philosophy does not observe things in isolation, but always fixes its gaze on the totality and unity of reality.[63] Given the challenge of freeing oneself of preconceptions in order to observe things as they actually are, the consolidation of philosophical and scientific knowledge could clearly not happen overnight. The history of philosophy is a long process of evolution guided by the principle of the tendency towards stability, in which human beings liberated themselves from the metaphysical notions of soul and substance, and slowly recognize experience as the only foundation of knowledge.

Petzoldt places Thales of Miletus at the start of this history of philosophical and scientific thinking. Thales had sought a first 'unitary and empirical understanding of the world', freeing himself from the old 'mythology' that looked for an 'explanation of the world in terms of factors that had not been experienced'.[64] In this sense, Thales can be seen not only as something of an 'empiricist' *ante litteram* but 'even as a positivist'.[65] However, the thinking of Thales contains all the usual errors: 'the representation of substance', 'the relationship between the two non-concepts [*Unbegriffe*] of the appearance and the thing-in-itself', as well as the logical error of '*pars pro toto*' that 'identifies the nature of the whole as that of one of its parts'.[66] These three errors are linked. People tend to seek out an *immutable substance* that is the bedrock of everything. They find it in a certain aspect or a *part* of the totality of things, which is therefore generalized to become a property of the *whole*, of substance. However, not everything can be identified using this single aspect or part of reality, so this leads to the idea that things *appear* to be different, even though *in reality* they are nothing other than manifestations of the same single and unchangeable substance.

Protagoras

Petzoldt also saw substantialist reasoning in those who came after Thales, such as in Parmenides and Heraclitus, who absolutized substance and change, respectively,[67] and in Democritus, who introduced the fatal mistake of a difference between primary and secondary properties.[68] Protagoras was the first to try to break out of this schema, and Petzoldt's portrayal of him in the *Weltproblem* makes him appear as an almost mythical figure.

> [Protagoras] was the first to understand change purely as a process, without some entity that undergoes change. [...] Nothing is absolute. Every entity acts continuously on other entities and is ceaselessly subject to the actions of other entities. Every entity exists and persists only in either doing or being done to, of action and reaction. If we take any property of any entity, it never exists in itself or for itself, but only in relation to other entities. [...] An 'entity' is not an absolutely

stable connection of 'properties,' but a relatively stable connection, and nothing else. These properties have no need of any substratum.[69]

It is clear that Petzoldt is projecting onto Protagoras the position held by Mach that 'Thing, body, matter, are nothing apart from their complexes of colours, sounds, and so forth – nothing apart from their so-called qualities'.[70] Indeed, Petzoldt goes so far as to say that 'Protagorean relativism holds that a property of an object exists only *in a functional connection* with other properties of the same object or of other objects',[71] thus attributing to Protagoras the concept of 'functional relation' developed by Fechner and later taken up by Mach and Avenarius. Perhaps aware of the strained nature of his interpretation, Petzoldt specifies that 'we do not know how far Protagoras and the other sophists in his circle took this relativism, how much they tested it, or the extent to which they succeeded in maintaining this general anti-metaphysical position with respect to individual concrete objects'.[72] At the same time, however, despite this apparent caution, Petzoldt does not shrink from identifying Protagoras as being the first to formulate 'the principle of relativism, the most significant philosophical position that has ever appeared since the foundation of science by Thales; in fact, the position that has remained the most significant one to this day'.[73]

Petzoldt does not stop at seeing Protagoras as the first relativist but goes so far as to attribute to him the same 'positivist' relativism that he himself advocates.[74] Petzoldt concedes that 'the failure of Protagorean thinking comes from the subjectivist twist it took', namely that 'there is no universally valid truth; everyone has his own measure of what is real, and another has another measure, and the same person even has different measures at different times; and each of these measures is in principle as valid as any other'.[75] However, Petzoldt also holds that 'it would be a great injustice to believe that Protagoras might have considered any conception as valid as any other'.[76] Indeed, Petzoldt adds that 'nothing was further from him than the foolish doctrine, which he continues to be held to some extent responsible for, whereby everything is valid for anyone who sees it as valid'.[77] Above all, according to Petzoldt, 'Protagoras never entertained the possibility of making the actually given world disappear, reducing it to a mere representation'.[78]

Petzoldt summarizes the real doctrine espoused by Protagoras thus:

> The rose 'has a scent' only through an olfactive organ. Honey is 'sweet' only when in contact with the tongue. A space is either warm or cold according to whether one comes from a warmer or a colder space. These predicates already implicitly contain their relationship to ourselves. And it is the same for all other predicates. [...] If then the world depends only on ourselves, our view of the world must also comply with the individual characteristics of each of us. The world of a blind person is completely different from that of one who sees, that of the short-sighted is different from that of the long-sighted. But *asking how the world actually is*, whether it is as the blind person conceives it, or the short-sighted person, etc. or in any other way, *is meaningless*, because it would mean once again removing the above-noted dependence of things on ourselves. Nothing can exist in itself, each thing is always in immediate connection with many other things, and thus

with all things. Not even a property as fundamental as shape can belong to a body independently of other properties. […] Just as nothing exists in itself and for itself, so the totality of things, the world around us, does not exist in itself and for itself, but only ever relative to a subject that perceives it. […] All perceptions of the world, all our intuitions about the world, *as they are equally actual so they are equally true*, in the same way that perspectives of an object seen from different points of view are absolutely equally valid and equally correct. If instead we direct our thoughts to how the world actually is with no reference to a subject, to how it is in itself and for itself, we can reach no truth at all, but only opinions. […] But here – and *this* is Protagoras' deepest conviction – opinions are not all equally true, but are all equally false, because they all have the same impossible and self-contradictory aim of grasping the world without a subject to grasp it.[79]

For Petzoldt, although Protagoras's relativism recognizes that the perception of the world, and our knowledge of it, is always relative to the individual that perceives it, it is not a form of subjectivism. This is because it rejects the notion that personal knowledge of the world of this kind provides only the appearance of how things are, in contrast to how they might be in themselves and for themselves. If everything is as it is relative to other things, then even the relationship between things and whoever perceives them is definitely real. Although Protagoras 'denies that perception via the senses is universally valid, he does not deny its lawfulness [*Gesetzmässigkeit*]'.[80] In other words, the structure of the world is such that it must *necessarily* be perceived in a certain way by an organism with certain organs that operate in a certain way. It is therefore real, true and necessary that it appears one way to a blind person, a different way to a short-sighted person, a different way to a long-sighted person and in yet another way to a person with unimpaired vision. This does not mean dismissing the impressions of these individuals as illusory falsehoods that are 'merely' subjective but instead recognizing that this is how things actually are.

Petzoldt thus attributes to Protagoras the idea that human cognition cannot be reduced to subjectivity because our relationship with the world is governed by necessary and objective laws. Moreover, Protagoras also sustained the validity of human knowledge from a different point of view. According to Petzoldt, it is unthinkable that a man of Protagoras' stature 'would have forgotten that logic demands that any knowledge that is generally accepted must meet certain conditions' in order to accommodate the 'psychological foolishness of holding that, in certain circumstances, any opinion can be judged as "true"'.[81] In other words, Protagoras did not overlook the fact that 'science is a *social construct*'.[82] Protagorean relativism thus went beyond recognizing individual experiences and the differences between them and should be understood as an invitation to focus on 'what is universally valid and connects the thinking of all sound minds [*das Denken aller Gesunden*]'.[83]

In conclusion, Petzoldt maintains that, from its first formulation by Protagoras, relativism is not the same thing as subjectivism or scepticism. Rather, relativism can and must be interpreted as a form of positivism that does not deny the possibility of knowing objects but rejects the notion that something like reality-in-itself exists and holds that the existence of objects relative to ourselves *is* in fact reality.

Even if Petzoldt sees identifying relativism with subjectivism and scepticism as an error, this error is understandable if one considers how difficult it is for individuals to satisfy the relativist demand of 'breaking with the representation of substance in all its forms', that is, with 'the most deeply rooted of preconceptions'.[84] As we feel a biological and psychological need to postulate an immutable substratum beyond our representations, a doctrine that rejects this notion appears to negate the soundness of our knowledge, whereas its objective is exactly the opposite, namely, to assert that our knowledge of reality is true and valid, in that there is no reality that exists in itself in an absolute and isolated manner.

As a result of this misunderstanding of Protagorean relativism as subjectivism and scepticism, human thinking missed the opportunity to embark on the road to true knowledge much earlier than it did.

> If this bright Greek mind, unhindered by darkness or phantoms, had persisted in the awareness of those who came after him, thus leading to further elaboration, he might have merged with the ethics of Jesus just as well as what actually took place on the basis of the absurd fantasies of Plato and Aristotle. He might have saved the Greece of Pericles, Phidias, Sophocles and Euripides both from the abyss of perverted sophism and from the mirage of the dazzling doctrine of ideas, and humanity would have been spared the tyranny of ecclesiastical dogma.[85]

Hume

Continuing his reconstruction of the history of philosophy, Petzoldt observes that Anaxagoras did not follow Protagoras's line but held on to the concept of substance and added a new error to it: the separation between the corporeal and the mental. Petzoldt emphasizes that for Thales and the early natural philosophers 'there was no absolute opposition between living and not living, between animate and inanimate', because 'substance itself was living, animate', in the sense that 'the living being was a *property* of matter'.[86] 'In Anaxagoras, on the other hand, the animate living being only belongs to a *part* of the world', that is, to the 'spirit', which is distinct from 'dead matter, incapable of movement, simply the stuff of which the world is made'.[87] These errors were perpetuated over the course of the development of philosophy in a variety of forms. Petzoldt notes that Plato and Aristotle got lost in their desperate search for substance, and observes that 'the philosophy of Plato and Aristotle is one of the principal factors, if not the principal factor, in the decline of Western science and technology, and thus in the onset of the Middle Ages'.[88]

Petzoldt then goes through the main protagonists of the history of philosophy, showing how their ideas were based more or less explicitly on the notions of substance and the soul, which was seen as the vital force in contrast to inanimate matter. He recognized, however, that some writers introduced valuable innovations. Descartes, for example, although he had postulated the principle of the immediate givenness of consciousness and held that the brain was the locus of all thinking, thereby creating the false problem of the opposition between the internal world and the external world,[89] was the first to formulate, thanks to this very opposition, the notion that 'the course of organic processes can be fully understood without reference to the soul, just as in the case of inorganic processes'.[90] This view was taken further by Spinoza, to whom we must attribute the 'great leap forward' of having 'eliminated the so-called interaction between the body and the mind [...], thus opening the way to modern thinking'.[91] Petzoldt credits Berkeley with dispensing with the distinction between primary and secondary properties, which 'is completely unjustified, given that I perceive both types in exactly the same way, and can make no distinction between their modes of existence'.[92]

However, the first to properly pick up the thread of the Protagorean line of thinking was Hume. Petzoldt credits him with having 'rediscovered the fundamental error of philosophical thinking, the error of the representation of substance'.[93] Before Hume, philosophers were still engaged in trying to explain the relationships between inanimate matter and the mind. This opposition was never under discussion, whether by those who derived everything from the mind (Berkeley), or by those who derived everything from matter (Hobbes), or even by those who held an intermediate position (Locke).[94] Hume disrupted this rigid framework by showing that 'thorough thinking can find in experience neither matter nor mind'. There is 'neither some kind of immutable substratum of complexes of sensations that we call things', nor any mental principle 'to which our perceptions and representations are bound'.[95] Moreover, Hume was the first to rid himself also of the 'old representation of causality', which 'alone would suffice to place [him] among the greatest thinkers of all time'.[96]

With respect to the almost mythical attributes Petzoldt assigned to Protagoras, his portrayal of Hume is more nuanced. Unlike Protagoras, Petzoldt saw Hume as having, at least partially, fallen back into subjectivism and scepticism, thus demonstrating that he had not freed himself completely of the representation of substance. The proof is Hume's identification of 'what is immediately given' with 'the content of consciousness, subjective experience, the internal world', which follows precisely the line of the philosophical tradition launched by Descartes.[97] As he starts from a subjectivist base, Hume asks how it might be possible to know reality, coming to the conclusion that 'reason cannot prove the existence of the external world, nor can the senses assure us of it', and therefore all that lies beyond our consciousness is merely an 'indemonstrable and arbitrary supposition'.[98] Subjectivism thus leads Hume to scepticism.

Nevertheless, Petzoldt considers this to be simply one aspect of Hume's thinking, which he sees as having a kind of internal fissure. On the one hand, there are his philosophical reflections, the 'theory', which induce him to adopt the subjectivist-idealist position whereby the content of the consciousness is the alpha and omega of human knowledge. On the other hand, there are his deep convictions, which are aligned with the 'realism of the common man'.[99] Indeed, Hume never truly questioned our knowledge of the reality of the world, at most considering how it could be justified philosophically. As he was unable to base it on reason or on the senses, he resorted to the imagination, seeing it as 'the ultimate judge of all systems of philosophy'.[100] This interplay between the two aspects of Hume's position was also at work in his conception of causality. Petzoldt emphasizes that Hume 'never believed or stated that an inviolable conformity with laws was not present in nature' but had merely shown that 'the intellect has no reason to presuppose it' and therefore that 'its roots should not be sought in the intellect but in the imagination'.[101]

Petzoldt concludes that 'Hume is not a positivist, however much he prepared the ground for positivism thanks to his compelling criticism of representations of substance and causality, his struggle against any kind of metaphysics, and the rigour with which he insisted on the role of experience'. At the same time, however, Hume 'is not a radical sceptic', as the concept of belief allows him to accept everything that is produced by the imagination.[102] The limitation of Hume's approach is that, 'having resolved the things of the "external world" and the past experience of the "internal world" into homogeneous fundamental elements', he should have 'then designated those elements as completely undifferentiated initially with respect to the internal and the external'. This would have led him to 'the unavoidable conclusion that the external world deserves exactly the same recognition as the internal world' as 'both worlds emerge from undifferentiated elements through reciprocal correlation and differentiation'.[103] In other words, Hume did not take the additional step taken by Mach and Avenarius (and Schuppe) towards what is today called 'neutral' monism, the doctrine that the fundamental components of the world are neither physical nor mental until they become differentiated as such.

Kant

If Hume picked up the thread left by Protagoras, Petzoldt sees Kant's position as a clear retrograde step in the evolution of thinking. Kant not only 'leaves Hume's idealist premise intact' as he 'constrains the possibility of scientific knowledge to the phenomenal world'[104] but his approach also revives substantialist thinking in that it asserts that 'the knowledge/experience […] must be understood as the result of two components', 'the unknowable thing-in-itself and the unknowable I-in-itself', which are nothing other than 'particular modifications of nature and the soul, of material and mental substance'.[105] Thus 'the actual reality that opens up before us is to be understood on the basis of something that Kant's own doctrine states is unfathomable by man'.[106] Instead of resolving the problems that Hume had left open, namely, 'the conflict between imagination and intellect', Kant had simply 'created a new and much deeper abyss between phenomenon and thing-in-itself'.[107]

According to Petzoldt, therefore, Kantian doctrine is anything but a basis for knowledge, as it rests on two absolutely unknowable pillars: reality-in-itself and the I-in-itself. Referring sarcastically to the slogan '*Zurück zu Kant!*' (Back to Kant!), which had become a rallying cry for the adoption of Kantianism as a theory of knowledge since Otto Liebmann's 1865 conference on *Kant und die Epigonen* (Kant and his Followers),[108] Petzoldt writes that 'a consequence of the famous "back to Kant" is that anyone who still hopes that mankind can achieve a fundamental knowledge of actual reality is met with sarcastic and pitiful laughter'.[109] Considering the frustrations Petzoldt experienced while at the Berlin *Philosophische Gesellschaft* (Philosophical Society), when he constantly came up against an audience consisting mainly of followers of Kant, these words are to be understood as autobiographical rather than rhetorical.[110]

This reference to Liebmann is not, however, the only passage in which Petzoldt comes up against nineteenth-century neo-Kantianism. A little further on, Petzoldt emphasizes that the attempt to sustain Kant's position and at the same time 'avoid transcendence' leads to the 'escape route of seeking out the requisite dispositions of the mind' (the a priori forms of knowledge) 'in the organization of the human brain'.[111] Although it is not cited explicitly, it is clear that Petzoldt is referring here to the reinterpretation of transcendental philosophy proposed by Lange in his *History of Materialism*. Lange believed that the conditions of possibility of experience could be identified with our 'psychological' or 'psychophysical organization', that is, the structure of the organs responsible for human cognition.[112] According to Petzoldt, the inevitable result of this position is that:

> the very *factor that is supposed to make the experience at all possible* is made to rely on *a part of this experience*, namely, the organization and functions of the brain. But if the lawfulness of events is first introduced into the world by the brain, one wonders how it was possible for organisms to develop in a way that resulted in the brain.[113]

In other words, if – as stipulated by Kantianism, including that of Lange – the necessity of the course of nature depends only on the subject and cannot be attributed to nature itself, when this subject is identified as a product of nature, as in the case of the configuration of the brain through evolution, one must ask how a reality devoid of necessity can lead to a system that requires necessity as a condition of possibility of experience. Petzoldt maintains that the only solution is to concede that reality itself contains within itself the necessity of events or, more precisely, *Eindeutigkeit*, and that this is why the brain, an outcome of evolution, is inherently subject to the requirement of *Eindeutigkeit*. However, this means conceding that 'the Kantian doctrine of the full lawfulness of everything that exists or happens is ultimately based on facts', thus renouncing the very essence of Kant's Copernican revolution.[114]

The interesting point of this discussion of Lange's position is that it reveals Petzoldt's intellectual development and his move away from his initial neo-Kantianism, in particular Lange's physiological reinterpretation of it. Petzoldt retains Lange's concept of a psychophysical 'organization' rooted in the functions of the brain as a valuable remnant of Kant's notion of forms of knowledge.[115] However, he also takes the decisive anti-Kantian step of attributing the necessity of nature to nature itself, and only secondarily to the conditions of possibility of experience embodied in the organization of the brain. Even though the main aim of Petzoldt's approach was to defeat Kantian subjectivism, his bitter diatribe recognizes two fundamental merits of the Kantian position. The first is that Kant had brought a certain degree of unity to the opposition between the material world and the mental world that typified dualism by including both subject and object within the phenomenon, albeit at the cost of postulating the unknowable thing-in-itself alongside the phenomenon. Kant's notion of the phenomenon leaves no room for the old opposition between objective reality, which only has primary qualities, and subjective perceptions, which display secondary qualities. For Kant, space and time, as well as sensations, belong to the reality-for-us that is the phenomenon, in which subject and object are in a perpetual relationship as indispensable components of all experience.[116]

The second merit of Kant's position, according to Petzoldt, is the view that 'the objects in phenomena are independent of specific acts of perception and more generally of the perceiving individual'.[117] Although Petzoldt emphasizes that he 'cannot concur with the reasons for which Kant made this conceptualization possible' – that is, the doctrine of the *I think* as a transcendental, impersonal and universal subject – it provides a solution to the problem of how we should consider objects in the world when we cannot perceive them, which is fundamentally the same position as the one that Petzoldt adopts. In Kant's view, or rather, in Petzoldt's take on Kant, 'objects – albeit only as phenomena – are still there for every individual even when they are not perceived and, in particular, everyone has the right to represent them as they would be if he was actually observing them'. This is a 'significant step towards a revalidation of the convictions of "common sense" [*gesunde Menschenverstand*]', that is, towards 'naive realism'.[118] Indeed, when Petzoldt later speaks of his own position, he identifies two points of contact with the common man: 'the object perceived is not just a complex of projected sensations behind which the actual object is hiding' and that 'the object perceived continues to exist even when we cease to perceive it',[119] and in particular it continues to exist 'with all of the properties that it had when I perceived it'.[120] Kant and Petzoldt thus are in agreement about the common man being correct when he thinks of objects as existing with the same empirical properties they have at the point at which they are experienced.

Our aim here is not to determine whether this is actually the position maintained by Kant (assuming it were even possible to provide an unambiguous interpretation of his views). Our focus is not on Kant himself but on how Petzoldt's interpretation of Kant contributes to an understanding of Petzoldt's own position. We will therefore focus a little longer on the question of the existence of objects beyond their empirical givenness. This is a key issue in Petzoldt's relationship with the Kantianism of the times, and a central theme of his reflections in the years between the writing of the second volume of the *Einführung* and the *Weltproblem*, the works that laid out his philosophical position. This is demonstrated by his correspondence with Mach in 1905, where Petzoldt returns several times to the problem of the 'nature of the existence of complexes of sensations when they are not being perceived'.[121] Moreover, the problem of the existence of things independently of our perception of them must have been particularly thorny in Petzoldt's relativist framework. If reality-in-itself does not exist, not even as a limiting idea, but everything exists purely as an ensemble of reciprocal relations, it becomes difficult to explain *whether* and *how* we should conceive of things when they are not in a relation with ourselves. Petzoldt does not hide from the difficulties stemming from his theory, but proceeds to analyse them systematically in the second volume of the *Einführung*.

It is clear that Petzoldt cannot assert that it is possible to know how things are beyond our own cognition of them, as this would mean returning to an absolute and metaphysical view of reality. However, nor can he assert that we do not know how things actually are, as that would mean a return to Kantianism. The result is therefore that:

> things *are* as they *appear* to us. Their *elements* are not molecules or atoms, but are red, blue, hard, soft, angular, oblong, round, fleeting, enduring, etc. […] They *are* exactly as we find them in different circumstances, that is, complexes of these elements.[122]

However, this view of the world, which follows Mach's line, leads to the following question: 'How can we conceive of the existence of unperceived things?'[123] If we believe that this means how the world is in itself and for itself, then 'the question is meaningless', because 'conceiving of the world independently of conceiving is an impossible undertaking and a self-contradictory one': 'a blind person cannot conceive of a world of colour' and 'a normal person can only conceive of the world in his own way'.[124] This would seem to require agreeing with the idealist, who denies the existence of a reality-in-itself and reduces every object to its conception by the subject. Indeed, Petzoldt himself acknowledges that an 'idealist' might object that Petzoldt is simply 'stealing his job, as all of [the idealist's] effort is actually directed at showing that the world cannot be thought of independently of how we represent it; and if it is thought of as dependent on how we represent it, then it is merely a thought world, and therefore only a representation'.[125]

However, Petzoldt agrees only with the first part of the idealist's reasoning, rejecting the final conclusion, as 'thinking of the world and creating a representation of it do not turn the world into a thought [*Gedankending*], a representation'.[126] In other words, the fact that we cannot think of the world independently of thinking does not mean that the world depends on thought, which is the idealist's erroneous conclusion. Petzoldt holds that it would only be possible to assert that the world is just a representation if 'it exists for as long as I think of it, while what I do not think does not exist at all'.[127] It would thus only make sense to maintain that the world is just a representation if one concedes that it disappears when I close my eyes. However, although Petzoldt and the idealist agree that the world depends on our representation of it, in that we cannot conceive of it independently of that representation, no one actually believes that it depends on that representation in the sense that our minds have the power to actually make it cease to exist.

In any case, as the 'idealist' cannot really maintain that the world is a representation *in this latter sense*, the line separating the subjectivist view and Petzoldt's view turns out to be as ephemeral as ever, as further demonstrated in Petzoldt's elaboration of his argumentation:

> We do not consider the world to be dependent on the thinking of the individual or of individuals. Or, more precisely, we do not consider it to be dependent on *acts* of thinking or on some *actual* thinking, but consider it to be dependent – in a purely logical sense – on *thinking in general* [*von dem Denken überhaupt*]. The idealist confuses the two, and the result is agnostic semi-solipsism.[128]

Petzoldt's position becomes indistinguishable here from the (neo-)Kantian doctrine of '*Bewusstsein überhaupt*' (consciousness in general), in the sense of a pure form of thinking in general, decoupled from any relationship with the empirical knowledge of any individual subject.[129] This notion became the object of the reflections of several writers at the end of the nineteenth century and the start of the twentieth,[130] including Emil Laas,[131] Heinrich Rickert[132] and Wilhelm Schuppe. In 1902, Petzoldt becomes acquainted with Schuppe, moving closer to his position and eventually placing him firmly alongside his mentors Mach and Avenarius as key figures in the new direction of human thinking, as we have seen and as we will now examine in greater detail.

The exponents of the new philosophical approach

Schuppe

Wilhelm Schuppe (1836–1913) had a lifelong academic career at Greifswald and was the prime exponent of the 'theory of immanence' or 'immanentism'.[133] He first laid out his approach in the hefty volume *Erkenntnistheoretische Logik* (Gnoseological Logic, 1878), and then provided a summary in the slimmer *Grundriss der Erkenntnistheorie und Logik* (Outline of the Theory of Knowledge and Logic, 1894). The fundamental idea of his approach was that the whole world presents itself to the consciousness and in it, but does so to consciousness in general, the impersonal '*Bewusstsein überhaupt*', rather than to individual consciousness. The whole of reality, every individual 'I' and the world, can be found within this general consciousness. Precisely because nothing exists beyond *this* consciousness, the Kantian concept of the unknowable thing-in-itself should be rejected. However, this does not mean a return to idealism, as consciousness in general does not coincide with individual consciousnesses. For Schuppe, the fundamental error of subjectivist idealism, the error which inevitably leads it into solipsism, was that of confusing the two senses of consciousness, thus turning the whole of reality into no more than a representation created *by an individual subject*.

As in the case of Avenarius, Schuppe's works were also noteworthy for their intractable style. This meant that they were ignored, or misunderstood as representing the very subjectivist idealism that Schuppe was combatting. This prompted Schuppe to write an open letter to Avenarius in 1893, which was published in his journal. It was designed to demonstrate the affinities between his own philosophy of immanence and the ideas expounded by Avenarius in *Der menschliche Weltbegriff* (The Human Concept of the World, 1891).[134] Avenarius based his empiriocriticism on the concept of the 'found' (*das Vorgefunden*), that is, on the fundamental original experience, which contains the 'principal coordination' between the I and the environment, not in the sense that one is the *finder* and the other the thing that is *found*, but in the sense that both are *found*.[135] Similarly, Schuppe elaborated his philosophy of immanence starting from the consciousness in general, which contains the relationship between the world and each individual I.

At the same time, however, Schuppe suggested that Avenarius had not focused adequately on the role of the *finding I* (*das vorfindende Ich*) required to keep together the two parties to the principal coordination in order to ensure that 'the I and the environment are components of *a single* experience'.[136] As it was 'impossible to have an experience without an experiencing subject, like a thought that no-one thinks or a feeling that no-one feels', Schuppe held it necessary to 'enrich the "principal coordination" with the *finding I* itself and its environment'.[137] However, Schuppe immediately made it clear that the *finding I* consists only of an 'abstract conceptual moment', and is thus not something actually distinct from the concrete individual I.[138]

Two replies to Schuppe's letter appeared in the following issue of the journal: a long polemical piece by Avenarius's student Rudolph Willy,[139] and a short note by Avenarius himself. In particular, Avenarius emphasized that 'either the I whose absence in my analysis Schuppe complains about is an empirical value (as an empirical relation), or it is not an empirical content'.[140] In the first case, Avenarius pointed out that this would mean that his description of the I is incomplete and needs to be supplemented, albeit without changing its substance. Avenarius also stated that if this was the case he would be happy to 'accept with gratitude this completion of the analysis'.[141] In the second case, however, as the *finding I* would be something superempirical, it would alter the natural concept of the world, as the empiriocriticist view is that the only content that forms a stable part of our world view is purely empirical content. Therefore, this second case would not only require the exclusion of the superempirical concept of the *finding I* from the description of the I but indeed also its complete elimination from our concept of the world.

In the years in which he established contact with Avenarius, Schuppe also tried to involve Mach in the setting up for the *Zeitschrift für immanente Philosophie* (Journal for Immanent Philosophy, 1896–9), a project he was working on with Richard von Schubert-Soldern. We know that Max Reinhard Kauffmann, a young scholar working with these two philosophers, went to see Mach in 1894 to invite him to get involved in launching the journal. Mach rejected the offer, however, citing the same reason he had used to reject Avenarius's repeated requests for him to collaborate on the *Vierteljahrsschrift für wissenschaftliche Philosophie*, namely, that he was a physicist not a philosopher.[142]

It is important to remember, however, that Mach was not familiar with Schuppe's work at this time. He only read Schuppe's works in 1902, recognizing a position that was so close to his own that he wrote Schuppe a letter.

> Esteemed colleague! Please forgive me for having only now become familiar with your *Grundriss der Erkenntnisstheorie und Logik*, and only through an argumentative note written by [Hans] Driesch. I have read your highly detailed book for the first time, in which each sentence demands consideration, and is worthy of consideration. My keen interest therefore drew me to your *Erkenntnistheoretischen Logik* and I have now also acquired *Das menschliche Denken* [Human Thinking]. The first thing that impressed me about your writings is the fact that even a natural scientist with no knowledge of philosophy is able to read them without first acquiring a specialized dictionary.[143] Secondly, I was overjoyed to see your approach to the philosophical clichés that physicists adopt, against which I have been battling for forty years in my own field. In general, I struggled to find anything that did not resonate with me, anything that I was unable to accept willingly. Even your abstract I, which some have found shocking, is in no way strange to me. […] I have reached the conclusion that we can orient ourselves scientifically only *in* the world, and not *above* [*über*] the world. And this positioning must clearly start with the I. At times, even solipsism seemed to me to be the only coherent perspective. However, I no longer believe that investigation *must* remain anchored to this point of departure.[144]

Mach sent Schuppe two of his works along with the letter, probably *The Science of Mechanics* and *The Analysis of Sensations*.[145] Taking the opportunity provided by the appearance of the fourth edition of the latter, Mach inserted a reference to Schuppe in the chapter entitled 'My position with respect to Richard Avenarius and other thinkers'.

The only difference with respect to the eulogy in the letter was on the topic of the I, with Mach writing that Schuppe's 'conception of the Ego constitutes a point of difference between us; but not a point on which it would be hopeless to reach an understanding'.[146] The link between Schuppe and Mach was reinforced further when Mach dedicated his *Knowledge and Error* (1905) to Schuppe, who returned the compliment in the second edition of his *Grundriss* (1910).

Returning to Petzoldt, his relationship with Schuppe began when Schuppe published *Der Zusammenhang von Leib und Seele* (The Connection between the Body and the Mind, 1902), in which he analysed the various positions of several of his contemporaries on the question of psychophysical parallelism. One of the positions examined was the one Petzoldt sets out in the first volume of the *Einführung in die Philosophie der reinen Erfahrung*.[147] Schuppe sent his work to Petzoldt, thus initiating a correspondence that became a lively exchange of ideas over the years, so much so that they also met in person.[148]

In particular, in *Der Zusammenhang von Leib und Seele* Schuppe criticized Petzoldt's proposal to dispense with the concept of the necessity of nature in favour of that of *Eindeutigkeit*. Schuppe noted that Petzoldt himself accepted the lawfulness and universal determination of all of nature, and thus its necessity. What needed to be dispensed with was only the naive and anthropomorphic notion of necessity that implied some kind of constraint imposed by a cause when it *acted* upon the effect. The concept of *Eindeutigkeit* was therefore in effect only a 'new and artificial expression'.[149] Schuppe thus held that the concept of causality should be understood only as: 'necessity = existence itself = simply "this is how it is"'.[150]

Once it is assumed that the necessity of nature does not entail the cause acting on the effect but only the recognition of actual lawfulness as evidenced by experience, for Schuppe the presumed problem of the impossibility 'of the body to *act* on the mind, or the mind on the body' was also eliminated.[151] The three cases of the connections between mental content, between physical phenomena and between mental content and physical phenomena only ever required the empirical recognition of actual connections and therefore necessary connections. Schuppe therefore did not reject psychical causality or psychophysical interaction, and criticized Petzoldt's position that mental processes did not univocally determine one another, but were entirely determined by the cerebral substratum. Indeed, for Schuppe, 'the subject immediately feels the causal connection between his thoughts, his mood, his feelings, desires and decisions'.[152]

In his correspondence with Mach, Petzoldt commented on the work that Schuppe had sent him: 'Although I don't come out well in it and my point of view has been misunderstood completely, I am truly very pleased to have made contact with Prof. Schuppe.'[153] The reason for this pleasure was discovering how close their positions were. While Petzoldt rejects Schuppe's concept of psychical causality, he agrees fully that the concept of the necessity of nature should not be refuted but should simply be reinterpreted as the actuality of nature. However, Petzoldt remains convinced that his concept of *Eindeutigkeit* is better suited for this purpose, as it explains not only connections that involve a temporal sequence but, more generally, any kind of connection, including even the dependence of mental phenomena on the physical phenomena of the brain.[154]

As well as espousing Schuppe's view of the necessity of nature, Petzoldt states that he also concurs with his rejection of the metaphysical concept of parallelism. Both hold that the difference between the mental and the physical does not correspond to what is *inside* experience and what is *outside* it, as is the case if one understands the mental as the internal domain of representations and the physical as the material reality that is external to it. Both Petzoldt and Schuppe saw reality and experience as one and the same thing, with both the mental and the physical lying *within* it. In short, given that empirical reality is this varied but fundamentally homogeneous complex of contents, these contents may be regarded as either 'mental' or 'physical', depending on the circumstances.

Petzoldt wrote 'Die Notwendigkeit und Allgemeinheit des psychophysischen Parallelismus' (The Necessity and Generalness of Psychophysical Parallelism, 1902) in reply to the criticisms of Schuppe and others. In the Preface he specifies:

> My position is fundamentally very close to that of Schuppe, even though he did not know it as it had not yet been expressed adequately in the large publication he considered, given that my more general considerations are in the last part of the volume, as yet unpublished […] Schuppe thinks that I am defending a parallelism that I find as senseless as he does. […] I see nothing fundamentally different in the brain with respect to sensations and representations, even though I entertain neither spiritualist nor materialist conceptions of actual reality.[155]

In the first volume of the *Einführung*, the one commented on by Schuppe, Petzoldt had indeed concentrated on the problem of the *Eindeutigkeit* of mental content in its dependence on the brain, leaving the more strictly philosophical question of how the relationship between the mental and the physical should be understood to the second volume. When he writes the second volume, Petzoldt therefore takes the opportunity to once again clarify his position with respect to that of Schuppe, elaborating on the ideas sketched out in 'Die Notwendigkeit und Allgemeinheit des psychophysischen Parallelismus'. Encouraged by the reinforcement of his personal relationship with Schuppe, Petzoldt goes beyond noting the similarities between their positions, and for the first time includes Schuppe as a key figure in the new approach, stating that 'Today we find ourselves at a turning point in philosophical thinking. Three men, independently of one another, and each following his own path, have opened up a new vision of the world: Wilhelm Schuppe, Ernst Mach and Richard Avenarius'.[156] The same idea is repeated in the *Weltproblem*, where the three are noted as 'the founders of positivism', which Petzoldt sees as 'historically necessary, logically inevitable, and – most probably, in its main points – definitive'.[157]

What made Schuppe deserving of being placed alongside Mach and Avenarius was that he had dispensed with the idea of the thing-in-itself. Schuppe holds that the very concept of things-in-themselves is 'impossible' because they cannot exist in space or in time, cannot have perceivable properties and cannot be the cause of our sensations.[158] However, this does not mean returning to 'Berkeley's subjective idealism', which only 'cancelled the "external" reality' that was supposed to be out of the mind, but 'without correcting the concept of mind'.[159] For Schuppe, the reality that is given in consciousness is neither the mere representation of the individual nor the manifestation of a supposed reality that is external to consciousness, but, in fact, *is* reality:

> The space that we are conscious of is one and the same space, and if its individual aspects and what is perceivable in it change in the content of the consciousness according to fixed laws, and therefore diverge reciprocally, there is nothing contradictory in this, nothing that is impossible or even just extraordinary, and it is utterly obvious that it is one and the same reality that is perceived by several people at the same time or at different times and that differences in perception – to the extent that they are validly connected to specific conditions that conform with the laws of this same objective reality – have the same existence as that which is perceivable by everyone.[160]

Thanks to this conception, Schuppe had taken the only path that allows Kantianism to be overcome through Kantianism itself by taking the unknowability of the thing-in-itself to its extreme consequences, rejecting this concept in favour of definitively equating reality with phenomenal reality. Schuppe had thus preserved only the valuable aspects of Kantian philosophy: the reunification of the physical and mental domains within consciousness and the affirmation that phenomenal reality is not 'mere' representation, as it does not depend on the single acts of representation of an individual subject, but on the impersonal *Bewusstsein überhaupt*.

A letter from Petzoldt to Mach confirms this interpretation of Petzoldt's appreciation of Schuppe's Kantianism:

> I find that there is just *one* glinting oasis in the metaphysical desert of Kantianism: the theory of the empirical reality of phenomena. If we take just all that is good in it, we approximate to Schuppe's position. (One day I must ask Schuppe whether this is the route he took to achieve his extraordinary liberation from Kantian mists).[161]

Schuppe's approach is therefore Kantianism revised and corrected, the bridge between Kant and relativistic positivism. However, further on in the letter we see once again what the limitations of Schuppe's position are: 'If we also dispense with the idea that the elements of the world are phenomena or the content of the consciousness, then we have truly undifferentiated elements and true positivism with no shades of idealism.'[162] That is, the role of the I and of consciousness in Schuppe's approach means that it is still weighed down by traces of subjectivist idealism. This prevents it from rising to the level of true positivism as represented by Mach and Avenarius. As we can see in Petzoldt's *Einführung*:

> Everything that constitutes the I is given in the same way that the components of the environment are given. Strictly speaking, the I does not find the components of the environment and itself [...], but both – the I and the environment – are found *in the same way*, given, experienced, or however one might want to describe *experience that has not yet been differentiated into subject and object*. Here we

reach a point where those who are willing to embrace the new conception of the world often run the risk of failing to free themselves of old preconceptions, all the more so when we consider that on this point one of our three leading lights, Schuppe, insists on maintaining a position that the other two find untenable. [...] Schuppe does not acknowledge the equality [*Gleichberechtigung*] of the I and the environment but instead still attributes a privileged position to the I. [...] The I finds itself and the environment; it is the finder and the found; simultaneously I as subject and I as object; and the *finding I* knows itself to be identical to the *found I*.[163]

Schuppe's position thus falls victim to a '*petitio principii*, as experience only requires an experiencer, thoughts a thinker, and feelings one who feels when experiences, when thoughts and feelings are already considered to be objects [...]. But this is in fact what must be doubted and denied on the basis of psychological analysis'.[164] What does psychological analysis teach us? How should we think of the dependence of experiences on the subject, if not as dependence on an experiencing I? The answer is: simply as a functional connection between specific empirical content. This allows us to obtain 'a solution to the problem that is in line with Schuppe's view'.[165] Schuppe holds that saying that I see something means: 'I (the I as subject) find a specific something that has a colour and simultaneously an eye and its system of nerves (the I as object)'.[166] Petzoldt, on the other hand, holds that the important point is the dependence of this something that has a colour on the eye and the system of nerves, whereas Schuppe's supposed I-as-subject can simply be ignored. This means that this '*functional* dependence becomes the scientific encapsulation of the lay expression 'I see with my eyes'''.[167] In other words, the statement 'I see the tree' means neither that 'the I-as-subject finds the tree' (the position of traditional philosophy) nor that 'the I-as-subject finds the tree and the eye and a certain functional relation between them' (Schuppe's position), but simply 'there exists a functional relation between the eye and the tree', with no need to resort to a subject that finds these elements or the relation that holds between them.

Petzoldt recognizes the value of Schuppe's 'constant firm reiteration of the "identical nature" of the I-as-object and the I-as-subject'. Nevertheless, 'this "identical nature" is incomplete, otherwise any distinction would be impossible'.[168] Therefore, if on the one hand 'Schuppe insists that the perceived world is understandable in itself as is', in line with 'the core and the essence of the philosophy of experience', on the other 'his concept of the *finding I* takes him out of the domain of experience'.[169]

To summarize, Petzoldt attributes a fundamental role to Schuppe in the history of philosophy for his elaboration of Kantianism, and for going beyond it. Kant was to be praised for sustaining the unity of phenomenal experience and the independence of such experience from the individual, but he still represented a retrograde step in the history of philosophy for retaining the concepts of the I-in-itself and the thing-in-itself. Schuppe preserves what is valuable in Kant, but also dispenses with the notion of the thing-in-itself, thus definitively identifying reality as phenomenal reality. However, Schuppe's concept of the *finding I* retains the old notion of subject, which casts an idealist shadow over the philosophy of immanence, preventing it from achieving the true positivism of Mach and Avenarius.

Mach and Avenarius

We have clarified what Petzoldt sees as the progress made by Mach and Avenarius. Along with Schuppe, they reject the notion of a reality that exists beyond experience, holding that it is experience that gives us reality. There is therefore no material world beyond what we experience, and the physical and the psychical, the I and objects, are on the same plane, the content of a fundamentally unitary and homogeneous experience. Unlike Schuppe, however, Mach and Avenarius avoid the error of regarding this experience as subjective, as something that is in the consciousness, or that is given to an I. Avenarius refers to the totality of such experience as the 'found', Mach as 'elements' (or sensations). The problem is that of trying to say what cannot be said, because if such experience is what embraces everything, it cannot be defined, as any definition must define something relative to something else. As we have seen, Petzoldt criticizes subjectivism thus:

> The world as a whole, the totality of what is found, what is originally and immediately given is neither inside nor outside, neither appearance nor thing, neither representation nor object, neither conscious nor subconscious, neither physical nor psychical, neither I nor not-I. It is therefore completely devoid of any logic to ask what the world as a whole is. The concept that might be used to define its essence would in fact lack the opposite concept, precisely because the task would be to define the *totality* of what is given.[170]

The important point is not to provide a name or a term but to emphasize that this 'original experience is *undifferentiated* with respect to the psychical or the physical'.[171] Therefore, when Avenarius defines experience as the 'found', and Mach defines it as 'elements', they should not be understood in a subjectivist sense, as, at least partially, Schuppe does.[172] Indeed, the main aim of Petzoldt's philosophical work is to shed light on the 'frequently misunderstood fundamental point' that identifying experience with reality does not mean reducing it to mere representation, but in fact means rejecting any such conception.[173]

> Nothing is further from our aim than wishing to deny the solidness, the concreteness and the actual reality of the world and its independence of perception. Quite the contrary: after all the idealist attempts to make actual reality disappear, and after all the materialist efforts to replace its brilliant colours and its vividness with a world that is grey, dark and dead, we aspire to return to a world that is as bright as the fresh mornings of our infancy and of young people.[174] Restoring to science the conception of the common man, showing that naive pre-scientific realism is salutary and reasonable (putting aside any animist contamination, of course), so as to bridge the abyss that continues to separate man from nature, this is our objective.[175]

Thanks to the 'complete triumph over the representation of substance', Mach and Avenarius had found 'the path to fundamental relativism', thereby 'rediscovering the long-hidden truth' first noted by Protagoras: 'there is no difference between our image of the perception of the tree and the perceived part of the tree itself, because, in our perception of it, we directly grasp the object as perceived'.[176]

Nevertheless, Petzoldt also notes a difference between his mentors that places Mach one step ahead of Avenarius. As he puts it:

> In Avenarius […] the I still plays a role that is too prominent in the theory. It is understood as the central party to the principal coordination, of the indissoluble union of what is called the I and the environment. This creates the impression that the I must be a *real* condition [*reale*] that is indispensable to actual reality [*des Wirklichen*].[177]

In the *Weltbegriff*, Avenarius had described the 'principal empiriocritical coordination' as the '*belonging-together and inseparability*' of the I-experience and the environment-experience in every experience that is realized', being the 'fundamental correlation and equivalence of both experiences, in that both, the I and the environment, belong to *every* experience and do so in the same sense'.[178] The idea that the world (or, to use Avenarius's term, the environment) cannot be given without the I is seen by Petzoldt as an excessive imbalance in favour of the subject, further demonstrated by Avenarius's conception of how reality should be understood when it is not being perceived.

Avenarius had addressed this issue in a note in his 'Bemerkungen zum Begriff des Gegenstandes der Psychologie' (Remarks on the Concept of the Object of Psychology, 1894–5), replying to the question of whether and how the prehistoric world could be thought of, before mankind existed. 'Investigating the periods of our current environment that preceded in time the appearance of man' would indeed seem to entail a need to 'conceive of an opposite party [the environment] with no central member [the I]', thus dissolving the principal coordination.[179] Indeed, according to Avenarius, '*if we concede that* this is the meaning, investigation would be illegitimate' and 'as senseless as asking what a component of the environment is like "in itself"'.[180] However, precisely because 'the questioner *cannot avoid mentally adding himself* [*sich hineinzudenken*]', the only possible meaning of the question is:

> How would the earth, or the world before the emergence of living beings, and in particular of man, be determined if I were to mentally add myself as observer? This is somewhat like asking how we should conceive of […] a different solar system if we could *observe it* from our earth using perfected instruments.[181]

Petzoldt does not concur with Avenarius here, as he holds that in this solution 'the individual I that poses the question, or the thought of this I, becomes a condition not only for *the act of thinking* about the as yet uninhabited earth but also for the justification of the belief in the existence of the earth at that time'.[182] In other words, Petzoldt agrees that we need the I in order to conceive of the world prior to the emergence of mankind, but not to postulate its existence. Even if the I did not exist the world would still exist, but it would not be conceived. It is thus an error to assume that the principal coordination between the I and the environment is such that if the I disappears then the environment also disappears, because the I is part of the world, not a precondition for it. The solution that 'avoids these errors' is therefore 'not to attribute to the I such a prominent position from a theoretical point of view'.[183] In fact:

> The best thing is not to extend the concept of the I unless this is necessary. What sense is there in assuming an I even when an individual is so immersed in dreamless sleep, or in the contemplation of a landscape or a masterpiece, or in listening to music, or in scientific work, or in practical tasks, that he becomes 'oblivious to himself'? The I-complex is not present in such moments, the physiological substratum is at rest.[184]

This conception clearly only has any sense if the idea of the *finding I* is dispensed with completely and if we assume that reality is a collection of 'undifferentiated' contents, neither physical nor psychical, as advocated by Petzoldt. All that then needs to be said is that contents are given in reciprocal relations, and that some of the contents that constitute the world are called 'I'. But as the I (one or several) is in fact part of this totality of contents in constant flux and in reciprocal relationships with one another, the I itself may indeed not be present. The fact that my knowledge of the world depends on me (on my brain) – in the sense that there exist functional relations such that certain changes in my nervous system determine certain changes in the contents that constitute the world – does not mean that the collection of contents that is the world must always and necessarily include something like an I, neither like the *finding I* of Schuppe, nor like the *found I* of Avenarius.

In this respect, Petzoldt harks back to Mach, whose conception is best suited to represent relativistic positivism. Indeed, in the *Einführung* Petzoldt cites almost the whole of Mach's famous passage on the 'unsavable I' (*unrettbares Ich*):

> The primary fact is not the ego, but the elements (sensations). [...] The elements constitute the I. *I* have the sensation green, signifies that the element green occurs in a given complex of other elements (sensations, memories). When *I* cease to have the sensation green, when *I* die, then the elements no longer occur in the ordinary, familiar association. That is all. [...] This content, and not the ego, is the principal thing. This content, however, is not confined to the individual. [...] *It is impossible to save the I.* [...] We shall then no longer place so high a value upon the ego, which even during the individual life greatly changes, and which, in sleep or during absorption in some idea, just in our very happiest moments, may be partially or wholly absent. We shall then be willing to renounce individual immortality, and not place more value upon the subsidiary elements than upon the principal ones. In this way we shall arrive at a freer and more enlightened view of life, which will preclude the disregard of other egos and the overestimation of our own.[185]

This dissolution of the I also explains Petzoldt's rejection of Kantianism.[186] Even if it is possible to salvage from the Kantian approach the doctrine of the reality of phenomena, what then becomes inadmissible is the Copernican revolution that had raised the *I think* to the status of the cornerstone of all reality. The difficulty of understanding and embracing relativistic positivism in fact derives from the entrenched notion that the I is the fundamental condition of the world, its alpha and omega.

It is important to bear this point in mind; otherwise one runs the risk of being thrown by the next part of Petzoldt's argumentation, which resurrects the concept of the soul (*Seele*). We know that the whole of reality cannot be defined, because the opposite concept is missing, and 'we only have names: the world, actuality, everything, what is given, what is found'.[187] Nevertheless, Petzoldt holds that we can use two different concepts to describe that totality from two different points of view: the concepts of nature and soul. In this sense, we must say that 'nature and soul are not two sides of a third reality but are in fact this third reality itself'.[188] Petzoldt elaborates further:

> Our aim is not to use the concept of soul as a counterweight to the concept of nature. When we observe what is found as univocally determined, we call it *nature*; when we take it in its immediate existence, one entity next to another and one after another, we call it *soul*. Nature is that which is physical, the object of physical consideration; the mind is what is psychical, the object of psychological consideration. The soul is not the internal part of nature, nature is not the external part of the soul. Rather, they are one and the same thing, observed from different angles. Nature is not matter in the sense of materialism, the soul is not the spirit in the sense of the spiritualist conception of existence. There is no psychophysical parallelism in the metaphysical sense. The physical and the psychical are one and the same thing.[189]

This reveals both Petzoldt's closeness to Schuppe's concept of *Bewusstsein überhaupt* and his distance from it. On the one hand, Petzoldt's concept of 'soul' goes back to the idea that to avoid a return to materialism the whole of reality can and must be included not only in the concept of nature but also in the concept of something that is conscious, whether it is called 'consciousness', 'mind', or 'thinking':

> We make the concept of the world's existence dependent on the concept of thinking about the world, we see both concepts as being in a *correlation*, so that one cannot exist without the other. Only what can be *thought* can *exist*; the unthinkable, that which does not conform to thinking, cannot *exist*. And *thinking* would no longer have any meaning if it was not counterposed to a *thinkable existence*.[190]

However, this 'mind' or 'thinking' becomes even more impersonal than Schuppe's *Bewusstsein überhaupt*, which was still the consciousness of the thinking individuals, of the *finding I*, albeit *in general*, that is, an abstract concept derived from these individuals. Petzoldt's concept of the 'soul', however, is impersonal in the sense that the I of each individual no longer exists as such but only is part of collections of contents. The soul that embraces all the intrinsically 'undifferentiated' contents that constitute the world surpasses and encompasses each individual human I, collectively, and those of sentient beings in general. As Petzoldt put it: 'the mind extends beyond the I'.[191]

To summarize, Petzoldt sees the thinking of Schuppe, Avenarius and Mach as the first modern attempt to dispense both with metaphysical substance (the notion of a reality-in-itself, independent of experience) and with the metaphysical soul (the I-in-itself, the subject as a correlate of every experience). However, Schuppe deviated from this positivist/relativist direction when he maintained the notion of the *finding I*, even though he had made this concept coincide with a simple conceptual abstraction based on the concrete I. Avenarius, on the other hand, had correctly rejected the idea of the *finding I*, stating that the I is only a given complex of the contents of empirical reality, but he erred when he made empirical reality dependent on the individual I being given. The only fully valid conception was therefore that of Mach, who maintained that it is elements that are primary, with the I being merely a potential, contingent and unnecessary complex of these elements.

This sheds light on the words that Petzoldt writes to Mach in one of his letters:

People describe me as a follower of Avenarius. But this is only partially true. At the same time as I feel a link with Avenarius I also feel a link with you. In fact, I was a follower of Avenarius,' but I can learn nothing new from him, whereas with you it is different. I continue to learn new things from you, and there is a long way to go before I manage to extract all of the gold that your writings contain so much of. Indeed, I fear that I will never cease to learn from you.[192]

Petzoldt's relativistic positivism

At the start of this volume, we noted that the starting point for Petzoldt's reflections was the quest for a unitary account of scientific knowledge, a quest that characterized the nineteenth century in general. Petzoldt initially embarks on this quest by adopting the neo-Kantianism that had emerged and spread in German-speaking scientific circles around the middle of the century, and was embodied in writers such as Helmholtz, Lange and Du Bois-Reymond. According to this view, science adopts a materialistic-mechanistic conception of the universe for methodological reasons, but this interpretation of the world cannot account for reality in itself and for itself. A number of questions, such as the essence of material reality, the essence of consciousness and the reciprocal relationships between them, must therefore forever remain beyond the scope of natural science. This idea ultimately legitimizes the adoption of approaches that are not materialistic-mechanistic in fields other than science.

The allure of this solution consisted of the fact that it appeared not only to meet the requirements of the science of the day – where the principle of the conservation of energy had reinstated the assumption of the closed causality of nature, extending its validity to all phenomenal fields, including living beings – but also to quell the concerns of those who did not wish to derive the whole of reality from mere matter and the concepts of physics. As in all forms of Kantianism, the system of knowledge was thus guaranteed to be unitary by a conceptual device based on an unknowable reality-in-itself and a subject that could resort to a range of ways of relating to it, thus allowing these ways to coexist without contradiction, all equally valid.

However, Petzoldt later rejects this approach for two main reasons: its agnosticism and its scepticism. Indeed, the Kantian approach retains the idea of a divide between human thinking and reality which means that we can never hope to grasp the actual nature of the world. This dissatisfaction with the Kantian position first draws Petzoldt to Fechner, whose approach was valuable for its highly unitary approach (provided that its metaphysical and even mystical traits were eliminated). Fechner saw the physical and the psychical, the organic and the inorganic, simply as different aspects of a single reality constantly striving to achieve states of ever-increasing stability. Under this view, mankind is not distinct from reality but is included in this process of a universal tendency towards maximum stability.

Petzoldt then moves from Fechner towards Avenarius and Mach, seeing in their economical conception of nature an opportunity to extend the principle of stability to human knowledge. The main thing that these two thinkers have in common is the biological interpretation of mankind's theoretical activity. If we accept psychophysical parallelism's tenet that every mental process is fully conditioned by the physiological substratum, we must consider human knowledge to be a specific case of the organism adapting to the environment. However, Mach and Avenarius had both rejected the metaphysical sense of psychophysical parallelism. Indeed, they rejected in general the traditional separation between the physical and the psychical as two distinct dimensions of reality or as two fundamentally different types of experience. Instead they emphasized the unitary and homogeneous nature of our experience of the world, which exists prior to any potential differentiation between subject and object, between physical and psychical, between material and mental.

Petzoldt therefore develops an elaborate philosophical approach that rests on two pillars, with Mach and Avenarius (and later Schuppe) reinterpreted in a radically anti-Kantian way. He rejects the transcendental tradition that considers the subject to be the main protagonist in cognitive processes but turns out to be a form of subjectivism with touches of scepticism. Petzoldt therefore postulates an interpretation of empiriocriticism whereby it is not a form of phenomenalism in which reality coincides with subjective experience but a form of positivism in which it is the experience of the subject that becomes part of reality.[193] Mach and Avenarius had shown that subject and object were not conditions of reality but components of empirical reality. The I is not the law-giver of nature but an organism that is located in nature and strives to survive using physiological means, including the cerebral activity on which cognitive processes depend.

Following the line laid down by Mach and Avenarius, Petzoldt thus states that reality is initially given as a series of contents that are 'undifferentiated', neither physical nor psychical in themselves, and are in constant flux in their reciprocal relations. These contents include sensations such as red, hot, sweet, round, audible and so on. This stream of contents, simply as given, is what Petzoldt calls 'the soul'. Things such as objects and the I are nothing more than complexes of these contents. No sentient I is required as a correlate of these sensations. Similarly, there is no substratum beneath the sensations and contents that constitute the object. The object *is* those sensations, those contents. It is of course possible to imagine that beings with organs and nervous systems unlike ours might have a different perception of the same object, corresponding to a different complex of contents, but our complex of sensations, as much as theirs, *is* that object, as every object exists only in some relation to other objects.

As Petzoldt puts it:

> If I think of the object as existing independently of my perception of it, I *must* fundamentally think of it with all of the properties with which I find it when I perceive it, as *consisting* of these properties. We can only think of things just as we find them and not as no-one finds them. We can only ever think of them from the perspective in which we find them, and not from a perspective from which we cannot think of them, or in general from no perspective at all. There is no absolute perspective and there is no absence of perspectives, there are only relative perspectives, and this holds for every point in time. We cannot disregard the relations in which things are with one another and with us. […] There is no contradiction in thinking of unperceived things as existing with perceivable properties, because in any case I think of them in *relation* to me. The contradiction would immediately arise if I wanted to attribute these properties to them as absolute properties, as independent of my own composition or that of beings like me. […] In general, it is absolutely impossible for me to think of actual reality in absolute terms.[194]

The problem with this perspective comes from our tendency to assume the existence of a fixed and unchangeable substratum: substance. Petzoldt does not claim that the substance of objects consists of a given complex of sensations, but that substance *does not exist*. To say that the objects of the world are complexes of sensations, as Mach did, does not mean that sensations are the substance of the world but that there is no substance beyond those sensations. We should not think of objects as typified *eternally* and *immutably* by certain perceptions but must think of them as actually typified by those perceptions in that context, in that moment, in that set of relations:

> If I think of things exactly as they are perceived, different for different individuals, and also different for different individuals in their ongoing existence, different for a deaf person, different for a blind person, different for some intelligence structured in a completely different way from mankind, where might the contradiction lie, or something that is unthinkable?[195]

This, then, is Petzoldt's relativism, based initially on the Protagorean view that the world is as it appears to each individual. However, as we saw when we examined Protagoras above, this is not a sceptical, subjectivist or solipsist relativism but a positivist relativism, in that this ever-changing stream of contents in reciprocal relationships has one fundamental characteristic: lawfulness, the necessity of nature, *Eindeutigkeit* or whatever one might wish to call it. The way these contents flow and change in their reciprocal relationships is not chaotic or disorderly, but is subject to precise conditions. The fact that each individual perceives objects in a different way is not the result of individual arbitrariness, is not mere 'appearance', or an error, but is the result of the lawfulness of nature. Or rather, what Petzoldt calls 'nature', the physical world, is in fact the totality of reality considered in terms of its lawfulness, in contrast to the concept of 'soul' noted above.

Each piece of content changes its relations to others according to fixed relationships that are *univocally determined*, in the sense that, given a certain set of conditions, of all possible cases that case *alone* must be realized. The same happens in the dependency relations that hold between certain contents and the set of contents that constitutes the human organism, and in particular the brain. A specific sensation necessarily follows from certain conditions, as the *only* actual sensation from among all possible sensations. The same holds not only for sensations but also for every psychical piece of content, from the simplest to the most complex, eventually arriving at the highest concept of all, the concept of the world:

> We conceive of the diversity of world views as lawful in that we think of it as dependent on the diversity of individuals. Our position can be understood easily if we compare it to the description of the relations of motion between celestial bodies. Just as there is no absolutely stationary system of coordinates for that description, so there is no absolutely normal intelligence whose perceptions can be considered to be absolutely correct. And just as each of these individual descriptions in itself provides a lawful picture regardless of whether one starts from the Earth or the Sun or Jupiter, and just as each of these views is in complete harmony with each of the others, so the world view of these individual intelligences that emerges on the basis of each individual's organization, is free of internal contradictions and in harmony with all of the others.[196]

Petzoldt's relativism does not entail a fragmentation of the world into an infinite number of conceptions of reality. The unity of the world is guaranteed by *Eindeutigkeit*, by the lawfulness that connects, by necessity, the relations between all of the components of the world. This is a *positivist* relativism in that it maintains the notion of a knowable reality as the fundamental schema within which the knowledge of individuals is located. However, this knowable reality is not what is *beyond* the knowledge of individuals, is not what is *absolute*, unconstrained by any relation, but is in fact the network of all of the relations that hold between real elements.

Relativism therefore not only does not prevent us from knowing the world but is a necessary consequence of how the world actually is. As there is no such thing as an absolute reality, and the world consists exhaustively of a set of relations, we can only know the world insofar as we adopt a relativist approach. Relativism is thus a *positive* knowledge of the world. If reality is a series of contents that are functionally interconnected in such a way that if one changes then another changes, then the goal of consciousness must be to capture the conditions under which every object appears in relation to other objects. As the physical world, nature, is nothing other than the set of all contents seen from the perspective of the lawfulness of their reciprocal relations, the task of a scientific understanding of nature is to observe the relations that hold between the elements of the world *sub specie univocitatis*, in terms of their univocalness.

The psychophysical theory of knowledge, which explains how knowledge of the world depends *univocally* on cerebral activity, is therefore to be understood as belonging to the domain of physics, of natural science. Perception is only a psychological process if we consider the simple flow of perceptual contents, but it is a physical process if we frame it in the context of the necessity of nature whereby certain perceptions must necessarily occur as a consequence of given conditions. The same holds for every activity that we call 'psychical'. There is nothing 'merely subjective' about mental phenomena. They are just as objective as gravitation, in that both are fully and perfectly univocally determined.

There is, however, another sense in which Petzoldt's relativism is positivist. It is indeed true that every individual has a particular conception of the world that is fully justified in that it does not emerge from nothing, by chance, but is univocally determined in its necessary dependence on certain cerebral conditions. However, this does not mean that Petzoldt legitimizes any and every position, abandoning the search for a way to assess the validity of knowledge. Truth has nothing to do with how reality is given to us, or how it is given independently of us. In the first case, everything would be equally 'true' or equally 'false', because no one can perceive reality differently from how they perceive it. In the second case, 'truth' would be unachievable and even contradictory, like the thing-in-itself. As Petzoldt writes, 'From a theoretical point of view, Don Quixote's conception of the world is just as valid as that of Sancho Panza. "Correct" is not a theoretical concept, but a practical concept.'[197]

Our practical needs are nothing other than those of the human organism as a biological entity that has evolved and preserved itself within nature. Our evolution is based on a nature where things do not happen by chance but on the basis of univocally determined relationships, as the neural systems that develop are exclusively those that are conditioned by what constantly happens in nature. If one thing happened one day and the next day the opposite happened, for no apparent reason, there would be no way to develop certain cerebral functions through repetition and reinforcement, leading to the evolution of the brain as we know it. However, as reality is subject to the principles of univocalness and stability, the brain evolves as an organ whose very existence is based on univocalness and stability. Our 'practical' imperative is therefore to focus on what is constant and stable, in the sense both of what is constantly given in nature and of what is common to all of the conceptions of reality of different individuals. However, what recurs in reality and in the conceptions of it by different individuals is the univocalness of relations. For example, I can observe a swinging pendulum from different positions, perceiving it differently each time, but this does not prevent me from noting, each time, the same univocal relation between kinetic energy and potential energy that is captured by the law of the conservation of energy. Or, although I and another individual may have different perceptions of the same event, we must concur that these different perceptions are univocally determined by certain specific psychophysiological conditions in our bodies. By focusing on what is constant, that is, on the univocal relations between the contents of the world, we can even contemplate the possibility of 'successfully developing theories of physics that are independent of the specific nature of our senses', that is, 'doctrines that would also be valid for beings with different senses, if we were able to translate them into their sensory perceptions'.[198] In other words, a human being and an intelligent alien being must concur on the univocalness of the world, and on the fact that their different conceptions of the world are univocally determined by their cerebral apparatus.

To summarize, we need to observe the world univocally because that corresponds to our brain's requirement to favour what is constant and uniform. However, the brain favours what is constant and uniform because that is how the world is. The 'practical' concept of truth therefore refers yet again to the univocalness of reality itself, which turns out to be the precondition and the purpose of all knowledge. The emphasis on the fact that truth is a 'practical' concept is therefore not a reduction to mere pragmatism, detaching human consciousness from an effective grasp of reality and resorting to validation criteria that are wholly internal to consciousness itself. On the contrary, it further reinforces the link between consciousness and reality. It is the univocalness of reality that makes univocalness a practical requirement of the brain and of the cognitive processes that are determined by it.

Since the knowledge that is conditioned by nature will become reinforced over the course of development because the related cerebral structures will be gradually reinforced by their greater use, the scientific conception of the world, which observes it univocally, is the inevitable endpoint of the route that mankind is progressing along. In other words, as reality is univocally determined, the brain has not only evolved over time in line with this property but its future evolution will also proceed in the same direction. Scientific progress is therefore inevitable as it is a necessity of nature, univocally determined by the workings of the brain in a univocally determined environment. However, just as scientific progress is a necessity, so is the triumph of positivistic relativism, which holds that all of reality is given as a series of contents whose reciprocal relations are in constant flux and subject to univocally determined relationships.

Even though Petzoldt frequently reaffirms that nothing is absolute, univocalness looks to all intents and purposes like an absolute in positivistic relativism, in that everything is given only in its reciprocal relations except univocalness itself, being the fundamental characteristic of all of reality.

Petzoldt's approach may at times appear to be so close to Kantianism as to be almost indistinguishable from it. However, it is in fact a clear overturning of it in that the necessity of reality that holds phenomena together is not just a subjective requirement, a precondition of experience, but is in the first instance a characteristic attributed to reality itself. Petzoldt's conception thus frees itself of the duality of reality-in-itself and empirical knowledge, offering a view of the world in which the whole of reality is a unitary and homogeneous set of contents in constant flux governed by univocally determined relationships. Any remaining trace of 'agnosticism' must also be swept away, as mankind and mankind's knowledge are not separated from reality by an insurmountable transcendental barrier, but are an integral part of reality. As human knowledge is conditioned by the brain, it arises from and ends with this univocalness that is the fundamental characteristic of reality itself.

It is clear that this approach, even though it is built on foundations laid by Avenarius and Mach, is a significant transmutation of their positions, and in particular of Mach's position. As we have noted several times, it is not simply that this realist interpretation of the anti-metaphysical empiricism of Avenarius and Mach is just one of its possible interpretations, given that other students and followers developed their ideas in a clearly phenomenalist direction. What is more important is that Petzoldt dispenses with the critical caution that is one of the characteristic traits of Mach's thinking. It is well-known that Mach ceaselessly repeated that he was not and did not want to be a philosopher, precisely because he did not want to create a rigid belief system that claimed to say something definitive. Instead, he saw his work as an unending historical and critical examination of scientific knowledge (as reflected in the title of his famous work on mechanics), a task destined by its very nature never to reach completion. As he writes in *Knowledge and Error*, replying to Hönigswald's *Zur Kritik der Machschen Philosophie* (For a Critique of the Philosophy of Mach, 1903):

> there is no Machian philosophy, but at best a scientific methodology and cognitive psychology and both are provisional, imperfect attempts, like all scientific theories. I am not responsible for a philosophy that might be constructed from this with the help of alien additions.[199]

The relevance of these words to Petzoldt's work is clear, to his construction of a philosophical system based on Mach's reflections. The inevitable unfinished nature that Mach proudly claimed for his approach disappears in Petzoldt's relativistic positivism, which is proposed as an approach that aspires to a definitive stability.

4

Petzoldt and Einstein's theory of relativity

Relativity before relativity

The preceding chapters outline the evolution of Petzoldt's thinking from his early position to the definition of the philosophical system he presented in his *Einführung in die Philosophie der reinen Erfahrung* (Introduction to the Philosophy of Pure Experience, 1900–4) and in *Das Weltproblem vom positivistischen Standpunkt aus* (The Problem of the World from the Standpoint of Positivism, 1906). Petzoldt continues his work over the following years, producing innumerable publications until his death. The main objective of this intense productivity is to promote Petzoldt's interpretation of the ideas of Mach and Avenarius in order to lead to the ultimate triumph of relativistic positivism.

In particular, Petzoldt's new work targets natural scientists, partly in the hope that they might be more likely to accept the ideas he was promoting, partly having failed to make an impact on traditional philosophers. Relativistic positivism had adopted the main results achieved by modern science in the previous century, such as the law of the conservation of energy and the full dependence of psychical content on the brain, so it could be promoted as a philosophy for scientists. Petzoldt therefore misses no opportunity to present his position not only at scientific conferences on chemistry or physics and in specialist journals but also in educational publications targeting a readership with a general interest in science.[1] Mach's death in 1916, following that of Avenarius in 1896 and that of Schuppe in 1913, leaves Petzoldt as the last surviving exponent of this position, thus reinforcing his role as its public advocate.

Petzoldt's promotional activities are, however, not limited to disseminating his ideas and those of his mentors to as broad a public as possible, but extend to the re-elaboration of these ideas in the light of the new scientific paradigm. In particular, the revolution sparked by Einstein's formulation of the theory of relativity is both a challenge and a resource for Petzoldt. On the one hand, the success enjoyed by Einstein's ideas required all philosophers to take the new paradigm into account and provide their own interpretation of it. On the other, it provides Petzoldt with the opportunity to stress the link between Einsteinian relativity and the ideas of his mentor Ernst Mach, a link that Einstein himself had acknowledged.

This aim of integrating the theory of relativity with his own philosophical framework is one of the most interesting aspects of Petzoldt's work in the first decades of the twentieth century. However, our analysis of this development should start *before* the theory of relativity, that is, with Mach's criticism of Newtonian physics and Petzoldt's discussion of it in his writings.

Mach's criticism of Newton

In 1883 Mach had published what would turn out to be one of his most acclaimed works, *The Science of Mechanics*. In its outline of the evolution of physics, the work questioned some of the cornerstones of Newtonian physics, such as the concepts of mass, inertia and absolute space and time, thus paving the way for Einstein's revolution in the early twentieth century. To understand Mach's criticism, we first need to outline some aspects of Newton's work.[2]

Newton introduces the definitions of the quantity of matter (mass), the quantity of motion, the inherent force of matter (inertia), impressed force and centripetal force at the start of the *Principia*. Then, in the Scholium, he discusses the absolute notion of space, time, place and motion. It was not in fact necessary to define these concepts as they are 'familiar to everyone'. However, what was necessary was to distinguish between their 'absolute', 'true', 'mathematical' versions, and those that are merely 'relative', 'apparent', 'common'.[3]

Newton knew that Galilean relativity meant that the laws of mechanics had to hold for any system that was stationary or in rectilinear uniform motion. He wrote, in Corollary V: 'When bodies are enclosed in a given space, their motions in relation to one another are the same whether the space is at rest or whether it is moving uniformly straight forward without circular motion.'[4] However, his view of the cosmos and his approach to physics prevented him from extending this principle to the point of eliminating the idea of absolute space as a reference frame for all motion. As we can see from the definitions that open the work, Newton's starting point was a body with a certain mass. This body innately has inertia, that is, 'the power of resisting', whereby it 'perseveres in its state either of resting or of moving uniformly straight forward'.[5] This approach clearly required the postulation of an absolute space with respect to which the body could be said to be stationary or in motion. In other words, if, as Newton holds, inertia is a property of a single body, then, if we imagine that the universe contains just one body, this body will have inertia. But as there is only this one body, it is impossible to establish whether it is at rest or in motion *relative to another body*, and it is therefore necessary to consider its state of rest or motion relative to absolute space.[6]

Once we accept absolute space, it follows that there are two types of motion: absolute, that is, the movement from one point in absolute space to another, and relative, where a body changes its position relative to another body. The problem is how to distinguish between absolute motion and relative motion, given that we know about bodies and relations between bodies, but we cannot perceive absolute space at all. According to Newton, 'absolute and relative rest and motion are distinguished from each other by their properties, causes, and effects'.[7] In particular, 'The effects distinguishing absolute motion from relative motion are the forces of receding from the axis of circular motion', that is, centrifugal force.[8]

Newton laid out several arguments to support this point, above all the famous thought experiment known as 'Newton's bucket'. A bucket of water is hung at the end of a rope and twisted until the rope is rigid. When released, the bucket begins to spin but the water does not, retaining a flat surface. Gradually, the motion will be transferred from the bucket to the water, which will also begin to spin. Consequently, centrifugal force will raise the water towards the rim of the bucket, and the surface of the water will become concave.

Newton uses this to demonstrate the difference between relative motion and absolute motion. At first, the water is in relative motion with respect to the bucket, as it is not possible to determine whether the water is rotating and the bucket is stationary, or vice versa. However, the absence of centrifugal force shows that the water is in fact stationary, meaning at rest with respect to absolute space, whereas it is only when it *actually* begins to spin that we see it rise towards the rim of the bucket, making the surface concave.[9]

Mach's counterargument to Newton consists of two fundamental points. The first is to show that absolute space cannot be detected in any way, making it an unusable concept and thus a metaphysical one. The second is to postulate that the forces that seem to require the assumption of absolute motion, such as inertia and centrifugal force,[10] in fact presuppose a relation between a body and the other bodies in the universe, and cannot therefore account for the behaviour of a body in absolute space.[11]

With respect to the first point, Mach notes that 'No one is competent to predicate things about absolute space and absolute motion; they are pure things of thought, pure mental constructs, that cannot be produced in experience'.[12] As we have no way of measuring the position of a body with respect to absolute space, it cannot serve as a frame of reference for the description of the motion of bodies. 'If we take our stand on the basis of facts, we shall find we have knowledge only of relative spaces and motions', which makes descriptions of phenomena using different frames of reference all equally valid.[13] For example, the Ptolemaic and Copernican systems must be considered 'equally *correct*' in all respects, as experience only tells us that celestial bodies change their positions relative to one another, and it is simply a convention which celestial body we consider to be at rest and which in motion.[14]

For his second point, Mach points out that 'even in the simplest case, in which apparently we deal with the mutual action of only two masses, the neglecting of the rest of the world is impossible'.[15] We have no way of knowing what would happen if there was just a single body in the universe. We do not know whether it would comply with the law of inertia, continuing to move uniformly along a straight line forever, or whether it would exert centrifugal forces if it was rotating. We do not know these things not only in the sense of having no other bodies to use as reference points to determine whether or not it is in fact in motion but also because we have no way of knowing whether the law of inertia and centrifugal forces belong to the motion of a body *in and of itself* or to its relationship with all other bodies. As Mach writes, 'When, accordingly, we say, that a body preserves unchanged its direction and velocity in space, our assertion is nothing more or less than an abbreviated reference *to the entire universe*'.[16]

In other words, if the law of gravity describes the forces that emerge in the relationship between two or more bodies (i.e. acceleration), then we cannot be sure that the absence of acceleration (i.e. inertia) is not itself also dependent on the reciprocal relationship between the bodies. This is what Mach means when he writes, 'the masses that in the common phraseology exert forces on each other as well as those that exert none, stand with respect to acceleration in quite similar relations. We may, indeed, *regard all masses as related to each other*'.[17] In later editions of this work, Mach takes these ideas further, arriving at what can be considered his 'research programme', which defines the task of the scientist as knowing 'the immediate connections of the masses of the universe' and developing 'an insight into the principles of the whole matter from which accelerated and inertial motions result in the same way'.[18] In his formulation of general relativity, Einstein will call this 'Mach's principle', that is, the idea that gravity and inertia are 'essentially the same' (*wesensgleich*) and 'completely determined by the masses of the bodies'.[19]

If inertial forces are determined by other bodies in the universe, the same must apply to centrifugal forces.[20] Mach therefore dismisses Newton's interpretation of the bucket experiment. For him, it 'simply informs us, that the relative rotation of the water with respect to the sides of the vessel produces no noticeable centrifugal forces, but that such forces are produced by its relative rotation *with respect to the mass of the earth and the other celestial bodies*'.[21] Conversely, to validate Newton's position it would be necessary to 'fix Newton's bucket of water, rotate the fixed stars, and then prove the absence of centrifugal forces', which is clearly impossible.[22]

Mach thus confutes Newton's position, pointing out on the one hand the uselessness of absolute space from a kinematic perspective, given that it contributes nothing to providing a description of the motion of bodies, and on the other the indispensability of referring to other bodies from a dynamic perspective, given that, whether forces appear or, as in the case of inertia, they do not, we must assume that the behaviour of a body is determined by the presence of other bodies in the universe. Both of these points therefore lead to the assumption of the fixed stars as a frame of reference or, to put it another way, to making explicit the fact that the fixed stars already serve as the implicit frame of reference for physics research.[23]

Neumann and Lange

Before turning to Petzoldt's position on absolute space, we must consider two other philosophers whose work is essential for a full understanding of Petzoldt's argumentation: the mathematician Carl Neumann[24] and the physicist Ludwig Lange[25] (not to be confused with the author of *History of Materialism*, Friedrich Albert Lange, whose work we referred to in an earlier chapter).[26]

In 1870, several years before Mach published the first edition of his *Mechanics*, Neumann had published a work entitled *Ueber die Prinzipien der Galilei-Newtonschen Theorie* (On the Principles of the Galilean and Newtonian Theory), in which he analysed a number of problems inherent to the classical formulation of the law of inertia. The proposition that 'a material point that was set in motion will move on – if no foreign cause affects it, if it is entirely left to itself – in a straight line and it will traverse in equal times equal distances' is in fact 'totally unintelligible', as it fails to define what this straight and uniform motion is relative to. For example, 'every motion that in relation to one heavenly body is rectilinear will appear curvilinear in relation to every other heavenly body'.[27]

Neumann concludes that it is necessary to preface the law of inertia, which is the first principle of the Galilean-Newtonian system, with the assumption that 'some specific body must be assigned in space to serve as the basis of our judgment, as that object in relation to which all motions are to be determined'.[28] Neumann referred to this object as 'Body Alpha'.[29] Neumann did not consider it to be a problem that this 'unknown body' located 'in some unknown position in space' remained to some extent indeterminate, as the Galilean relativity found in Corollary V of Newton's *Principia* means that 'the Body Alpha can be replaced by some other Body Alpha, provided that this other body is assigned a motion that is rectilinear and of constant velocity with respect to the first body'.[30]

Despite being somewhat curious, this concept of a Body Alpha was an advance on Newton's position. Although Neumann retained the concept of absolute motion, he rejected that of absolute *space*, replacing it with an absolute reference point, or rather, an absolute reference body. This freed him from the metaphysical conception of absolute space as a kind of container of the universe, recognizing that motion must always be seen as the movement of one body relative to another. Neumann holds that absolute motion cannot be thought of as a change of location within an empty space, but as the possibility to refer 'all changes of place' 'to one and the same object'.[31] Moreover, in his recognition of the possibility of substituting Body Alpha with any other object that moves uniformly relative to it, Neumann moved closer to 'postulating the existence of an infinite set of dynamically equivalent "inertial frames of reference"'.[32]

As well as criticizing the lack of a spatial reference point in the Newtonian law of inertia, Neumann pointed out that a definition of the temporal parameter was also missing. The law stated that a body covers equal distance *in equal time* unless acted upon by a force, but without explaining how this should be understood. Neumann endeavoured to clarify the concept of time by noted that we always measure intervals of time with reference to the motion of certain bodies (the position of the sun, for example). This means that we have no way of knowing whether a body *in isolation* proceeds at a constant velocity, and always need at least *two* bodies, with one serving as a temporal reference point for the other. The statement that the motion of a body is straight and uniform must therefore be expressed with greater precision, as follows: '*Two* material points, each of them left to itself, move in such a way that the equal paths of one of them always correspond to the equal paths of the other.'[33] This conception is a precursor of Mach's formulation that temporal parameters are always spatial references to the position of bodies whose movement serves as a measure of time.

Neumann, like Newton, also provided a thought experiment to validate the thesis that the concept of absolute motion cannot be dispensed with.

> Let us assume that among the stars there is one which is composed of fluid matter and is somewhat similar to our terrestrial globe and that it is rotating around an axis that passes through its center. As a result of such motion, and due to

the resulting centrifugal forces, this star would take on the shape of a flattened ellipsoid. We now ask: What shape will this star assume if all remaining heavenly bodies are suddenly annihilated (turned into nothing)?[34]

Neumann's answer is that its shape would remain unchanged, because, contrary to what Mach would propose a few years later, 'centrifugal forces are dependent only on the state of the star itself; they are totally independent of the remaining heavenly bodies'.[35] Neumann considered the opposite view to be absurd, where the star would suddenly take on the shape of a sphere when the rest of the universe disappeared, as it could no longer be considered in motion relative to some other object, and thus, being stationary, could not exhibit any centrifugal force. It must therefore be conceded that absolute motion exists, and to define absolute motion we need a Body Alpha.

Even though Neumann's work preceded the first edition of *The Science of Mechanics*, Mach did not include any reference to it until the fourth edition in 1901, where he concedes that Neumann's example of the fluid star provides 'the most captivating reasons for the assumption of absolute motion'. Nevertheless, he rejects the argument, seeing it as 'too free a use of intellectual experiment', in that 'it is not to be antecedently assumed that the universe is without influence on the phenomenon here in question'.[36]

Explicitly picking up on the work of Neumann and Mach, the young Ludwig Lange published a series of works in 1885 and 1886[37] in which he set himself the task of 'finding a fully valid substitute' for the Newtonian concept of absolute space, which 'is a phantom that should never be made the basis of an exact science'.[38] Lange recognized Neumann's success in eliminating the notion of absolute time by replacing it with a 'inertial timescale' (*Inertialzeitscala*), that is, the notion that the 'the fundamental timescale of dynamics is to be defined through the motion of a point *left to itself*'.[39] Lange therefore poses the question of 'whether it is possible to eliminate also absolute space by a similar procedure', using an 'inertial system' as a reference point to determine the location of bodies.[40] If an 'inertial timescale could be defined through one single point left to itself' then, according to Lange, a 'the three-dimensional inertial system can be defined through three points left to themselves'.[41] Lange thus produced the first definition of an 'inertial frame of reference', a concept that would become an integral part of physics in the years to come. Lange first established that 'for three points P, P', P'', moving arbitrarily relative to one another' it is always possible 'to construct a coordinate system', or rather, 'infinitely many coordinate systems, in relation to which these points move rectilinearly'. He then defined an inertial system as 'a coordinate system of the kind that in relation to it three points P, P', P'', projected from the same space point and then left to themselves – which, however, may not lie in one straight line – move on three arbitrary straight lines G, G', G'' that meet at one point'.[42]

Lange later observed that, as it is always possible to construct a coordinate system for three points such that they move in an inertial manner, a construct of this kind is a mere *geometrical* convention, whereas something more concrete from the point of view of *physics* can be stated about the inertial motion of a single point starting from a fourth point. Having defined an inertial system using three points, Lange therefore reformulated the law of inertia as follows: 'In relation to an inertial system the path of an arbitrary fourth point, left to itself, is likewise rectilinear.'[43]

Mach did not become aware of the publication of Lange's works immediately, and makes no mention of them in the second edition of the *Mechanics*, which was published in 1889. Passages addressing Lange's theory of the structure of inertial systems did not appear until the third edition in 1897. Here, Mach praises the young scholar for his work, which he considers 'one of the best that have been written' in the growing debate on the law of inertia, both for his 'methodical movement' and for 'careful analysis and study, from historical and critical points of view, of the concept of motion', which has yielded 'results of permanent value'.[44]

In particular, Mach recognizes that 'the law of inertia can be referred to such a system of time and space coordinates and expressed in this form', adding that Lange's approach 'especially appeals to my mind, as a number of years ago I was engaged with similar attempts'.[45] However, Mach also specifies that he had 'abandoned these attempts, because I was convinced that we only apparently evade by such expressions references to the fixed stars and the angular rotation of the earth' as the unavoidable space-time coordinates of physics.[46] Indeed, as we have seen, Mach saw the fixed stars not only as a possible frame of reference for the kinematic description of motion but also in dynamics, as we must assume that they are a precondition for the force of inertia. Mach's response to Lange was therefore that it is 'quite questionable, whether a fourth material point, left to itself, would [...] uniformly describe a straight line, *if the fixed stars were absent*'.[47]

The law of inertia and the relativity of motion

Petzoldt joins the debate on the interpretation of the law of inertia in 1895, in 'Das Gesetz der Eindeutigkeit' (The Law of Univocalness). In particular, he takes a position on two fundamental issues: the validity of the law, that is, whether it should be considered an empirical proposition or an a priori principle, and whether it is necessary to postulate absolute space as a frame of reference for the inertial motion of a body.

On the first point, Petzoldt splits the law of inertia into several statements, all related to the principle of *Eindeutigkeit*. In the first place, the law tells us that a body cannot modify its state of rest or motion spontaneously, that is, in the absence of any conditions that determine such change. Petzoldt thus harks back to an earlier argument of Euler's, who had based the law of inertia on the principle of sufficient reason.[48] The law of inertia would thus have the same validity as the principle of *Eindeutigkeit*, which – as we have seen – Petzoldt regards as a condition of possibility not so much of experience (as Kant did) as of our existence. Petzoldt therefore writes, 'we cannot imagine that a body would change its state "by itself" unless we want to give up [*aufgeben*] on ourselves'.[49] The law of inertia is thus neither empirical in the sense of deriving from a specific experience or series of experiences nor purely a priori, but rather, it depends on human evolution in an environment in which everything that happens is univocally determined.

In the second place, for Petzoldt the law of inertia contains a 'conditional' proposition:[50] *if* the motion of a body is in a straight line and uniform, *then* it proceeds in this motion unless some external force intervenes to change its state. This means that the law of inertia is not strictly speaking an empirical proposition but has the same validity as the 'principles of geometry, which, as postulates, can be neither "confirmed" nor disproved by experience'.[51] Indeed, just as geometrical figures are idealized constructs with respect to the shapes found empirically in objects, so uniform motion in a straight line is an ideal abstracted away from the actual motion of bodies.

In both cases, the law of inertia states that, in the absence of other factors that lead to a change of state, a body at rest remains at rest and a body in uniform motion along a straight line remains in uniform motion along a straight line. Petzoldt also considers another case, however. The law of inertia states that a body proceeds in uniform motion along a straight line 'after a single push' (*auf einen einmaligen Anstoss*). In the first two cases, the law of inertia tells us what happens *in the absence* of a means of determination, *in the absence* of change, but in this case it specifies what happens when there is a means of determination (such as a push), that is, when the state of the body changes.

Introducing a 'push' – albeit a 'single' push – into the law of inertia seems somewhat curious, given that, by definition, this law pertains to a body upon which *no force* acts. In any case, even here Petzoldt splits the issue into two parts. The first questions why we assume that the body in question follows a trajectory along a straight line. The second questions the assumption that the motion is uniform and has a constant velocity. With respect to the straight trajectory, Petzoldt states that this is the one that is determined univocally, in the specific sense that it is the exceptional case that stands out against the infinite number of possible cases. There is in fact an infinite number of lines that link two points, but only *one* straight line.[52] To say that the body follows a straight line 'after a single push' is therefore to some extent valid a priori, as the principle of *Eindeutigkeit* allows us to *know in advance* that the case that will be realized is the one that is in some way unique. The question is more complex when it comes to uniform motion, in that we could, for example, think of the motion of the body as accelerating uniformly 'without compromising the principle of *Eindeutigkeit*'.[53] Petzoldt therefore concludes that 'in this case it must be the individual experience, that is, the experiment, that univocally determines thinking'.[54]

In the 1901 edition of *Mechanics*, Mach examines this passage in 'Das Gesetz der Eindeutigkeit' and describes Petzoldt's position on whether the law of inertia should be considered 'an axiom, a postulate or a rule':

> Petzoldt deduces the law of inertia in part only from experience, and regards it in its remaining part as given by the law of unique determination. I believe I am not at variance with Petzoldt in formulating the issue here at stake as follows: It first devolves on experience to inform us what particular dependence of phenomena on one another actually exists, what the thing to be determined is, and experience alone can instruct us on this point. If we are convinced that we have been sufficiently instructed in this regard, then when adequate data are at hand we regard it as unnecessary to keep on waiting for further experiences; the phenomenon is determined for us, and since this alone is determination, it is uniquely determined. [...] I believe I am right in saying that the same fact is twice formulated in the law of inertia and in the statement that forces determine accelerations.[55]

Although Mach tries to downplay the differences between his conception and Petzoldt's, it is clear that his notion of determination is decidedly weaker. Mach holds that statements that every event must be univocally determined, including therefore the law of inertia itself, are simple empirical generalizations, obtained through induction. For Petzoldt, on the other hand, the principle of inertia and that of *Eindeutigkeit* derive from experience in the sense that they have been etched into us by evolution, and their absence would be a threat to our preservation. Petzoldt himself states, in a footnote in *Das Gesetz der Eindeutigkeit*, that he 'cannot agree with Mach when he writes that "all forms of the law of causality spring from subjective impulses, which nature is by no means compelled to satisfy"', as 'our very existence demonstrates that a "necessity" of this kind undoubtedly exists'.[56]

In *Das Gesetz der Eindeutigkeit*, Petzoldt also takes a position on the relationship between the law of inertia and absolute space, accepting 'the objections of Mach and Neumann' with respect to the need to 'clarify what is meant by rest, straight trajectory and uniform motion'.[57] Petzoldt states that 'there is no need to introduce a "Body Alpha"', as suggested by Neumann, as it is sufficient to 'see ourselves as references bodies'.[58] Petzoldt holds that this is what already happens 'tacitly', as even when 'we try to eliminate mentally all bodies expect the one whose motion we are considering, this cannot include our own body', as we cannot avoid 'thinking of ourselves as in some kind of spatial relationship with the body in question'.[59]

Petzoldt therefore states that he 'fully agrees with Mach' that 'rotation does not provide any *greater* validation of the assumption of "absolute" motion than it does for translation', adding, however, that 'what is relative' is for him 'the relation to ourselves and not to the fixed stars', in contrast with Mach's position.[60] Given that 'we always think of ourselves relative to the bucket of water', it is not the case that 'we actually need the fixed stars to think of its rotation'.[61]

It is therefore clear that Petzoldt does not understand that Mach's use of the fixed stars as a frame of reference is not simply about the 'relativity of representation', that is, the standpoint from which a phenomenon is observed and described. It is also about the 'relativity of bodies', that is, the fact that 'physical processes are determined by the relations of bodies with each other in the world'.[62] Appealing to ourselves, to our own bodies as a frame of reference implicit in any assessment of the motion of objects, might be an alternative to using the fixed stars as a tool in the kinematic description of motion, but not to the hypothesis that the masses of the universe have dynamic effect, as conditions of the force of inertia, unless one also wished to maintain that the centrifugal forces in the water in Newton's bucket, and even in Neumann's fluid star, are in fact determined by our own bodies.

Mach noted Petzoldt's 'rejection of absolute motion', but in the third edition of *Mechanics* he also added: 'I cannot understand how all the *physical* difficulties involved in the present problem can be avoided by referring motions to one's own body'.[63] The problem is not simply that of having a *geometrical* reference point against which changes of position can be assessed but to recognize that inertia and centrifugal force involve the *physical* dependency of a body on the other bodies in the universe. Moreover, referring to the tenets of *The Analysis of Sensations*, Mach reiterated that a key feature of physics research into dependency relations is precisely that 'abstraction must be made from one's own body, so far as it does not exercise any influence'.[64]

The long-distance debate with Mach

Following the publication of the third edition of *Mechanics*, which contained the references to 'Das Gesetz der Eindeutigkeit' noted above, Mach sends the volume to Petzoldt with a request to redact a new edition, probably fearing that his poor health might prevent him from doing so himself.[65] After re-reading the book, Petzoldt replies to Mach in a letter that goes into great detail on where their views converge and where they diverge. He thanks him for referencing his own position in the volume, and makes it clear that they occupy 'the same ground' with respect to their 'conception of mechanics' and to their 'conception of the difference between physics and psychology', namely that 'they do not differ in terms of what they investigate, but in the direction from which they approach it'.[66]

We cannot be certain, however, that at the time of writing this letter, in 1901, Petzoldt had already developed his own vision of the difference between the perspective of physics and that of psychology, as this was only laid out in the second volume of the *Einführung*, in 1904. As we noted at the end of the second chapter, Petzoldt follows Mach's line in the book, maintaining that the two disciplines do not differ in terms of the object of research, but of the 'direction from which they observe things and processes'.[67] At the same time, Petzoldt distances himself from Mach in seeing research into the dependency upon the body as part of physics rather than of psychology. For Petzoldt, the study of physical connections is in general the study of the relations of univocal determination that are given in reality, even where one of the two terms of the dependency relation is the bodily individual, whereas psychology addresses the simple and disorderly sequence of empirical representations as they are given.

Although we cannot be certain of Petzoldt's thinking in 1901, there is no doubt that the conception of the difference between psychology and physics outlined in the *Einführung* is fully consistent with the view that physics can use the human body as a reference frame for the description of the motion of objects. Despite his declaration that he and Mach occupy the same ground, it is therefore clear that Petzoldt distances himself from Mach when he attributes to physics a perspective that starts with the individual, both in the sense of the psychophysiological study of the organism and that of using one's own body as a reference frame for the kinematic description of motion. For Petzoldt, both cases involve univocally determined relations and therefore pertain to physical knowledge of the world.

In his commentary on the new edition of Mach's *Mechanics*, Petzoldt also states that he has 'become even more convinced' of Mach's thesis that the law of inertia is simply a version of the hypothesis that forces are accelerations, whether positive or negative, and therefore that in the absence of forces a body maintains its state of rest or of constant velocity.[68] However, even more importantly, Petzoldt 'concedes' that '*from the perspective of physics* the issue is only ever to describe the movement of a body relative to the Earth or to the fixed stars'.[69] This suggests that he recognizes that one's own body cannot be a condition for the law of inertia or for centrifugal forces *in this sense* (i.e. from a dynamic perspective). However, Petzoldt adds that 'with all due respect' for Mach's approach to these issues, he 'cannot concur with his view that [...] the *physics* question exhausts *all* the questions inherent in the law of inertia'.[70] In particular, although Mach's framework eliminates all '*physics* problems', it leaves unanswered some 'gnoseological' (*erkenntnistheoretisch*) questions.[71] Indeed, according to Petzoldt, 'issues such as Newton's bucket, or Neumann's celestial body, are no longer *physics* problems' but problems of the theory of knowledge.[72]

Petzoldt therefore disagrees with Mach about Neumann taking his thought experiment too far by distancing himself from experience to the point of trying to imagine what might happen if the whole universe disappeared apart from one celestial body. Petzoldt emphasizes that 'without assumptions of some kind it is impossible to go beyond what is immediately given' and it is therefore not easy to determine where and how 'to draw the line between the legitimate and the illegitimate use of thought experiments'.[73] As long as 'the laws of thinking and the necessity of *Eindeutigkeit*' are respected, therefore, thought experiments should be considered admissible'.[74]

Petzoldt not only defends Neumann's fluid star hypothesis but adopts it himself, proposing a slightly different version. If we imagine removing all celestial bodies *one at a time*, there should be no changes in the shape of the fluid star rotating on its axis. It would remain a flattened ellipsoid. When we come to the last remaining celestial body, what would happen 'when the last infinitesimal remnant of the last fixed star disappeared'? Should we perhaps assume that 'the celestial body in question would suddenly take on the form of a sphere'? Petzoldt's reply is that any such thought is 'untenable', as it goes against the '*continuity* of the process' of gradual elimination.[75]

For Petzoldt, 'the resolution of this paradox', whereby the *continuous* process of removing bodies from the universe should eventually change, *all of a sudden*, the shape of the star into a sphere, does not lie in the fact that 'one cannot do without at least a part of the fixed stars, however miniscule', as Mach appears to say when he accuses Neumann of being too cavalier in his use of thought experiments.[76] Indeed, 'the abstraction was stretched too far; not with respect to the fixed stars, however, but with respect to one's own body'.[77]

It should be noted, however, that, contrary to what Petzoldt suggests, Mach held that the fixed stars could not be disregarded *as a whole*. It is therefore debatable whether he would have accepted Petzoldt's alternative version of Neumann's thought experiment. Mach would probably have refuted the assumption that the gradual removal of celestial bodies had no effect on the shape of the star, or at least rejected it as indemonstrable. Confirmation of this comes in the passage in the *Mechanics* on Newton's thought experiment where Mach writes: 'No one is competent to say how the experiment would turn out if the sides of the vessel increased in thickness and mass till they were ultimately several leagues thick'.[78] That is, not only do we not know what would happen if the whole universe disappeared but we cannot even formulate hypotheses about gradual changes in the masses of the universe (in the sense of their increase, but also – by analogy – of their decrease).

Several arguments are presented, albeit in a somewhat confused manner, in Petzoldt's reasoning that the unremovable element in Neumann's thought experiment, as well as in Newton's, is not the fixed stars but one's own body.

Firstly, Petzoldt casts doubt on the claim that our body 'has no influence'. We have seen that Mach's reply to Petzoldt noted that the peculiarity of physics is that 'abstraction must be made from one's own body, *so far as it does not exercise any influence*'. Petzoldt replies that 'by the same token, if not with even greater justification, when you say "but it is not to be antecedently assumed that the universe is without influence on the phenomenon here in question", one can likewise assert "That the I is without influence, etc. etc."'.[79] Petzoldt thus retracts what he had written at the start of the letter: '*from the perspective of physics* the issue is only ever to describe the movement of a body relative to the Earth or to the fixed stars'. He now attributes to one's own body not only the role of a geometrical reference point for the description of motion but also in terms of dynamics. His intention to reintroduce the role of the body *even from the perspective of physics* is confirmed by the passage in the letter where, having pointed out to Mach that he himself asserts that it is acceptable to abstract from one's own body, only so far as 'it does not exercise any influence', Petzoldt adds that in fact 'this is often not the case, even in physics research [...]. In particular, in mechanics the body is no less a *physical object* than any other celestial body, both as a whole and in terms of its parts'.[80]

We have already noted the problems faced by this argument, namely, the fact that reference to the stars as a *physical* condition for inertia and centrifugal force is not about the presence of any celestial body whatsoever but about the *masses* of the universe. The claim that the human body could act as a substitute *for the masses of the entire universe* is therefore a decidedly risky hypothesis. In any case, in partial justification of Petzoldt's position, it is important to note that this aspect of Mach's view became much clearer once Einstein had made it prominent, referring to it as 'Mach's principle'. It is therefore plausible that Petzoldt had not fully understood that Mach's position established a relationship between inertia and the masses of the universe.

The second line of argumentation found in Petzoldt's letter to Mach is perhaps more interesting from a philosophical point of view. Petzoldt uses his revised version of Neumann's fluid star to show the difference and at the same time the similarities between *physics* problems and *gnoseological* problems. The analysis of the effect of the gradual removal of the bodies of the universe on the shape of the rotating star can be seen as a matter for physics. However, as the experiment proceeds, at a certain point we reach the problem of the disappearance of the last body. Petzoldt sees this as a gnoseological problem, that is, the problem of whether we can know something that is *absolute*. The question arises of whether it is possible to imagine that the entire universe contains just one object. In such a case, we may in fact simply be forgetting about ourselves, and more generally about the fact that registering the presence of that single object requires the existence of a nervous system that is equipped with organs of perception and is able to establish some sort of relationship with the object.

This is what Petzoldt means when he writes: 'Why, therefore, can I not play the role of the fixed stars, not in a physics problem, but in a gnoseological problem that is linked to the physics problem by means of a continuous mental process?'[81]

> I cannot, of course, disregard this [my body] with impunity with respect to ultimate questions. But here we have, in fact, an ultimate question, that is, the gnoseological separation of mechanics from the rest of our knowledge, and the

not insignificant issue of whether using abstraction remains valid even when, as we proceed along a *continuous* path, we go beyond the constrained field of physics, or whether these are discontinuities that we must put a fence around; whether the laws of rotation – which we must see as independent of any part of the fixed stars and of any combination of such parts, and which we instinctively deem cannot be under the influence of any celestial body – can be maintained up to the limits of what is thinkable. Is it not completely natural that particular sciences may be faced with limiting questions, questions that challenge us to overcome or remove these limits that are in fact purely arbitrary?[82]

The *physics* issue ends up straying into the theory of knowledge when the issue of the effects of bodies on one another – as per the hypothesis of the gradual disappearance of the bodies – becomes the *gnoseological* issue of the thinkability and representability of bodies *in themselves*. Neumann's experiment does not make us ponder just the effect that the masses of the universe have on one another (as Mach believes), or just on the requirement of an absolute reference point for the description of motion (as Neumann believes), but takes physics to its limits, to the question of whether physics is able to represent bodies and their motion *in themselves*, absolutely, independently of any observer. An affirmative answer, as we have just seen, would entail 'the gnoseological separation of mechanics from the rest of our knowledge', which we know is always *relative*.

In Petzoldt's relativistic positivism, to say that physics is allowed to 'disregard our body' – albeit only 'as long as it has no effect', which may be uncertain – means that it can be considered an absolute science. Mach would thus end up legitimizing the very exceptionalism of physics that he had spent his whole life battling, that is, the idea that physics is the only science able to go beyond how the world is given to human beings and thus to grasp true reality as it exists beyond sensory experience.

From a gnoseological point of view, as Mach himself always maintained that physics had to keep its feet on the ground, so to speak, staying as close as possible to the concrete empirical data on which it is based, he more than anyone else must recognize that it is not possible to disregard the human body, which is the fundamental variable in all our relations with the other bodies of the universe. This is why Petzoldt turns Mach's own writings against him, citing two passages: the assertion in the *Mechanics* that, once we have fixed our gaze on the specific, 'we should not omit, ultimately, to complete and correct our views by a thorough consideration of the things which for the time being we left out of account';[83] and the following passage from *The Analysis of Sensations*:

> A magnet in our neighbourhood disturbs the particles of iron near it; a falling boulder shakes the earth but the severing of a nerve sets in motion the whole system of elements. Quite involuntarily does this relation of things suggest the picture of a viscous mass, at certain places (as in the ego) more firmly coherent than in others.[84]

Petzoldt thus concludes that 'it is easier to disregard the fixed stars than the I', as 'we undoubtedly remain closer to original experience when we abstract away all celestial bodies than when we abstract away our own body'.[85]

Let us recapitulate Petzoldt's arguments. (1) In general, physics disregards the observer's body when focusing on relations between bodies, but we cannot assume *in principle* that our body has no *physical* effect. In other words, the fact that our body has no physical effect can either be the result of empirical investigation or a provisional assumption, but in the latter case Mach's rule applies: 'we should not omit, ultimately, to complete and correct our views by a thorough consideration of the things which for the time being we left out of account'. (2) In terms of geometry, the description of the motion of a body always needs another body as a reference point. Using our own body for this purpose has the advantage that we always visualize spatial relations from our own point of observation, as shown by the fact that even when we consider 'the Earth moving round the Sun we usually imagine ourselves observing the scene from outside the solar system rather than from our actual location within it'.[86] (3) From a gnoseological perspective, Petzoldt's relativistic positivism states that we can never disregard our own body, that is, our nervous system and its relation to our environment, even when it comes to the science of physics. Indeed, as physics is nothing other than knowledge of the relations between the elements of the world *sub specie univocitatis*, it cannot and must not ignore the fact that our knowledge of the world is univocally determined by our cerebral substratum and by its relations with the world, which are likewise univocally determined.

From all three perspectives – physics, geometry and the theory of knowledge – we cannot disregard ourselves, our own body. The body can and must therefore serve also as a reference point in physics, both to maintain the link with concrete empirical data and to avoid any claim that physics research can lead to knowledge that is *absolute*.

These initial reflections on physical knowledge already contain the two pillars of Petzoldt's philosophical position: relativism and *Eindeutigkeit*. The first states that all of the components of reality are in reciprocal relations with one another, and – as we are also part of this reality – we can only conceptualize reality from our own point of observation, starting with the relations that hold between our body (and in particular our brain) and the rest of the world. The second, *Eindeutigkeit*, states that every relation between the components of the world, including ourselves and our brains, is univocally determined, in that, under the conditions given, the actual case is the only case that occurs from among the infinite number of possible cases.

Unfortunately, we do not know how Mach addressed Petzoldt's objections as his replies have been lost. Petzoldt does not return to this topic in his next letter, in which he simply informs Mach about his preliminary work on a new edition of the *Mechanics*. The only passage where there is some hint of a reference is where Petzoldt writes that he intends to 'leave the text unchanged' and to 'avoid any personal references in the argumentation, always remaining objective', as for him this is part of 'the ethics of writing'.[87] It may therefore be the case that Mach had asked Petzoldt whether he intended to add, perhaps as footnotes, the objections he had raised in his letter, and had received an appropriately reassuring reply.

In any case, there is no doubt that Mach must have replied to Petzoldt in one of his letters, countering his arguments, as Petzoldt is clearly referring to Mach's observations in his next letter when he writes: 'I have not yet reached a conclusion about the law of inertia. I have not yet found the time to reflect carefully upon the matter'.[88]

It is plausible to surmise that Mach responded to Petzoldt's philosophical and gnoseological arguments by shifting the focus to the *physics* issue of the dependence of inertia on the totality of the masses of the universe, or at least formulating the debate in terms of a purely scientific problem. This would be not only unsurprising but also consistent with Mach's approach, in that he was always somewhat wary of edging towards speculation of a more clearly philosophical nature. The most concrete evidence for this hypothesis, however, is the fact that over the following years Petzoldt decides not to formulate a position on these issues publicly or in his correspondence, having realized that he did not have the knowledge required to debate physics questions with Mach on an equal footing.

In a letter of the following year (1902) about reading the second edition of Mach's *Prinzipien der Wärmelehre* (Principles of the Theory of Heat, 1900) and his own work on the publication of the second volume of the *Einführung*, Petzoldt provides the following admission:

> I am impatient to finish [the *Einführung*] so that I can turn to theoretical physics, above all to become fully familiar with the third hundred pages of the *Wärme*. Then I will turn to Maxwell. I have horrendous lacunae and would need two years to get up to speed. If only I could undertake my studies over again from scratch! [...] In any case I think I am on the right track, even if perhaps in your eyes I am too constructive [*konstruktiv*].[89]

As well as acknowledging his own lack of relevant expertise, Petzoldt's words hint at a criticism that might have been levelled at him by Mach, that of being too 'constructive', where this term is clearly to be understood in the sense of a tendency to construct philosophical systems, in contrast to Mach's critical and cautious approach.

There is a return to the theme of his 'lacunae' in physics in a 1904 letter sent after visiting Mach. Petzoldt writes:

> I was delighted to find once again that we agree on the key points of our conception of the world. But I also carried away with me a somewhat oppressive awareness. I noted with renewed awe the vastness and depth of your knowledge, including of recent scientific advances, and my ignorance is an embarrassment to me. I have even come to the conclusion that my life so far has been spent writing too much and reading too little. I should have devoted more of my free time to acquiring a fundamental knowledge of physics and mathematics. I am no more than a Gretchen to your Faust.[90]

A 1906 exchange is also relevant. After resuming work on the *Mechanics* in preparation of his teaching duties, Petzoldt writes to Mach asking for clarification of some points where the mathematical proofs provided in the volume are unclear to him, in particular in the section on 'The Laws of Impact', where Mach analyses some of Galileo's experiments in a dynamics framework.[91] As before, Mach's reply is not available, but he must have provided abundant evidence that it was Petzoldt who was in error. Indeed, Petzoldt wrote to Mach to thank him for his detailed explanations and stating that 'having thought long and hard about the matter, I am now fully convinced of the correctness of your conception and of your mathematical analysis'.[92] Petzoldt also returns once again to the issue of his lacunae, reaffirming his intention to eliminate them:

> I am lacking in the practice of the practical application of principles. It has been my intention for some time now to devote myself to Fuhrmann's collection of exercises, or to some similar volume, but I have not yet had the time. Perhaps this coming winter will finally be the right occasion.[93]

This brief overview of the correspondence between Mach and Petzoldt reveals how Petzoldt's enthusiasm for the scientific advances of the day, including in the field of physics, comes up against his rather basic knowledge of the subject matter. This must be borne in mind when evaluating his forays into physics and will become even more evident when Petzoldt turns to Einstein's theories.

Petzoldt's critique of Lange

Having promised himself many times to deepen his knowledge of physics, Petzoldt finally had to find the right moment to embark on this task. In August 1907 he writes to Mach: 'Over recent years and especially last year, I have worked extensively on inertia and motion, and hope to be able to set down my views on the matter in writing shortly'.[94] The work in which Petzoldt expounds his ideas, 'Die Gebiete der absoluten und der relativen Bewegung' (The Domains of Absolute and Relative Motion), is published the following year in Wilhelm Ostwald's eminent journal, *Annalen der Naturphilosophie*, and consists of an elaborate critique of Lange's theory of the construction of inertial frames of reference for the description of motion.

Unlike his first forays into physics in 'Das Gesetz der Eindeutigkeit', Petzoldt's new work is based explicitly on the relativistic positivism he adopted at the start of the twentieth century. The work notably opens with these words:

> Man is himself part of the world, part of *his* world, the only world that is available for his perusal and his thinking. He is unable to attain any vantage point *above* this world, and can only observe every item from his original and invariable position. This is why there is only *relative* knowledge. In particular, this is true for the investigation of nature, of which we have knowledge only thanks to our senses, only in a functional dependence on our senses.[95]

In particular, Petzoldt's aim is to analyse the relationship between this 'broad sense' of relativism or, more precisely, relativism in a philosophical sense, and the 'narrow sense' of relativism where it has the more specific meaning pertinent to the 'problem of motion'.[96] It is important to bear in mind that, even though Einstein had published his first work on relativity two years earlier, Petzoldt's use of the term 'relativism' is unrelated to Einstein's theories, and is simply the hypothesis that 'motion without a frame of reference is unthinkable, and "absolute" motion is a meaningless concept'.[97]

Petzoldt goes on to distinguish between two different aspects of the issue of the relativity of motion: 'the *psychological* description of the immediate perception of the processes of motion', and their '*scientific* description'.[98] He asserts, of the first problem, that 'there can be no doubt that absolute motion does not exist with respect to immediate perception: here motion is entirely relative' as 'we cannot abstract away from ourselves, from the observer'.[99] Moreover, in this case 'the relationships of motion are fundamentally reversible [*umkehrbar*]', so that what is initially perceived as being at rest can later be perceived as in motion, and vice versa.[100] This is demonstrated by the fact that one can learn to alternate between perceiving oneself as in motion and the world as at rest, or oneself at rest and the world as in motion, even while walking.[101]

Somewhat surprisingly, however, Petzoldt does not hold that this concept of relativity can be transferred *sic et simpliciter* to the scientific domain:

When we are concerned with the motion of bodies independently of ourselves, in their univocal connections with the motion or states of other bodies, that is, to identify the natural laws of motion, a completely different question arises. Here, the immediately experienced visual space is replaced by a metric space (generally Euclidean) which is never experienced and cannot even be represented, a pure thought object [*Gedankending*], a mere concept. We univocally project our experience, our visual space, onto this concept [*auf ihn bilden wir unsere Erfahrungen, unseren Sehraum eindeutig ab*].[102]

In particular, 'only complete abstraction away from the subject makes metric space possible, as the space of natural science', since 'the visual spaces of individuals differ one from another', whereas metric space is constructed specifically so that it is 'the same for everyone'.[103]

Petzoldt thus picks up on some of the themes of Mach's discussion of space, first presented a few years earlier in *Knowledge and Error* (1905), where Mach distinguished between '*physiological* space' and 'conceptual *metric* space'.[104] Petzoldt's debt to Mach's book is confirmed in their correspondence. In a letter written in the year in which the book was published, Petzoldt tells Mach that he is reading the chapter on space and emphasizes how 'appealing' he finds 'the explanation of geometric space as a concept', even highlighting his emphasis with an exclamation mark. In particular, Petzoldt states that he 'concurs in large part' with Mach, specifying, however, that he finds 'the term "concept" insufficiently precise', in that it would be preferable to speak in terms of something that is '*like* a concept', given that metric space 'albeit not perceptible or representable, is nevertheless real [*wirklich*], in that it is the only thing that allows the physiological spaces of different individuals to be linked in a non-contradictory manner'. In this sense, it should therefore be considered 'real and independent of me' in the same way as 'the complexes of elements' that constitute the 'things' of the world.[105]

Going back to 'Die Gebiete der absoluten und der relativen Bewegung', the difference between perceived space and space constructed for the natural sciences means that it is necessary to recognize that 'relativism in the narrow sense', that is, in the sense of physics, 'is in no way an inevitable consequence of the relativistic conception of the world'. The discussion of whether absolute motion is possible from the point of view of physics and geometry must therefore be based on different grounds than simply asking whether we can perceive or represent the motion of a body *in itself*, and the issue cannot be resolved in haste or labelled meaningless.[106]

Petzoldt's position is that the scientific definition of absolute motion can only be 'the motion of a body in metric space, where we consider such motion *independently of the presence of any other body* with the exception of specific coordinates [*von den besonderen Bahngestalt abgesehen*]'.[107] Three questions arise: whether this concept of motion is 'logically valid'; if so, whether it is also 'physically valid'; and, if so, whether it is also 'useful' (*Zweckmässig*).[108]

With respect to logical validity, Petzoldt holds that 'there is nothing in the concept of space as defined in natural science that requires this space to contain anything'. Indeed, the space itself and the bodies located within it are two distinct and 'correlative' concepts. Moreover, as motion does not pertain to space but to things, space must be seen as 'an indispensable *condition* of motion'.[109] In purely logical conceptual terms, we can speak of a single body in motion in abstract metric space, even if such motion cannot be *represented*.[110] The answer to the first question is therefore that the concept of absolute motion does not lead to any conceptual contradiction and is therefore to be considered logically valid.

With respect to the physical question regarding absolute motion, Petzoldt returns to Neumann's thought experiment: if we could remove every other body from the universe, leaving only the fluid star rotating on its own axis, then

> if it were legitimate or even necessary to think of the appearance of centrifugal forces as independent of the presence of other bodies, which has so far not been decided, [...] then it would be legitimate to describe or to define the state of that body as absolute rotation.[111]

Continuing his long-distance conversation with Mach, Petzoldt defends Neumann's thought experiment, conceding that his argument has nothing to do with 'facts' but only with 'representations'. However, these are not 'arbitrary representations, but representations that are linked via certain abstractions, in a completely natural way, to the facts observed'.[112] These reflections can therefore be seen as instances of the 'adaptation of thoughts to facts and to each other' that Mach himself holds to be the correct way to proceed in the pursuit of scientific knowledge.[113] Petzoldt recognizes, of course, that Mach would 'reject' these arguments as 'nobody can know how a solitary body in cosmic space would behave'. However, he remains unconvinced by the view that here we would come up against an insuperable 'limit of experience', because this objection appears to hark back to the kind of 'mystical' thinking that constantly resurfaces in human history.[114]

Neumann's thought experiment tells us, therefore, that absolute motion is not to be seen as an outlandish concept in physics, because if centrifugal forces depended on the motion of the body (which is neither confirmed nor denied) rather than, as Mach would have it, on the body's relationships with other bodies, then the *physical* concept of absolute motion would not be meaningless, at least for rotating bodies.

Petzoldt then turns to inertia and how it is addressed by Newton and by Lange. Petzoldt notes that Newtonian physics splits the motion of bodies into two components: the first, inertia, is absolute and belongs to the body itself, while the second, gravity, is relative and depends on the body's relationship to other bodies.[115] As a body would move uniformly and along a straight line even if there were no other bodies in the universe, under this view absolute motion does indeed exist, as motion in absolute space. As Petzoldt puts it:

> By means of this independent component, the body, if taken into consideration in isolation, would exhibit uniform rectilinear inertial motion, which – apart from time – would involve only the body and metric space.[116]

Petzoldt later notes that Neumann, Mach and Lange dispute Newton's assumption that 'the concept of inertial motion can be used *in physics* without defining a coordinate system or a temporal scale for the motion'.[117]

Petzoldt's account of Lange's solution for constructing a frame of reference to define inertial motion emphasizes the central role played by the 'phoronomic principle' (i.e. the kinematic principle) which states that 'whether three (or fewer) points should be considered to be moving along a straight line depends only on which coordinate system is being used'.[118] As we have seen, Lange's position is that 'for three or fewer points, rectilinear motion is a mere *convention*'[119] as 'for three points P, P′, P″' that are *moving arbitrarily relative to each other* it is always possible to construct a coordinate system in relation to which these points move rectilinearly'[120] and therefore 'the spatial content of the law of inertia only has any meaning in terms of a fourth point'.[121]

Petzoldt defines Lange's proposal as 'wrong' (*irrtümlich*) in no uncertain terms. Even if it purports to hold for points that are moving 'arbitrarily', which is necessary if one wishes to assert that their rectilinear trajectory is a mere convention, it in fact presupposes a series of conditions that constrain their motion.[122]

> There is no doubt that Lange did not wish to impose any limitation to the three imaginary geometric points in motion. His phoronomic research would otherwise have been meaningless, as he could have started with physical points 'left to themselves'. However, despite having the opposite intention, Lange in fact limits the freedom of movement of the geometric points, albeit tacitly and without realising that he is doing so.[123]

Lange assumed that the motion of the points is continuous and uninterrupted, that they do not reverse or oscillate and so on. In particular, he thought of them as in reciprocal motion, and therefore not moving independently of one another, either drawing indefinitely further away from one another or drawing closer to the point of meeting before beginning to draw away again indefinitely.[124] Petzoldt suggests that

Lange was in fact *already* thinking of the three points as in inertial motion, which means that it is not true that 'the spatial content of the law of inertia only begins with a fourth point not acted upon by any force'.[125] Petzoldt therefore concludes that it is mistaken to assert that we can speak of inertial motion only in terms of a fourth point. It is equally mistaken to assert that the definition of the frame of reference *precedes* the definition of the principle of inertia, as Lange holds. In fact, we already think of the three points as moving inertially; that is, we think of their motion as *absolute*, exactly as stated by Newton.

If Lange were correct, three points not acted upon by any force could move in an arbitrary manner, or rather, nothing could be said about their motion in terms of *physics*, given that any trajectory could be attributed to them, by convention. There would therefore be no independent component of the motion, meaning independent of the presence or behaviour of other points. However, Lange's inability to speak of the behaviour of the single point without surreptitiously attributing to it the inertial motion that in theory has not yet been defined means that it cannot be denied that inertia actually belongs to the properties of the single point.[126] In other words, Lange merely demonstrated that we need three points *in inertial motion* to define a fourth point in inertial motion. This would mean that he is not actually defining rectilinear motion in order to *then* be able to introduce the law of inertia, but is using the law of inertia to define rectilinear motion.

To summarize, Petzoldt asserts that on the one hand we have Newtonian physics, which starts from the absolute motion of a single point that moves with a rectilinear uniform motion in metric space, and on the other hand there are the 'relativists' such as Lange, who strive to demonstrate that the motion of a body without reference to other bodies is unthinkable. However, as the relativists in fact tacitly assume that *every single point* moves in a uniform and rectilinear manner, Newton is correct, and inertial absolute motion is the cornerstone of physics. As Petzoldt writes, absolute motion 'disposes of [*beseitigt*] the arbitrary coordinate system required by Lange, as without this [absolute motion] he would be unable to speak of motion in general'.[127] The result is that 'a rectilinear trajectory can be considered an ultimate fact that cannot be reduced further or a component that needs no further explanation in the construction of actual motion within space as conceived in natural science'.[128]

Petzoldt concedes that there is something 'disquieting about basing a theory on something that can never be experienced and cannot even be thought', such as the absolute motion of a body, but he maintains that:

> this disquiet vanishes with the realization that the metric space of natural science that every natural fact must be related to, and therefore the geometric shape of objects […], are no less beyond the limits of experience and are in fact concepts justified only by their function in the description of natural events.[129]

For this reason, 'there is no obstacle to allowing the concept of absolute motion' and above all 'there is no obstacle to basing a theory of physics on this concept'. Indeed, any 'resistance' to doing so must be seen as the result of a 'confusion between metric space and visual space', in that it is only in visual space that everything is relative and it is not possible to speak of absolute motion.[130] Petzoldt therefore modifies the position he had laid out in 'Das Gesetz der Eindeutigkeit':

> The assertion that we ourselves are the frame of reference that can never be put to one side, even in thinking, and that Neumann's rotating body can therefore never be thought of in absolute terms ('Gesetz der Eindeutigkeit,' 192 ff.), refers to visual space and not to metric space. I was clearly not fully aware of this distinction at the time.[131]

Petzoldt therefore concludes that the concept of absolute motion is valid in physics research, thus replying in the affirmative to the second question he had set himself.

Turning to the third question, on the usefulness of this concept, the issue takes on a new form. According to Petzoldt, 'there is no doubt that we must in general prefer a relativist position over an absolutist one', as the former is the only one that can provide us with a 'picture that is more faithful to the facts, that is closer to the visual space of actual experience'.[132] Indeed, we must not forget that the development of knowledge proceeds until 'no conceptual structure that is closer to the facts is possible. As this would mean that the state of maximum economy of thought or maximum stability had been achieved'.[133] Newtonian physics is thus superseded not by Lange's position but by Mach's use of the fixed stars and the angle of rotation of the Earth as a spatiotemporal frame of reference, since this represents 'the simplest and most natural relativistic perspective, and the one that most closely matches the facts observed hitherto'.[134]

For the first time, Petzoldt appears to understand the purely *dynamic* role, and not just the kinematic role, attributed by Mach to the fixed stars in his conception of inertia. In his early works, Petzoldt had not grasped that Mach did not see inertia as dependent on the presence of *just any* reference body, but on the *masses* of the universe. Petzoldt now writes that 'the reference to the fixed stars already contains the law of inertia', as 'inertial motion is the logical correlate of acceleration', and if '*the concept of mass* contributes to the description of processes of motion, then it is absolutely impossible to dispense with the fixed stars as a reference system'.[135] Petzoldt thus finally recognizes that the crux of the problem is not so much the need to have a body to which motion can be related, but the dynamic role of the *masses* of the universe. He therefore declares that he 'agrees unreservedly with Mach' that the *dynamic* concept of mass lies at the heart of physics.[136]

The only limitation of Mach's position is that it sees inertia as a property of the relationship between the masses of *our* universe rather than as a property of bodies *in general*, whereas 'the human search for knowledge focuses on the discovery of *immediate* connections and the formulation of *general* laws'.[137] We would therefore only be able to decide whether Mach or Newton is *truly* correct if we could actually discover how a solitary body behaves, or if we could show that inertia is indeed the result of the interaction of masses. This would allow us to further generalize Mach's assumption about the relationship between the fixed stars and inertia, formulating a *general law* on the relations between the masses of bodies and not simply a description of the relations between the masses of our universe.

If we could use crucial experiments to resolve the matter, 'the outcome of such experiments could decide the fate of the Newtonian system and of the concept of absolute motion'. In the case of a negative result, we should not hesitate to set it aside, 'especially as we would only be getting rid of a conventional thought object, of a definition'.[138] If an experiment might one day succeed in demonstrating that inertia does *in fact* depend on the relationship between the bodies of the universe, then on that day the relativist theory of motion would no longer have to be considered only provisionally 'useful' (pending further knowledge on the laws of the universe and the dynamic properties of masses), but its usefulness would have become *stable*. We know, of course, that Petzoldt saw *stability* as the mark of any true scientific knowledge.

From what we have laid out above, it is clear that 'Die Gebiete der absoluten und der relativen Bewegung' is atypical of Petzoldt's intellectual development, above all for its favourable reception of Newtonian physics and the concept of absolute motion, which explicitly conflicts with the views he had initially expressed. This apparent incongruence can be explained by thinking of this work as the result of a conflict between the two fundamental principles of Petzoldt's philosophy: relativism and *Eindeutigkeit*.

Petzoldt's openness to Newton's conception of inertia derives from the fact that it is impossible for us to think that the motion of a solitary body *is not univocally determined*, that a physical point can move in any arbitrary manner of its own accord, thereby allowing the trajectory we attribute to it to be considered a matter of pure convention, as Lange would have it. This is why Petzoldt writes that the 'fatal' element in Lange's theory is 'its indeterminacy, the absence of laws'.[139] In contrast, Newton's position of basing physics on the body's inertial motion establishes that even the motion of a solitary body is univocally determined.

For Petzoldt, relativism can never mean the absence of determination, that is, plurivocalness. The value even of Mach's form of relativism, with its reference to the fixed stars, is not that *we cannot know how a body would behave in the absence of the other masses of the universe* (i.e. in the negative, as the absence of determination, or at least the suspension of judgement), but is the notion that inertia is *univocally determined by the masses of celestial bodies* (i.e. as *positive* knowledge, like a law of nature, albeit one that remains to be demonstrated).

Einstein's theory of relativity and relativistic positivism

Einstein and special relativity

At the start of the twentieth century, physics was grappling with a number of problems in electromagnetics and optics. Electromagnetism had taken a great leap forward with the publication in 1873 of James Clerk Maxwell's *A Treatise on Electricity and Magnetism*, in which he proposed a series of equations describing the fundamental laws governing electromagnetic interaction. However, a number of experiments suggested that Maxwell's equations were not invariant with respect to every inertial frame of reference, and were thus inconsistent with the fundamental principle of Galilean relativity. Physicists therefore began to wonder whether Galilean relativity was valid only within mechanics, or whether, despite its extraordinary efficacy, Maxwell's system of equations contained some kind of error, or whether there might be some way to reconcile the two through a transformation system different from the Galilean one.[140] The first to solve this problem was Hendrik Antoon Lorentz. The Lorentz transformations preserved Maxwell's equations between a frame of reference at rest and one in uniform rectilinear motion with respect to it by attributing a different temporal metric to the latter, which Lorentz referred to as 'local time'. However, Lorentz did not maintain that there was an actual difference in the flow of time between the two systems, and saw local time merely as a mathematical artefact whose validity consisted in its usefulness in performing calculations.[141]

In optics, the nineteenth century had seen the temporary abandonment of the corpuscular theory of light of the Newtonian system, and the ascendance of the wave theory formulated by Thomas Young and Augustin Fresnel, which was supported by Maxwell's discovery that electromagnetic waves propagated at the speed of light. The conviction that a wave required a medium through which to propagate had therefore led to the hypothesis of a luminiferous aether through which light waves could travel. Various experiments were devised over the years to try to determine the behaviour of this aether and thus of the light that travelled through it. In particular, the objective was to determine whether the aether was completely at rest, as maintained for example by Lorentz, or whether it was dragged by the objects that travelled across it, as sustained by Heinrich Rudolf Hertz.

In 1851, to study the drag effects of a medium on the speed of light, Hippolyte Fizeau devised an experiment in which an apparatus of tubes and mirrors sent light through water moving in opposite directions. Earlier experiments by Fizeau and Foucault had shown that the speed of light through a medium like water was determined by the speed of light in a vacuum and the refractive index of the medium. If the aether was dragged by the motion of the water, observations from the stationary laboratory should show a faster speed of light when it was travelling in the same direction as the water and a slower speed when it was travelling in the opposite direction. In contrast, if the aether remained at rest, the motion of the water should have no effect on the speed of light. Fizeau's experiment produced ambiguous results, as the speed of light did change according to the motion of the water, but it did not add up to the speed of the water according to Galilean principles of the composition of motion. In particular, the change in the speed of light was consistent with the Fresnel drag coefficient, which depends on the refractive index of the medium. This means that the change in the speed of light could not be attributed to the effect of a partial dragging of the aether by the water, as the refractive index depends on the frequency of the light wave, so that the speed of two rays of light of different colours is affected differently by the motion of the water. To account for the effect, it would have been necessary to hypothesize the existence of *different* aethers for each frequency of light, with each dragged differently by the motion of the water. Any such hypothesis was, of course, not tenable. Fizeau's experiment therefore appeared to show that the aether was *not* dragged by the motion of bodies and was completely at rest, as maintained by Lorentz.[142]

In an attempt to demonstrate the presence of a luminiferous aether that was completely at rest, in 1887 Albert Abraham Michelson and Edward Morley set up an experiment with an interferometer to detect the 'aether wind' that should have been caused by the motion of the Earth through the aether. Their method was to use a system of mirrors to split a single beam of light into two perpendicular beams. The speed of one of the two beams should show a phase difference according to whether the motion was with or against the aether wind, that is, according to whether or not the sum of their speeds equalled the speed of the Earth moving through the aether. Contrary to expectation, the Michelson-Morley experiment showed that the speed of light did not change according to the orientation of the interferometer, and was not dependent on its direction relative to the motion of the Earth. This result contradicted Galilean relativity, whose transformations should have led to the speed of light adding to or subtracting from the speed of the Earth relative to the aether. The only other possible explanation for this negative result was that the aether was dragged by the motion of the Earth, with the Earth therefore remaining at rest with respect to it, but this option conflicted with the results of Fizeau's experiment.

An explanation of the Michelson-Morley experiment was proposed by Lorentz, whose position as an exponent of the theory of a stationary aether required him to account for the negative result of the attempt to verify the presence of an aether at rest. Lorentz hypothesized that the motion of a body relative to the aether produced a contraction of that body along the longitudinal axis of the motion. The reduction in the speed of the beam of light travelling against the aether wind would therefore not register, as it would be compensated for by the reduction in the distance it had to travel, given that the relevant arm of the interferometer would have been shortened by its motion relative to the aether. Lorentz therefore amended his system of transformations, adding a relation between the speed of a body and its contraction in length.

Thanks to Lorentz, time and the length of a body (a spatial parameter) became variable values dependent on the state of motion or rest, thus opening the way for the relativization of the dimensions of space and time later proposed by Einstein. Nevertheless, as well as considering 'local time' to be a mere expedient device in calculations, Lorentz continued to place the concept of an aether at absolute rest at the centre of his conception, maintaining the idea of a privileged frame of reference. Moreover, Lorentz attributed to the aether itself the physical property of causing bodies that move relative to it to contract.[143]

In 1905, Albert Einstein published *On the Electrodynamics of Moving Bodies*, in which he addressed the issues we have been considering.[144] His solution was based on two fundamental principles: the principle of relativity and the principle of the constant speed of light. The first states that the laws of physics remain the same for any non-accelerating frame of reference. This extended the Galilean principle of relativity beyond mechanics to the whole of physics, including electromagnetism and optics. The second states that light travels at a constant speed in any frame of reference, irrespective of whether it emanates from a body in motion or at rest. This meant that it was no longer necessary to attribute to light a speed relative to the aether but simply a speed relative to any inertial frame of reference. The concept of the aether could therefore be dispensed with completely.

The key aspect of Einstein's theory (or rather, one of the many elements worthy of note) was that making the laws of physics the same in any frame of reference required not only the speed of light to be constant but also the relativization of other factors, such as time and space. In consequence, whereas Galilean relativity stated that the speed of a body was a value that was dependent on the frame of reference, in the sense that, for example, what is in motion from one perspective is at rest from another, in Einstein's theory of relativity time and space also became relative values. This meant that in two frames of reference in uniform motion relative to each other, the same laws of physics apply and light will have the same speed, but the dimensions of space and time will differ. *The same body* will therefore have a certain size if measured in one frame of reference and a different size if measured in another, and *the same interval* will have a certain duration if measured in one frame of reference and a different duration if measured in another. The most important point, however, is that the system of transformations proposed by Einstein, which were largely based on those proposed by Lorentz, guaranteed the possibility of determining how these measurements of space and time differed between different frames of reference.

Petzoldt and Mach address Einsteinian relativity

Only a few specialists addressed Einstein's work when special relativity was formulated in 1905. It took about five years for the new theories to attract the attention of thinkers from outside the field of physics, thanks also to the explicative publications that started to appear. The general public's awareness of Einstein's theory really took off after 1919, when the solar eclipse of 29 May allowed the deflection of starlight by the mass of the sun to be measured, providing experimental confirmation of Einstein's theory of general relativity. The spectacular experiment splashed Einstein across the front pages of the press, launching the popularization (or vulgarization) of the theory of relativity.[145]

Petzoldt's first reference to the theory of relativity is in a letter to Mach dated 22 September 1910, in which he writes:

> I have not yet worked much on Einstein's theory of relativity as outlined by Classen.[146] However, Einstein fundamental idea seems truly excellent. I wonder, however, whether he has completely freed himself from the absolute. For example, I do not see why the speeds of light c and c' must be the same. It strikes me that, from a gnoseological [*erkenntnistheoretisch*] perspective, it makes no sense to compare c with c', given that they are the ultimate measuring references (as Classen states in the fourth from last section of his paper), and yet I must assume that there is a relation between the two clocks that c and c' are dependent on. However, this relation can only be established on the basis of the speed of light. Is this not perhaps a vicious circle? I hope to be able to study the original work soon.[147]

Petzoldt thus begins to assimilate Einstein's theories, albeit indirectly, in 1910. The blow dealt to the concepts of absolute time and space, and the emphasis on the relativity of their dimensions with respect to any frame of reference, were obviously attractive to a 'relativist' like Petzoldt. However, it is clear that he had not yet fully understood Einstein's theories, declaring his perplexity at the fact that the speed of light remained the same in different frames of reference.

It may be the case that Mach had asked Petzoldt for his opinion of Einstein's work, given that he had already had the opportunity to come across relativity after the publication of Hermann Minkowski's famous lecture *Space and Time* (1909), in which Einstein's theories were described by representing space-time as a four-dimensional space.[148] We know that Mach asked the physicist Philipp Frank to help him understand the mathematical aspects of the work. Frank sent him a paper outlining Einstein's and Minkowski's theories in a more accessible way the following year.[149] Moreover, in the second edition of his *Die Geschichte und die Wurzel des Satzes von der Erhaltung der Arbeit* (History and Root of the Principle of the Conservation of Energy, 1909), Mach added a footnote in which he states that his discussion of the concepts of space and time not as autonomous entities but as forms of the reciprocal dependence of events proceeds in the direction of the *principle of relativity*, referring to Minkowski's paper.[150]

The footnote must be interpreted as Mach's first response to the criticisms of his work by the physicist Max Planck. Planck had published his *Die Einheit des physikalischen Weltbildes* (The Unity of the Physical World Picture) in 1909, attacking Mach's concept of economy and the reduction of knowledge to sensations. Planck considered such views to be dangerous to scientific progress, as scientists need to believe that what they are studying is real and must eliminate anything that is subjective, like sensations.[151] In 1910 Mach wrote a response to Planck, reiterating that recent works by Lorentz, Einstein and Minkowski proved the fruitfulness of his approach to matter, space and time.[152]

Further confirmation of the 'strategic' purpose of Mach's footnote on the links between his conceptions and the theory of relativity is provided by the fact that Mach himself sent Einstein a copy of the second edition of his work on the conservation of energy. Mach's covering letter has been lost, so we do not know what he wrote to Einstein when he sent him the work, but it is revealing that Einstein's reply explicitly takes Mach's part in his dispute with Planck: 'You have had such an influence on the epistemological [*erkenntnistheoretischen*] views of the younger generation of physicists that even your current opponents, such as, e.g. Herr Planck, would undoubtedly have been declared to be "Machist" by the kind of physicists that prevailed a few decades ago.'[153]

Returning to Petzoldt, we see that he updates Mach on the progress of his study of the theory of relativity in June 1911:

> I am delving deeper into the principle of relativity, one step at a time. I find it ever more interesting that people do not go beyond the opposition between appearance and reality, as in Berg,[154] whose exposition I find excellent, and greatly preferable to those of Classen or Gruner.[155] You write that you feel that there is still something missing from the principle of relativity from a gnoseological point of view.[156] I agree; this derives from the fact that the proponents of the principle do not yet see things freely enough, in spite of it being clear that in other ways they have no preconceptions. The claim that the speed of light is constant still seems to me to play a strangely '*naïve*' role. It is clear that it is constant in different space-time frames only due to a deformation, and the principle must assume this. But I see

mathematics as more advanced in the foundation of geometry than theoretical physics. Perhaps Minkowski should not have established a relation only between space and time, but should also have included the speed of light? Perhaps I have not yet fully understood the matter, but in any case I find the recent developments in theoretical physics as well as in mathematics extraordinary. This puts an end to any absolutism and apriorism. Old Protagoras rises from his grave; a resurrection that is followed by the effusion of the holy spirit of relativism.[157]

We can see that Petzoldt's enthusiasm for the new theory, even without a full understanding of it, comes with some perplexity about the speed of light being constant.[158] In contrast with the first letter, however, here Petzoldt tries to rationalize it (incorrectly) as a kind of mathematical postulate, or an expedient device in calculations, rather than as a physical fact.

In the year that this letter was written, Petzoldt includes an initial reference to the 'relativization of space-time by Einstein and Minkowski' in the new edition of the *Weltproblem*, emphasizing that it cannot be held inconceivable by anyone who had already adopted Mach's deconstruction of Newtonian physics.[159] This note was clearly a way of taking a stand in defence of Mach and against Planck's continuing attacks. Indeed, when he sends the new edition of the work to Mach, Petzoldt takes care to let him know that he has taken a position on the theory of relativity, emphasizing that this new footnote might be of interest to him.[160]

The following year, Petzoldt got in direct touch with Einstein, who was teaching in Prague at the time. Although the relevant letters are no longer available, Einstein is one of the signatories on the announcement of the foundation of the *Gesellschaft für positivistische Philosophie* (Society for Positivist Philosophy) in Berlin, of which Petzoldt was the main promoter. We can therefore surmise that Petzoldt had written to Einstein to inform him of the project and to ask for his permission to include his name as one of the signatories.

The same year, 1912, saw the publication of 'Die Relativitätstheorie im erkenntnistheoretischen Zusammenhange des relativistischen Positivismus' (The Theory of Relativity and its Gnoseological Connection to Relativistic Positivism), Petzoldt's first paper on the theory of relativity. This was a transcription of a talk Petzoldt gave at the *Deutsche Physikalische Gesellschaft* (German Physical Society) on 8 November 1912. In this brief ten-page article, Petzoldt lays out the common ground between his own relativistic positivism and Einstein's theory of relativity, analysing in greater detail what he saw as the links between Mach's conception and the recent developments in physics to which Mach had referred. In particular, as foreshadowed in the 1911 letter, Petzoldt saw the fundamental convergence of the two positions in the rejection of the opposition between 'being' and 'appearing', a rejection that is the main teaching of any form of relativism.[161] There is no such thing as a thing-in-itself, a special frame of reference, an absolute space and time that are 'true' reality with respect to which everything that is relative is 'mere' appearance. Everything that is relative must be seen as actually existing, as the fact that different perspectives lead to different accounts is not a contradiction but is in fact a necessary consequence of univocally determined relations.

The same notions appear in another work published by Petzoldt in the same year, under the entry 'Naturwissenschaft' (Natural Science) in the ten-volume encyclopaedic work *Handwörterbuch der Naturwissenschaften* (Dictionary of Natural Science, 1912–15). Apart from being the second place in which Petzoldt focuses on the links between the theory of relativity and relativistic positivism, this work is significant in that it demonstrates the prestige Petzoldt had acquired in scientific circles. It is noteworthy that he was assigned the task of drafting such an important entry – one that was to set the general philosophical frame for the whole encyclopaedia – in such an ambitious project for the Fischer publishing house, with contributions from eminent scientific figures of the day. Moreover, it is equally significant that Petzoldt was able to use the space allocated to him to express his personal vision of relativistic positivism as a necessary step in the development of human thinking.

The two works published in 1912 make Petzoldt the second philosopher to publish works on the theory of relativity, after the neo-Kantian Paul Natorp, who had addressed it two years earlier in his *Die logischen Grundlagen der exakten Wissenschaften* (The Logical Foundations of the Exact Sciences, 1910). Petzoldt himself mentions Natorp's work, noting that it only contains an 'apparent adoption' of Einstein's principle of relativity, while 'retaining an absolute framework' disguised within the field 'of pure mathematics and logic'.[162] Indeed, Natorp held that the theory of relativity confirmed the Kantian conception that *empirical* space and time are relative, which is precisely why it is necessary to presuppose pure (absolute, a priori) space and time as the foundation of the intelligibility of empirical space and time.[163]

Petzoldt saw Natorp's neo-Kantian interpretation of the theory of relativity as an attempt to find a compromise with Einstein's theories, a compromise that was not bold enough to fully embrace the radical nature of the rejection of absolute time and space. In contrast, relativistic positivism not only wholeheartedly adopted the results of the theory of relativity but had to some extent anticipated them. According to Petzoldt, 'Einstein's theory takes a step in the direction indicated by Mach' in that 'it makes the concepts of physics – in particular those that concern the shape of bodies and the temporal sequence of events, or the rate of a clock – more emphatically dependent on the observer's perspective […] than had ever happened in mathematical theories of physics'.[164]

We will turn to Petzoldt's interpretation of the theory of relativity in the next section, continuing for the moment with the evolution of his relationship with Einstein and the theory of relativity.

After the publication of his 1912 talk on relativity, Petzoldt continues to interest himself in Einstein's theories, following their development closely and commenting on them to Mach. His engagement with the scientific circles of Berlin allows him to keep abreast of ongoing progress, updating Mach continuously. In 1913, Petzoldt tells Mach that he has spoken with Max von Laue, who had recently been awarded the Nobel Prize in Physics for his research on X-rays and was very active in the analysis and development of Einstein's ideas. He told Petzoldt that Einstein was preparing a work on gravity in which he takes up Mach's idea that centrifugal forces are dependent on relative rotation.[165] This was in fact when Einstein was publishing the *Entwurf* (Outline of a Generalized Theory of Relativity and of a Theory of Gravitation), which – as the

German title suggests – contained the first 'draft' of general relativity, with gravitational processes described as space-time curvature using field equations.

As soon as he receives Petzoldt's letter, Mach contacts Einstein again, sending him the most recent edition of the *Mechanics*. Einstein responds by sending him the *Entwurf*. In his covering letter, Einstein emphasizes that his own theory provides 'a splendid confirmation of your [Mach's] brilliant investigations on the foundations of mechanics, in spite of Planck's unjustified criticism' as it substantiates the view 'that *inertia* has its origin in some kind of *interaction* of the bodies', as Mach had hypothesized for some time.[166]

At the start of the following year, Petzoldt publishes a lengthy piece entitled 'Die Relativitätstheorie der Physik' (The Theory of the Relativity of Physics, 1914) in the second volume of the journal of his Society for Positivist Philosophy. The work is an introduction to the theory of special relativity, starting with the conflicting Fizeau and Michelson-Morley experiments, and a critique of Lorentz's interpretation of the contraction of bodies in motion. As before, Petzoldt frames Einstein's theories within the broader context of relativistic positivism as based on Mach's work. However, the piece makes no mention of Einstein's new theories as contained in the *Entwurf*, even though they would have reinforced the links between relativity and Mach. Indeed, Petzoldt's correspondence reveals that he had not yet had an opportunity to study Einstein's latest work. In a letter in early March, Petzoldt writes to Mach to say that he had only been able to skim the *Entwurf*, which looks to him like a 'development' (*Weiterbildung*) of Einstein's earlier theory, and that he is planning to devote himself to it in the autumn.[167] In the same letter, Petzoldt thanks Mach for having sent him a letter of Einstein's, telling him that he had made a copy of it for his own use before returning it. Commenting on the contents of the letter, Petzoldt writes, 'I am very pleased to see Einstein's recognition of the importance of the gnoseological [*erkenntnistheoretischen*] perspective. One cannot overstate the importance of the fact that such an exceptional physicist shares your point of view'.[168]

It is usually held that the letter from Einstein that Petzoldt is referring to is one of the two sent to Mach in 1913, as they are the last two of the four that their correspondence consists of. However, the reference to a recognition of the affinity between Einstein's and Mach's gnoseological perspectives means that it cannot be excluded that it might be the 1909 letter, in which, as we have seen, Einstein tells Mach that he 'had such an influence on the epistemological views of the younger generation of physicists'. Indeed, in both of the 1913 letters, Einstein emphasizes the importance of Mach's work to the 'foundations of mechanics', referring to Mach's criticisms of Newtonian physics rather than to Mach's more general gnoseological position. It might therefore be the case that Mach saw the letter in which Einstein refers to the influence of his 'epistemological views' as of greater interest to Petzoldt, and that he therefore forwarded that one to him.

In any case, Einstein's approval is not limited to Mach's work. He writes to Petzoldt in 1914 with comments on his 'Die Relativitätstheorie der Physik':

> Esteemed Colleague! I read your comments on relativity theory in the *Zeitschrift für positivistische Philosophie* with much pleasure. From it I see with astonishment

that you are closer to me in your understanding of the subject, as well as with regard to the sources from which you draw your scientific convictions, than my true colleagues in the field, even as far as they are unconditional supporters of relativity theory.[169]

Although Einstein does not specify the nature of this closeness, it is clear from the rest of his letter that he shares Petzoldt rejection of the opposition between 'appearance' and 'reality' and thus his view that the relativistic effects of motion on space and time are not merely what 'appears' in a given frame of reference but are indeed reality. Indeed, Einstein refers to his new research, writing that 'there is no distinction between a "real" gravitational field and an "apparent" gravitational field produced through the acceleration of the reference system'.[170] Einstein closes the letter saying that he would like to meet Petzoldt to discuss such matters of shared interest. Indeed, in 1914, Einstein moves to Berlin to take up the Chair of Physics at the city's university, as well as the position of director of the new Kaiser Wilhelm Institute for Physics.[171] This made it possible for them to meet, as Petzoldt was now teaching at the Technical University of Berlin as well as at a college in the suburb of Charlottenburg.

Einstein's arrival generated so much interest in the city that the Vossische Zeitung', a Berlin daily, asked Einstein to write a short article to outline his theory. In the piece, Einstein presents the relativity of simultaneity as 'the most important and also the most controversial' aspect of his theories, adding that 'it is impossible to enter here into an in-depth discussion of the epistemological [*erkenntnistheoretischen*] and *naturphilosophischen* assumptions and consequences' of this principle. However, for 'those who want to familiarize themselves with a more detailed substantiation and justification' Einstein recommends reading two works: *Physikalisches über Raum und Zeit* by the physicist Emil Cohn,[172] and Petzoldt's 'Die Relativitätstheorie der Physik'.[173]

Einstein and his physicist friend Paul Ehrenfest paid Petzoldt a visit in June 1914. Petzoldt must have given Einstein one of his books, probably the second edition of the *Weltproblem*. A few days later, Einstein writes to Petzoldt: 'I have just finished reading your book with great interest, from which I gather with delight that I have long shared your convictions.'[174]

Over the following years, before Petzoldt's posting to the front during the Great War, they continued to meet in Berlin's scientific circles and elsewhere. We know that Petzoldt attended talks given by Einstein,[175] and they probably met at the capital's scientific events, and perhaps even at each other's homes.

Einstein's support of Petzoldt went beyond some word of appreciation given in private. It included concrete attempts to further Petzoldt's academic career by backing his aspirations to take up a chair in philosophy. In 1914, Einstein wrote a letter to his cousin and future wife Elsa, informing her that he had 'sent a very warmly phrased letter of recommendation to the minister' in support of Petzoldt, and that he was sure that 'it will not misfire'.[176] Then in 1918, the mathematician Georg Helm writes to Einstein asking him to write a new letter of support for Petzoldt, as the chair of philosophy in Dresden had become available. A decade earlier, Helm, together with Mach, Schuppe and the physiologist Ewald Hering, had tried to secure this position for Petzoldt.¹⁷⁷ Einstein acceded to Helm's request to support Petzoldt's application. We know this from a letter to Helm in the following year that closes with Einstein asking whether his 'recommendation was any help'.¹⁷⁸

On Mach's death in 1916, Einstein wrote a eulogy for the journal *Physikalische Zeitschrift* in which he emphasized once again 'the greatest influence' of Mach 'on the epistemological orientation of natural scientists'.¹⁷⁹ In particular, Mach had shown that 'concepts that have proven useful in ordering things can easily attain an authority over us such that we forget their wordly origin and take them as immutably given', storing them in the 'treasure chest of "absolutes" and "*a priori*"'.¹⁸⁰ Mach, as 'epistemologist', had 'paved the way' for the theory of relativity, fostering it 'directly and indirectly'.¹⁸¹ Einstein even states that 'It is not improbable that Mach would have hit on relativity theory when in his time – when he was in fresh and youthful spirit – physicists would have been stirred by the question of the meaning of the constancy of the speed of light'.¹⁸² Einstein also touches on the much-discussed problem of Mach's sensualism, pointing out that his claim that the aim of science is 'striving for order among individual elementary experiences "sensations"', had led to a 'sober and careful thinker' like Mach being 'mistaken for a philosophical idealist and solipsist' by 'those who are less familiar with his works'.¹⁸³ When Petzoldt returned from the front at the end of the Great War, he resumed his contact with Einstein. He wrote to Einstein in 1919 to ask for some clarification of the problem of the rotation of a rigid disc whose circumference should be smaller than that of a disc at rest as a result of the Lorentz contraction, a problem also known as the Ehrenfest paradox, after the first person to raise the issue.¹⁸⁴ After a brief exchange of letters in which he tries to explain the problem, Einstein suggests that the only way to explain the matter is to meet 'face to face'.¹⁸⁵

In 1920, after the experiment that confirmed the theory of general relativity had turned Einstein into a celebrity, Petzoldt writes to him again, requesting a meeting in order to discuss 'epistemological aspects of relativity theory'.¹⁸⁶ The experiment had made Einstein's theories a talking point in philosophical circles, in particular stimulating intense debate at the Halle meeting of the *Kant-Gesellschaft* on 29 May 1920. Einstein had been invited to the conference by the society's president, Hans Vaihinger. He had initially accepted, and his participation was noted on the invitations. He declined later, however, having received a letter from Max Wertheimer, the initiator of *Gestaltpsychologie*, warning him of the significant risk to his scientific reputation of stepping into a trap laid by philosophers who were confident that their sophistry could demolish his theories. Indeed, the conference programme included a prominent role for the philosopher Oskar Kraus, one of the main critics of Einstein's theories.¹⁸⁷

Moritz Schlick, whose 1915 paper had made him the other main philosopher supporting the theory of relativity,¹⁸⁸ was also unable to be present at the Halle meeting. Petzoldt was thus the only attendee in the role of defence attorney for Einstein against Kraus and the other philosophers who opposed his theory. Some time later, Petzoldt writes to Einstein to let him know about the outcome of the conference, pointing out the confusion of philosophers about the theory of relativity, and in particular about its experimental proofs, but also about the interest it was generating in young scholars. Returning to the proposals about dialogue between philosophers and scientists that had enlivened the foundation of the since defunct

Society for Positivist Philosophy, Petzoldt closes his letter to Einstein on the note that that 'full clarity on all these problems will only be possible if physicists and philosophers discuss them together'.[189]

Einstein replies that he is in favour of such a meeting, 'provided only those people join whom we ourselves invite', showing that he had taken on board the recommendation to be wary of attending gatherings of philosophers. Einstein suggests that Petzoldt should visit him at home.[190] We do not know whether Petzoldt succeeded in organizing the meeting with Einstein and a few 'select' philosophers and scientists, as the 1920 letter is the last of their correspondence to have survived. It may be the case that Einstein, now that he had been propelled to the position of a celebrity invited to speak at the most prestigious universities in the world,[191] no longer had the time or the inclination to engage with a philosophical milieu that often appeared to be more interested in using the theory of relativity for its own ends than in actually trying to understand the recent developments in physics. It might also be the case that the posthumous publication of Mach's *Prinzipien der physikalischen Optik* (Principles of Physical Optics) in 1921 made Einstein distance himself from Mach's approach, including that of his main living representative, Joseph Petzoldt. Indeed, the Preface to this work contains Mach's famous disavowal of the theory of relativity:

> I gather from the publications which have reached me, and especially from my correspondence, that I am gradually becoming regarded as the forerunner of relativity. I am able even now to picture approximately what new expositions and interpretations many of the ideas expressed in my book on Mechanics will receive in the future from the point of view of relativity. [...] I must, however, as assuredly disclaim to be a forerunner of the relativists as I withhold from the atomistic belief of the present day. The reason why, and the extent to which, I discredit the present-day relativity theory, which I find to be growing more and more dogmatic, together with the particular reasons which have led me to such a view [...] must remain to be treated in the sequel.[192]

Thanks to the painstaking research of Gereon Wolters,[193] we now know that these words are the apocryphal work of Mach's son, Ludwig, who had shifted his position towards that of the mathematician and philosopher Hugo Dingler, an opponent of the theory of relativity. Ludwig Mach wanted to counter the pro-relativistic interpretation of his father's work that Petzoldt had promoted over those years and explicitly in the Appendix to the most recent edition of Mach's *Mechanics*.[194] There was clearly no reason for Einstein and his contemporaries to question the authenticity of the Preface to the *Optik*, so Mach's repudiation of the theory of relativity became a given in philosophical and scientific circles, gaining ground over the years and becoming a commonplace of the history of philosophy that is still in circulation today, despite the many rebuttals. A quick glance through the reviews that appeared when the work was published shows that Mach's text itself was seen as a senile work of meagre originality and attracted little attention, whereas the rejection of the theory of relativity by one who had until then been seen as its prime precursor, thanks in part to Petzoldt, immediately caused a furore.[195]

The publication of the fake Preface added Mach, albeit posthumously, to the increasingly vocal ranks of Einstein's opponents. From the 1920s onwards, the scientific and philosophical objections to the theory of relativity became enmeshed with political objectives linked to the anti-semitism that was increasingly plaguing Germany, and Europe in general. This turned into a true and proper propaganda campaign against Einstein and his theories, spurred by far-right political activists and culminating in the compendium volume *100 Autoren gegen Einstein* (100 Authors against Einstein, 1931).[196] One can only imagine Einstein's dismay when, in addition to the hostile general climate, the Preface to the *Optik* expressed a rejection of his theories by one of the thinkers who had influenced him most, and who had been a role model for him up to that point. This disappointment led Einstein to disavow Mach's influence almost completely. When he was asked about his stance on Mach's ideas at a conference in Paris in 1922, he replied:

> *From the logical point of view there does not seem to be much relation between the theory of relativity and Mach's theory.* For Mach there are two points to distinguish: on one hand, there are things that we cannot budge: these are the immediate facts of experience on the other hand, these are concepts that we can, on the contrary, modify. Mach's system studies the relations existing between the facts of experience; the ensemble of these relations, for Mach, is science. *This is a false standpoint here all in all*, what Mach had made was a catalog, not a system. *As good as Mach was as a mechanician, he was a deplorable philosopher.*[197]

These words clearly show a full turnaround with respect to his eulogies of Mach up to that point.

Ludwig Mach's falsification irreparably damaged the tapestry that Petzoldt had patiently woven over many years to link Einstein's theories with Mach and relativistic positivism. Nevertheless, Petzoldt continued to defend the view that Mach's work was the *erkenntnistheoretisch* groundwork for the theory of relativity and the theoretical framework within which it received its truest sense. In particular, Petzoldt tried to attribute Mach's apparent unprecedented rejection of Einstein's ideas to his poor health in the last years of his life, begging Ludwig Mach 'not to toss these matters into the already painful debate on the theory of relativity', especially in the light of the fact that 'if your father was alive today and in good health' he would have kept well away from those who had transformed 'a scientific debate into the lowest form of anti-semitic propaganda'.[198] For his part, in his post-1921 writings Petzoldt advocated preventing the 'unfavourable opinions expressed in the *Optik* on the current theory of relativity' from sending into oblivion Mach's great achievements in all his previous works and his battle against the concepts of absolute space and time.[199]

Petzoldt's interpretation of special relativity

Petzoldt produced a large number of works on special relativity over the years. We will focus on 'Die Relativitätstheorie der Physik', published in 1914. This is the most comprehensive exposition of Petzoldt's views on special relativity, in particular when compared to other more summary or polemical works.[200] Moreover, as noted above, this work is of particular interest as the one that Einstein lauded in 'Vossische Zeitung'.

From the opening pages, Petzoldt expounds his view that the theory of relativity is a '*Weltanschauung*' whose great importance is that it is 'located on the principal path of the development of human thinking', that is, the path that had already led to 'the elimination of the last remnants of the old representation of substance' and to 'overcoming the opposition between what is "real" and what is "apparent"'.[201] In particular, according to Petzoldt there are 'two theories of relativity', or at least 'two phases in the theory of relativity': the first, represented by Mach, saw the relativization of inertia and of centrifugal forces, and a preliminary relativization of space and time; the second, with Einstein, resulted in the relativization of 'the shape of bodies' and the 'temporal order of events'.[202]

As we have seen, Mach saw inertia, centrifugal force, and temporal metrics not as absolute properties but rather as relations that express the 'reciprocal functional dependency between phenomena'.[203] Petzoldt adds that Mach's criticism of mechanistic physics also anticipated the relativization of the dimensions of bodies. In classical physics, touch was seen as the sense par excellence, the only sense able to provide the true dimensions of a body, its actual constitutive material, and therefore, definitively, its substance. The fundamental form of interaction between bodies was therefore seen as impact, that is, physical contact between bodies. This, however, failed to explain any action at a distance, such as gravitational interaction, for example, and subsequently electromagnetic phenomena. Mach turned the problem with this approach around with a proposal that involved the relativization of the concept of the body:

> But where is a body? Is it only where we touch it? Let us invert the matter: a body is where it acts. A little space is taken for touching, a greater for hearing, and a still greater for seeing. How did it come about that the sense of touch alone dictates to us where a body is?[204]

Basing his thinking on 'not just physics and the history of physics, but also the physiology of the senses and the theory of knowledge', Mach arrived at the idea that the old 'concept of mass must be reformulated as a pure "expression of what is factual"', going beyond the mechanistic conception that sees mass as a quantity of matter, of tangible substance.[205]

Once we free ourselves from the traditional concept of the material body, it is also necessary to reconsider our knowledge of physical phenomena, going beyond the mechanistic materialism that frames all knowledge of nature as the study of matter in motion. This new science must therefore start by recognizing that it is impossible to know 'absolute truth', 'what the world is like in itself and for itself, with no reference to the nature of the entity that has cognition and volition'.[206] According to the relativistic positivism, of which Mach can be considered the initiator, science must set itself the 'task of determining the relationships between man and his immediate or remote environment, forming a system of conceptual reactions that allows the attainment of equilibrium, a stable relation, an ultimate connection with the flow of complexes of stimuli'.[207]

Starting from this 'theory of knowledge based on the physiology of the senses', Mach's relativity states that 'it is only possible to speak of relative motion', in that 'there is no observation and no means – whether optical, electromagnetic or mechanical – that can allow us to *perceive* absolute motion'. Therefore, to say that 'one of two bodies in relative motion to each other must be in absolute motion means relapsing into metaphysics, attempting to express something that cannot be experienced' and is therefore 'devoid of meaning' or 'usefulness' for physics.[208]

According to Petzoldt, Einstein was able to resolve the puzzle of the conflicting Fizeau and Michelson-Morley experiments, succeeding where all other physicists before him had failed, precisely by seeing the problem from this Machian perspective. Indeed, it was thanks to Mach that Einstein was able to reformulate the question in novel terms, freeing himself from the assumption of the existence of absolute space. As Petzoldt writes, 'already standing on the ground prepared by Mach, Einstein took that experiment in a quite natural and unbiased way as a new confirmation of his relativistic conviction'.[209]

Petzoldt recognizes that Lorentz had proposed a solution based on the 'hypothesis of a contraction' of the body. However, as Lorentz considered 'the contraction to be dependent on the *absolute* motion of the body' without evidencing 'any property that might allow this absolute motion to be identified', his conclusion is 'a physics concept that is completely devoid of content', that is, a concept that is 'pure metaphysics, indistinguishable from the natural science of Schelling and Hegel'.[210] Petzoldt pointed out that Einstein, in contrast, had developed his theories using a relativistic perspective from the outset. Therefore 'the relativistic perspective is essential to the theory of relativity', and in this sense relativity is not just a theory of physics but 'a principle, a fundamental proposition, a specific way of seeing the facts of physics, a conception of nature, and, ultimately, a *Weltanschauung*'.[211] The theory of relativity tells us that 'it is in general impossible to arrive at an absolute theory whose formulae are consistent with the facts', because 'any adequate mathematical theory' can only 'be applied to a specific frame of reference'.[212]

Just as Protagorean relativism cannot be reduced to the simple proposition that everything is different for everyone, thus rendering knowledge of the world completely arbitrary, so the theory of relativity does not simply state that the same body or time interval changes according to the frame of reference. It also states that 'among the descriptions' of a phenomenon undertaken in different frames of reference there is 'a univocal connection' that establishes how a body or a process *must necessarily* figure in a given frame of reference.[213] This means not only that every single description undertaken within *one* frame of reference is univocally determined but also that there exists a univocal determination when moving *between one frame of reference and another*, and thus that 'we can change our perspective and get different views of the "same" process'.[214] In other words, as we observe a given phenomenon within a single frame of reference we can calculate how it *must* figure if observed within a different frame of reference. The key point of the theory of relativity is not only that it is necessary to describe the world within a specific frame of reference but that it is possible to move from one frame of reference to another according to univocally determined relationships, thanks to the laws of transformation. The theory thus states not only that every description is in relation to a frame of reference but also that all frames of reference are in relation to one another, and therefore that reality does not consist of the absolute, in what exists in itself and for itself, but is given by this network of relations.

The link between Einstein's theory of relativity and this more general relativistic *Weltanschauung* is demonstrated, according to Petzoldt, by the interpretation of Lorentz's contraction. The contraction of a body in motion can only be interpreted properly by adopting a fully relativistic perspective. It is not to be understood as a *physical* effect, as the *dynamic* result of the action of forces generated in a body in motion, but should be interpreted in Mach's sense, simply as the description of a functional relation of the form 'if x, then y'.[215] Petzoldt writes that 'this contraction is noted, established and described as a mathematical function of motion, univocally dependent on velocity v, univocally determined by this velocity'.[216] This is therefore the appropriate interpretation of Minkowski's famous statement that 'the contraction is not to be looked upon as a consequence of resistances in the ether, but simply *as a gift from above*, as an accompanying circumstance of the circumstance of motion'.[217]

For those who had already adopted Mach's 'phenomenological physics',[218] in which atoms, forces and masses do not exist but are simply concepts that serve to describe the fact of the *connection between phenomena*, there is nothing mysterious about the existence of a functional relation between the speed of a body and its length. Only someone still using the mindset of mechanistic physics, in which material bodies and forces acting upon them exist, might find it perplexing. In mechanistic physics, a body must have a determined size, and if this size changes it must be because some force has acted upon the body to cause the contraction. Phenomenological physics, on the other hand, is not shocked by the idea that a body changes its length without a cause – that is, without a force that acts upon it, but only as a function of a changed frame of reference – because it had long previously rejected the metaphysical concept of cause, understood as something that acts upon a body and had redefined the concept as the mere presence of functional relations between empirical facts.[219] Petzoldt therefore goes so far as to state that Mach's position on the theory of relativity not only 'poses no problem' but is in fact 'self-evident' (*Selbstverständlich*).[220]

We see that Petzoldt considers the theory of relativity to be something of a generalization of the fact that 'any "body", from any point of view, "appears" in a different perspectival shape'.[221] Indeed, even in the field of perspective there is no privileged observation point from which we get the 'true' shape of the body, and with respect to which all the others are *mere* 'appearance'. Rather, *all* the points of view provide the 'true' shape of the body in relation to their specific observation point, as there exists a univocal relation between the observation point and the shape that the body *must* have when observed from that point. Moreover, the laws of prospective geometry permit us to calculate the shape that a body *must* have from a different observation point. Similarly, in the theory of relativity:

> the equations can be used to univocally correlate any point of a frame of reference to the point of a frame of reference that is moving in a straight line and at a uniform velocity with respect to it, that is, the space of one can be 'represented'

univocally on the basis of the space of the other. [...] There is nothing else *above* these spaces, and there is no nature *beyond* them. Any individual observer uses, for his own purposes, *his own* representation [*Darstellung*], *his own* description of nature, of his own space, and recognizes the corresponding space of any other observer as equally valid and puts his own representation into a univocal relation with the representations of others, which he constructs on the basis of his own representation.[222]

Petzoldt, therefore, does not concur with those who hold that the theory of relativity is intrinsically counterintuitive, unrepresentable, *unanschaulich*.[223] Projective geometry shows that we are quite familiar with picturing a body as simultaneously having different shapes from different points of view. What we cannot picture is the attempt to unify several different perspectives into one, trying to imagine a body having *several* different shapes *from a single point of view*. If a body is measured in two frames of reference in relative motion to each other, the frame of reference in which A is at rest will provide a given length, while in the frame of reference in which A is in motion a shorter length will be obtained and A thus is seen as contracted. According to Petzoldt, to assert that body A can simultaneously have two or more different lengths, depending on which frame of reference it is measured against, is in no way counterintuitive, just as we do not consider counterintuitive the fact that when two people observe each other at a distance each could say that they are of normal size while the other is shrunk. All the apparent contradictions manifest themselves only when we omit the frame of reference and try to unify the two points of view in an attempt to obtain an absolute perspective, which leads us to wonder how it can be possible for a person to be large and small at the same time, or how a body can be both contracted and uncontracted.

The issue of the *Anschaulichkeit* of Einsteinian relativity does not just address the problem of the difficulty of grasping or explaining the theory but also the epistemological issue of the status of our scientific knowledge.[224] For Mach and Petzoldt, the mathematical forms in which our knowledge of physics is expressed, which permit extreme levels of abstraction, must not occlude the fact that their ultimate value consists of their function as descriptions of empirical facts. The assertion that even the theory of relativity is *anschaulich* is therefore a reminder that it, too, is about our lived experience, the empirical perception that a scientist cannot avoid starting from. As Petzoldt reminds us, 'there is no valid proposition in the natural sciences that is not based on observation through the senses'.[225]

Petzoldt sees this as the point where two *Weltanschauungen* collide: the relativistic one, which includes the modern theory of relativity, and the one derived from Plato and Kant.[226] According to the latter, we can never grasp 'true reality' through the senses, which only provide us with 'appearance': the phenomenon and not the thing-in-itself. In the former, on the other hand, there is no such thing as *absolute* reality apart from how it is given in a specific frame of reference. Therefore our senses do not give us just mere appearance but provide us with reality relative to the frame of reference of which we (our nervous systems) are the centre. Indeed, not only do our senses not deceive us in general, showing us what appears to be instead of what is, but the very concept of 'sensory illusion' is illogical. Even in so-called illusions there is no 'error', but only ever an ensemble of univocal functional relations that determine how reality must *necessarily* be perceived in its relationship with my sense organs.[227]

Einstein's theory of relativity, with its assertion of the indispensability of a frame of reference and the equivalence of all frames of reference, is an essential part of the relativistic *Weltanschauung*. It would therefore be a perversion of its whole spirit and content to try to interpret it along Platonic or Kantian lines, forcing it into the real/apparent schema. A theory is not independent of the epistemological frame within which it is formulated or interpreted. Some philosophical positions support and foster scientific advances; others block them or slow them down. Petzoldt therefore states that 'physics, chemistry and biology have today reached a point where it is no longer irrelevant to their valid development which *Weltanschauung* researchers adopt'.[228] The '*Weltanschauungen* that are in decline', like the Platonic/Kantian one, are unable to promote scientific development, whereas 'positivistic physics, mathematics and epistemology' show a 'strong convergence' on relativism.[229]

Just as psychophysiology showed Mach that there is no supposed reality beyond sensory appearance, and that our sensations constitute what is real, so the theory of relativity tells us that there is no reality in itself without a frame of reference, and that all frames of reference give us actual reality and not merely its appearance. For Petzoldt, this is the heart of the *erkenntnistheoretisch* links between Mach and Einstein, similarly to the links between the experimental psychophysics of the late nineteenth century and the new theoretical physics of the early twentieth. As he wrote concisely in the third edition of the *Weltproblem*, 'The mission of the theory of relativity [...] is to make the relativism of the physiology of the senses [*sinnesphysiologischen Relativismus*] the foundation of the natural sciences'.[230]

Petzoldt's objections to the theory of relativity

As well as the enthusiasm that Petzoldt demonstrates for the theory of relativity, in a tone that will become ever more emphatic over the years, his works also contain explicit or implicit objections to some aspects of it. As Klaus Hentschel emphasized in his monumental volume on the philosophical reception of Einsteinian relativity, the misinterpretations (*Fehlinterpretationen*) by philosophers were characterized by taking from Einstein's position everything that was consistent with their own thinking and dismissing anything that did not lend itself to appropriation of this kind.[231] Petzoldt was no exception. Indeed, he was probably one of the most striking examples of this kind of relativity *à la carte*.[232]

As can be gleaned from his correspondence with Mach, Petzoldt's main fear was that some form of the absolute might lurk within the folds of the theory of relativity. This probably came from his desire to emulate his mentor, setting himself the task of applying to Einsteinian physics the same historical/critical approach that Mach had deployed so successfully to Newtonian physics. However, if Mach's analyses were based on a deep knowledge of the subject, Petzoldt's blunders on issues such as the rotation of a rigid disc and the Twin Paradox show that his familiarity with Einstein's ideas was decidedly superficial.

The constant speed of light was one of the first points that puzzled Petzoldt. Unlike in his correspondence with Mach, Petzoldt's writings do not go so far as to characterize this aspect of Einstein's theory as a return to the absolute. However, the way he addresses this issue reveals his doubts.

In 'Die Relativitätstheorie der Physik', 'the principle of the constant speed of light' is defined as 'a supposed natural law', that is, as 'a hypothesis whose validity lies in its success', in 'serving to provide the most thorough description possible and the one that is most consistent with the many functional connections that have been established'.[233] Going into greater detail, Petzoldt recognizes that the value of a proposition of this kind is that it allows the relativization of the measurement of time, in that the assumption that the speed of light is constant leads to the conclusion that the clock of a frame of reference in motion must necessarily run more slowly. Petzoldt therefore concedes that the assumption that the speed of light is constant is an essential component of the theories of Einstein and Minkowski, and not simply an aspect that could be modified or dropped with no impact on the rest of the theory. However, this does not mean that this assumption might not be superseded in the future, perhaps when the theories of Einstein and Minkowski are revised. What will endure, apart from their historical role, is their relativization of space and time. In other words, Einstein's equations and Minkowski's geometrical models, which are based on the assumption that the speed of light is constant, might be supplanted by other theories of *physics*, but what will remain is their *erkenntnistheoretisch* significance, that is, the service they rendered to relativism.[234]

According to Petzoldt, the provisional nature of the theory of relativity as a theory of physics is evidenced by the relationship between facts and hypotheses in any scientific explanation. A theory is based on hypotheses that serve to provide a description that approximates as closely as possible to all the facts ascertained up to that point. In the case of Einstein's theory, the new facts that had to be taken into account were the Fizeau and Michelson-Morley experiments. The assumption that the speed of light is constant fulfils this task perfectly. However, Petzoldt considers that it cannot be decided at this point that no new facts will emerge – from new experiments, for example – that will require this assumption to be dropped, with new hypotheses formulated to describe the new facts:

> To conclude that speeds greater than the speed of light will never be verified in nature, or that they are in general impossible, would mean completely disavowing the relationship between theory and fact. We cannot ask more of a theory than to encompass in a univocal connection all of the facts that are known at the point at which it is formulated. [...] It is only experience and not theory that can decide whether a temperature of absolute zero exists, and what it is, and whether there exist bodies that have a speed greater than that of light.[235]

The reference to absolute zero in this passage refers to Mach's discussion of it in *Principles of the Theory of Heat*. Mach criticized the various concepts of absolute zero proposed by physicists over time, showing that these were not the result of measurement but the expression of the hypotheses on which each based their theory of heat.[236] Petzoldt is clearly attempting to replicate his mentor's line of argumentation, adapting it in a somewhat slavish manner to the new debates in physics, striving thus to continue Mach's critical analysis of scientific knowledge.

Another aspect of Einstein's theory of relativity that Petzoldt feels justified in correcting, if not criticizing, is the invariance of the laws of nature. Petzoldt obviously does not want to question the assumption of *Eindeutigkeit*, which states that the connections that exist in nature must be describable using univocal functional relations. He remains firm in his conviction that, in physics, all knowledge must express the necessary connection between phenomena, to the extent that he states that 'if one of the consequences of the theory of relativity was the foregoing of empirical constraints [*Gebundenheit*] in the sequence of events, it would not be a tenable theory'.[237] In particular, all physics knowledge consists of equations stating that – given certain conditions, certain variables – certain other conditions, certain other variables, will occur. In the case of the theory of relativity, a change of frame of reference leads to different values in the measurement of space and time, in line with univocally determined relationships, while the laws of nature and the speed of light do not change. Petzoldt claims, however, that it must be possible to develop systems of equations in which, when the frame of reference changes, the measurements of space and time remain unchanged and it is the laws of nature that change (in a univocally determined way):

> The theory of relativity *demands* the laws of nature to be independent of the reciprocal state of uniform motion of the frames of reference, and therefore leads to the abandonment of the fixed shape of bodies and the *congruity of the clocks in different frames that are in reciprocal uniform motion*. By the same demand, however, it would be possible to retain the congruity of shape and of clocks in frames that are in reciprocal uniform motion and thus let the 'laws of nature' change from one frame to another, or one might imagine some combination that lies between these two extremes. No experiment could resolve this issue.[238]

Petzoldt's attempt to replicate a famous argument of Mach's, such as that of the equivalence of the Ptolemaic and Copernican systems, is again evident here. Just as both of the astronomical models provide an equally valid description of the motion of the planets of the solar system, simply keeping different values fixed, one considering the Earth to be static and the other the Sun, so models other than the Einsteinian one must be possible, changing which parameters are to be kept constant and which are allowed to change. Making space and time invariant would not mean resurrecting the absolute but would rather produce a different relativization, one in which it would be the relationships between the variants and the invariants of Einstein's theory of relativity that would be relativized. Petzoldt holds that the theory of relativity overstates the distinction 'between spatiotemporal determination on the one hand and the "laws of nature" on the other', as both are nothing other than 'equally valid means of determination, members of the same group of concepts, namely, the physical means of determination used ubiquitously in physics equations'.[239] This means that the 'laws of nature' are not *absolutely* invariant, as they are simply the set of means of determination that are deemed invariant within a given model. They are in fact invariant *relative* to the model adopted, given that different means of determination will serve as 'laws of nature' in a different theory.[240] In the case of Einsteinian relativity, the distinction between invariant 'laws of nature' and spatiotemporal determination 'does not therefore derive from the facts themselves, but only from history', that is, from the accidental development of the theory, and it could therefore be superseded by alternative models or exist alongside them.[241]

Another aspect of the theory of relativity that Petzoldt sees as problematic is Minkowski's position on space-time. Petzoldt agrees that, as the theory of relativity attributes to each observer their own frame of spatiotemporal reference, it represents a step towards experience, and therefore towards a physics that is as close to the facts as possible and thus as stable as possible. Although Petzoldt does not identify *visual* space with the *geometrical* system of spatiotemporal coordinates used in physics, he sees 'this individualization of what used to be general space as something closer to the physio-psychological fact that everyone has their own *visual* space and that this is in a univocal relation with the visual spaces of others'.[242]

However, Petzoldt notes that 'an absolute space that contains *everyone's* experiences is unthinkable, in that each body within that space would have to have an infinite number of different shapes simultaneously'.[243] He is therefore wary of Minkowski's position on space-time, namely, the 'application of a four-dimensional Euclidean space to provide a unitary account of the infinitely large number of spatiotemporal systems of the theory of relativity'.[244] Even though Petzoldt praises Minkowski's geometrical model as a 'powerful creation' that represents a 'completion of the original theory that is indeed indispensable (now that we have it)', he emphasizes that we must never forget that 'it is just a conceptual system'.[245]

In fact, Minkowski's model does not envisage just *one* four-dimensional space-time represented by the coordinates x, y, z, t but an innumerable number, to the extent that what he calls 'the world' is 'the *multiplicity of all thinkable x,y,z,t systems of values*'.[246] He also explicitly states that according to the theory proposed by himself and Einstein 'we should then have in the world no longer the space, but an infinite number of spaces'.[247] We can therefore infer that Petzoldt had misunderstood Minkowski's position, given that he always speaks of it as if the fundamental property of the four-dimensional model proposed by Minkowski were that of providing 'a *unitary* account of the infinitely large number of cases', just as projective geometry provides the structure of the three-dimensional shape of a body based on the innumerable views of it from different perspectives.[248] Another possibility is that Petzoldt was misled by Minkowski's emphasis of the concept of the 'world' as the exhaustive set of 'world-points' (i.e. of every point of space at a point in time).[249] Indeed, one can easily imagine that Petzoldt, a relativist, was instinctively averse to steps such as the one Minkowski's takes when he defines as a 'postulate of the absolute world (or the world-postulate)' the assertion that 'only the four-dimensional world in space and time is given by phenomena'.[250] Even though Petzoldt does not cite the expression 'postulate of the absolute world', he does cite the rest of Minkowski's statement, specifying that he does not agree with the idea that 'a four-dimensional world is *given* by phenomena'.[251] He concedes that Minkowski 'probably did not mean this in a metaphysical sense', but points out that from the perspective of relativistic positivism no geometrical space can be 'given', irrespective of whether it is three-dimensional or four-dimensional, as these are merely '*conceptual* tools of thinking'.[252] Minkowski's closing words must have sounded even more alarming to Petzoldt's ears, where he speaks of the 'pre-established harmony between pure mathematics and physics'.[253] Indeed, statements of this kind ran the risk of placing centre stage once again the metaphysical conception of the natural sciences that Mach had fought against all his life by holding that the laws of nature, equations and geometrical models are not true reality beyond the appearance provided by the senses but tools whose purpose it is to provide a concise description of experience.[254]

Relativity and the physics and physiology of the senses

We have seen that Petzoldt's interpretation of the theory of relativity included an attempt to relate the new physics with the physiology of the senses. We will now examine this aspect in greater detail, as it is precisely the relationship between sensory experience and scientific knowledge that is one of the most controversial points of Mach's theory of knowledge, as well as being one of the issues that criticism of Petzoldt's interpretation of relativity was directed at. This issue is elaborated in greater detail in works written later than the ones we have examined so far.

Petzoldt's writings contain a twofold connection between the theory of relativity and the study of sensations. Firstly, as Mach shows, the origins and significance of *every* item of scientific knowledge are based on sensations, including therefore physics and, more recently, the theory of relativity:

> You may think what you will of the essence of sensations, you may consider them to be appearance or actual reality, in any event they are the only basis and the ultimate criterion for the laws elaborated in the natural sciences. […] Whether you note the position of the hand of a clock or a set of scales, the indicator of a thermometer, of a voltmeter, of an ammeter, the position of a telescope with respect to celestial coordinates, or more generally any indicator of any scientific instrument of measurement, whether you count the swings of a pendulum or a metronome, or the beats of a pulse, whether you compare the pitch of two sounds, *we are always dealing with the observation and evaluation of complexes of sensations*, or at least of our memory of them. Einstein calls them coincidences, we call them: *coincidences of sensations*.[255]

Petzoldt's refers here to the concept of 'coincidence' first introduced by Einstein in 1916 in his presentation of general relativity.[256] Einstein generalized the principle of the invariance of the laws of nature with respect to the frame of reference, extending it to also cover systems that are not in reciprocal rectilinear uniform motion: 'The general laws of nature are to be expressed by equations which hold good for all systems of co-ordinates, that is, are co-variant with respect to any substitutions whatever (generally co-variant).'[257] Einstein adds:

> That this requirement of general co-variance, which takes away from space and time the last remnant of physical objectivity, is a natural one, will be seen from

the following reflexion. *All our space-time verifications invariably amount to a determination of space-time coincidences.* If, for example, events consisted merely in the motion of material points, then ultimately nothing would be observable but the meetings of two or more of these points. Moreover, the results of our measurings are nothing but verifications of such meetings of the material points of our measuring instruments with other material points, coincidences between the hands of a clock and points on the clock dial, and observed point-events happening at the same place at the same time. *The introduction of a system of reference serves no other purpose than to facilitate the description of the totality of such coincidences.* […] As *all our physical experience can be ultimately reduced to such coincidences*, there is no immediate reason for preferring certain systems of co-ordinates to others, that is to say, we arrive at the requirement of general co-variance.[258]

Even though Einstein's line of argumentation clearly shows Mach's influence, Petzoldt adds a further twist, turning coincidences of material points into 'coincidences *of sensations*'. Petzoldt thus projects Mach's position onto that of Einstein in order to assert that it was thanks to Mach's influence that Einstein did not 'see a world of absolute corporeal elements, but a world of "sensations", of coincidences of sensory impressions, an aesthesiophysiological multiplicity'.[259]

Petzoldt also uses the concept of 'coincidence' to reformulate the theory of special relativity:

The observers of two systems in reciprocal motion note that *simple* coincidences of sensations (observations of rulers and clocks) differ one from another, but that there is complete agreement on the reciprocal correlations of *pairs* of coincidences (that is, on the 'laws of nature'). But even these divergences are univocally determined (by means of the Lorentz transformations) and therefore they in turn obey the 'laws of nature'. Thus, even in 'simple' coincidences we are ultimately dealing with pairs of coincidences or of complexes of coincidences, which should be associated with the velocities of the systems *as well as with the organization of the sensory elements of the nerves and the central nervous system.*[260]

This passage shows how Petzoldt justifies going from 'coincidences' to 'coincidences *of sensations*'. The pure datum, the simple unrelated datum, does not exist in knowledge, whether it be a spatial determination, a temporal one, a point of matter or even a coincidence in Einstein's sense. Only relations exist, or coincidences. Every datum thus presents itself in a network of relations, coinciding with other data. However, the human brain is also part of this network of relations. Therefore, if the data of the natural sciences are coincidences, as Einstein maintains, these coincidences are *in turn* given as coinciding with certain cerebral states, and in this sense they can be called coincidences of sensations. In other words, if, for example, science considers the coincidence between the hands of a clock and the figures on its face to be a *simple* coincidence, it is in fact abstracting away from the fact that this *simple* coincidence is actually part of a *complex* of coincidences of which the nervous system and the sense organs are also a part. That is, it is abstracting away from the fact that in the end this coincidence is a coincidence of sensations.

However much Petzoldt emphasizes the closeness between Einstein's position and Mach's, it is important to note that relativity is no different to any other theories of physics with respect to its links to sensory data. It makes no difference whether or not Einstein introduces the figure of the observer, whether or not he understands this observer in a literal sense, whether or not he speaks of coincidences, because *in any case* relativity, *like any other piece of knowledge*, derives its meaning from sensations, and is based on data that are not located, so to speak, in a vacuum, but are in a functional relation to the brain. Newton's physics was no more and no less founded on sensations than Einstein's, but Einstein – thanks to Mach's influence – is simply much more aware of this than Newton was.

From a different point of view, however, the theory of relativity does in fact enjoy a unique status which allows it to establish a special relationship with the physiology of sensory perception. Petzoldt notes that both set themselves against the mechanistic view of the world, which makes them natural allies. As Petzoldt states:

> [The mechanical conception of nature] consists of the fact that all natural processes – physical, chemical and biological – can be traced back to mechanical processes that take place between impalpable and invisible atomic particles the size of electrons or smaller. The stage for these processes is infinite three-dimensional space, completely homogeneous throughout and subject to Euclidean laws of geometry, and known as Euclidean space. This counts as absolute reality, completely independent of the human body. Another equally absolute fact is time, running in a completely uniform manner, independently of any measurement, observation or clock. Space and time make motion possible, which is therefore equally absolute, and in its actuality independent of the frame of reference and of the observer. [...] Absolute space, absolute time and absolute motion are common to all of the innumerable versions of the mechanical conception of nature, versions that differ only in the structure of the substratum of this motion. Moreover, they all concur that sensations are merely subjective and can tell us nothing about actual reality.[261]

As the theory of relativity puts to one side the concepts of absolute space, time and motion that characterize mechanistic physics, it 'definitively eliminates the mechanistic conception and establishes a significant link between physics and biology', and in particular between physics and the physiology of sensory perception.[262] Indeed, both help to show that reality 'replicates in different forms from one individual to another' and therefore that 'the actual world of mankind is an aesthesiophysiological and biological world, and not the hazy world of mechanism'.[263] The physiology of sensory perception and biology share with the theory of relativity the view that absolute reality in itself and for itself does not actually exist, and that what exists is an environment that is in a constant relation with the individual, or rather, as many environments as there are individuals that relate to it, given that every being relates to the surrounding world from his own specific point of view, with his own specific needs and his own specific cerebral schemata.

In any case, Petzoldt does not hold that the theory of relativity converges with the physiology of sensory perception because it introduces the figure of the observer, understood *literally*, as the individual that receives a series of sensory impressions. Petzoldt in fact recognizes that 'whenever one speaks of an *observer* one only means the *connection* with his frame of reference or his coordinate system'.[264] There is therefore no reason to fear that 'an extraneous psychological or "mental" element might be introduced by the multiplicity of observers'.[265] What the theory of relativity and the physiology of sensory perception have in common is rather the awareness that any description is always made relative to a specific frame of reference. With the theory of relativity, the frame of reference is the four-dimensional coordinate system at rest or in motion. With the physiology of sensory perception, the frame of reference is the individual nervous system including the brain and its peripheral sense organs. In other words, what the two scientific approaches share is the awareness that there is no such thing as a *unique* description of nature, as mechanism would have it, because only 'the multiplicity of descriptions allows the creation of an adequate "picture" of natural events'. Scientists are 'like architects or engineers who cannot construct a building based on a single drawing but need to use floorplans, vertical sections and cross sections of all kinds'.[266]

To summarize, Petzoldt sees two links between the theory of relativity and the physiology of sensory perception. The first is not specific to the theory of relativity, as it concerns the fact that *every* piece of knowledge, including physics, cannot but be based on the data of sensory experience. Nevertheless, Petzoldt attributes a greater awareness of this to Einstein, given his Machian background. The second link, however, is specific to the theory of relativity, in that the theory is based on the recognition of the frame of reference as an essential means of determination for the physical description of the world. Even though the theory of relativity and the physiology of sensory perception *do not focus on the same frame of reference*, as the frame consists of spatiotemporal coordinates for the former and of the cerebral system for the latter, both reject any *absolute* knowledge of the world that is outside any frame of reference, and therefore reject a mechanistic view of the world.

5

Criticism of Petzoldt's interpretation of relativity

Petzoldt was active within a broader context of philosophical debate on the interpretation of the theory of relativity. It is therefore no surprise that his views on the links between Einstein's theories and relativistic positivism were attacked by the exponents of other philosophical directions who were also coming to terms with the recent development in physics. In particular, three eminent thinkers of the day laid criticisms at his door: Ernst Cassirer (1874–1945), Moritz Schlick (1882–1936) and Hans Reichenbach (1891–1953).[1] A detailed analysis of the debates between Petzoldt and each of these would merit independent research that is beyond the scope of the current volume. We will therefore focus on their disagreements on the specific issue of relativity, providing a concise account of how the different interpretations of Einstein's theories emerge as a litmus test for wider philosophical divergence on the theory of knowledge.[2]

Cassirer

In 1921, Cassirer published *Zur Einsteinschen Relativitätstheorie* (*Einstein's Theory of Relativity*), in which he provided his own interpretation of Einstein's theory, namely, that it confirmed the Kantian intuition that space and time were not objects or things, but conditions of possibility for the construction of objectivity on scientific grounds.[3] Cassirer took issue with the interpretation offered by Petzoldt, who was one of the main targets of the work even where he was not mentioned explicitly. The core of the argument was that as 'there can be no doubt that all our knowledge begins with experience', one should not wonder '*whether*' the theory of relativity 'has issued from experience but merely as to how it is based on experience'. That is, the question is whether it is necessary to go back to the neo-Kantian '*critical* concept' of experience, or to the '*sensualistic* concept' based on Mach and advocated by Petzoldt.[4] Neo-Kantianism and Machian sensualism were thus the two fundamental alternatives in the interpretation of physics theories in general and of the theory of relativity in particular.

In Chapter 3, we saw that Petzoldt's views were often so close to Kant's as to overlap completely, despite all of the criticisms raised by him against Kant and Kantianism. Similarly, Cassirer criticized Machian positivism but acknowledged the points of contact between his own position on cognitive processes and that of Mach. For example, in the earlier *Substanzbegriff und Funktionsbegriff* (*Substance and Function*) he asserted that 'the critical interpretation' derived from Kant 'is in complete agreement' with Mach's 'modern empiricism' where Mach maintains that the object of knowledge, the thing that is permanent, is not substance but 'mere relations'.[5] The disagreement would be rather on the 'logical meaning' of this permanence, where Mach sees it as 'a *property of sensuous impressions, immediately inhering in them*', whereas critical philosophy sees it as 'the result of intellectual work, by which we gradually transform the given according to definite logical requirements'.[6]

It is therefore no surprise that Cassirer did not distance himself from Petzoldt as much as his criticisms of him might lead one to think. It is in fact necessary to distinguish two arguments in his discussion of Petzoldt's interpretation: his attack on relativism and his attack on sensualism. Only the second of these appears to actually hit home, while the first rests on a misunderstanding of Petzoldt's thinking that masks the fact that the two thought along substantially the same lines.

In his work on Einsteinian relativity, Cassirer writes:

> The principle of relativity of physics has scarcely more in common with 'relativistic positivism', to which it has been compared, than the name. When there is seen in the former a renewal of ancient sophistical doctrines, a confirmation of the Protagorean doctrine that man is the 'measure of all things', its essential achievement is mistaken. The physical theory of relativity teaches not that what appears to each person is true to him, but, on the contrary, it warns against taking appearances, which hold only from a particular system, as the truth in the sense of science, i.e. as an expression of an inclusive and final law of experience. The latter is gained neither by the observations and measurements of a particular system nor by those of however many such systems, but *only by the reciprocal coordination of the results of all possible systems*. The general theory of relativity purports to show how we can gain assertions concerning all of these, how we can rise above the fragmentariness of the individual views to a total view of natural processes.[7]

Cassirer expresses himself along similar lines in lectures delivered as his work on relativity was being redacted, attacking Petzoldt's view of Einstein's theory as a confirmation of the Protagorean principle of man as the measure of all things:

> If there is no *unique* space and *unique* time for all perceiving subjects, but there are as many different spaces and times, as many different frames of reference, as many different observers, the shared synthetic *unity* of 'experience' is lost; and, it seems, only *single* perceptions remain. Each system now measures phenomena using *its own* metrics and for each system *its own* measurements are true. We are up against the principle of ancient Sophism, the Protagorean principle that Plato opposed so strenuously in the *Theaetetus* […]. 'Truth' is valid only with respect to the point of view of the individual who expresses it; there is no cogent "general" reality, but rather 'for each person, that person's perception is true'. […] This would mean that the modern principle of physical relativity, like the old principle of philosophical

relativity, would seek to tell us that the world, objective reality, is as it appears to each person; the perceptions of one subject would have no greater "objective" validity that those of another.[8]

We can see that Cassirer condemns Petzoldt's Protagorean interpretation of relativity, but misses the fact that Petzoldt never conceived of relativism in a sceptical solipsistic sense. We know that Petzoldt held that Protagoras himself would never have wanted to fragment reality into an infinite number of equivalent positions. What is missing in Cassirer's encapsulation of Petzoldt's position is the role of *Eindeutigkeit* as a necessary link within each individual experience and between all individual experiences. Although Petzoldt holds that there is no such thing as 'unique space and unique time' common to different observers, this does not mean that 'only *single* perceptions remain', because the fundamental point is that *single* perceptions form a *univocally determined complex*. It is *Eindeutigkeit* that provides 'the reciprocal coordination of the results of all possible systems' invoked by Cassirer.[9]

In contrast to Cassirer's position, Petzoldt holds that 'the sensory component of every perception is in no way simply subjective, secondary, unusable in the construction of a comprehensive theory of knowledge [...], but is a lawfully determined fact of nature [*ein volles Naturgegebenes, gesetzmässig Bedingtes*], and therefore objective and absolute'.[10] For Petzoldt, seeing the theory of relativity in the light of relativistic positivism does not mean indulging in the banality of 'everything is relative', as 'nothing is further from the truth than to maintain that "relativistic positivism" recognizes nothing as absolute and proclaims that "everything is relative"'.[11]

It is therefore not surprising that Petzoldt complained bitterly in a letter to Reichenbach about the fundamental misunderstanding of his work that was apparent in Cassirer's criticism:

Cassirer preaches to the converted. What manner of positivist would deny lawfulness! I have written a work entitled *Das Gesetz der Eindeutigkeit*! Cassirer rails against positivism because he finds harking back to Protagoras exceedingly bothersome. But he has clearly not read my *Weltproblem* carefully. Much of the futile harshness of the argumentation would vanish if one had first understood the opponent correctly (often only the presumed opponent).[12]

If Cassirer's criticism of Petzoldt's relativism seems off-target in that it attributes to him a position that he had never maintained, the debate on the relationship between relativity and the sensory substratum is more to the point. The key issue, indeed, is not whether a lawfulness exists or not but whether a lawfulness of this kind can be traced back to the level of the senses, as purely empirical, or whether, rather, it is transcendental. Petzoldt's reply to Cassirer points out that even within the theory of relativity 'the univocal connection – the connection between sensory coincidences and other sensory coincidences – must already be *findable* in each single system'.[13] Cassirer, on the other hand, maintains that lawfulness can never emerge from the purely individual plane of mere sensation, but necessarily requires some distancing from sensory perception by means of an elaboration that can serve as the basis for objective universal knowledge. Addressing Petzoldt's 'phenomenalism' elsewhere, he writes that Petzoldt must recognize that 'a series of perceptions must be completed conceptually'. However, when Petzoldt maintains that 'no fundamentally new content emerges in this completion', he does not understand that 'with this first step beyond "immediate" perception, the sensory perspective has already shifted imperceptibly towards a conceptual one', as the requirements 'of *Eindeutigkeit*, of continuous connection, of the closed causality of a series of experiences' are '*logical* requirements'.[14] As we have seen, Petzoldt refuses to accept that this elaboration is logical, a priori or transcendental, and frames it as a fundamentally *biological* requirement based on the functioning of the brain, which evolved the way it did in a stable and univocally determined environment.

According to Cassirer, it is precisely because Petzoldt starts from this incorrect account of the conceptual elaboration that is required in order to rise above the perspective of mere sensation that he ends up misunderstanding Einstein's work:

> It can and must be demanded that the multiplicity and diversity of the sensuous data here appearing can be united into a universal concept of experience. [...] This concept the theory of relativity no longer represents in the form of a picture but as a physical theory, in the form of equations and systems of equations, which are covariant with reference to arbitrary substitutions. *The 'relativization', which is thus accomplished, is itself of a purely logical and mathematical sort.* By it the object of physics is indeed determined as the 'object in the phenomenal world', but *this phenomenal world no longer possesses any subjective arbitrariness and contingency.* For the *ideality* of the forms and conditions of knowledge, on which physics rests as a science, both assures and grounds the empirical reality of all that is established by it as a 'fact' and in the name of objective validity.[15]

In other words, it is not sensory experience that provides the lawfulness that is required to achieve objectivity in the natural sciences. Quite the contrary, it is this lawfulness that is the condition for the possibility of experience. Cassirer thus reaffirms the sense of Kant's Copernican revolution even in the face of the new developments in physics, rejecting Petzoldt's claim to be able to construct a theory of knowledge based entirely on sensory experience and to justify the Einsteinian revolution on these grounds.

Cassirer therefore rejects Petzoldt's argument that the fact 'that Einstein refers gratefully to the decisive stimulus he received from Mach' proves that there is an inherent link between Mach's theory of knowledge and the theory of relativity.[16] Indeed, Cassirer maintains that 'a sharp distinction must be made between what Mach has accomplished as a physicist in his criticism of Newton s fundamental concepts, and the general *philosophical* consequences he has drawn from this achievement'.[17] While there might be a link between Mach's work as a scientist and Einstein's theories, 'there is no necessary connection between the theory of relativity and Mach's philosophy'.[18] Indeed, the 'constructive force' that lies at the heart of Einstein's revolution demonstrates that 'the system of physical knowledge is distinguished from a mere "rhapsody of perceptions"'.[19]

Schlick

Moritz Schlick's training as a physicist under Max Planck meant that he had no difficulty in being recognized as one of the best-qualified interpreters of the theory of relativity, in particular thanks to his *Raum und Zeit in der gegenwärtigen Physik* (*Space and Time in Contemporary Physics*), which appeared in 1917 in the journal *Naturwissenschaften*, and was published as a successful monograph immediately afterwards, appearing in four editions between 1917 and 1922. Alongside this main work on the theory of relativity, Schlick published a large number of articles and reviews designed to make Einstein's theories accessible to a lay audience and to engage with others active in the ongoing philosophical debate.[20]

As Schlick himself noted in 'Kritizistische oder empiristische Deutung der neuen Physik?' (Critical or Empiricist Interpretation of Modern Physics?),[21] one of his aims was to show that deciding between the sensualistic empiricism of Mach and Petzoldt on the one hand and the criticism of Cassirer and the other neo-Kantians on the other was in fact a false dilemma. There was a third way, one that recognized the need to go beyond sensations but without relapsing into the Kantian doctrine of synthetic a priori judgements. Schlick was thus able to use Cassirer's arguments against Petzoldt and Petzoldt's against Cassirer, offering his own 'critical realism' as a solution that overcame the limitations of both of these positions.

Schlick's criticisms also went beyond the theory of relativity, addressing more generally the different conflicting philosophical conceptions that interpretations of Einstein's theories were based on. In his *Allgemeine Erkenntnislehre* (*General Theory of Knowledge*) of 1918, Schlick devoted a large section to a critique of the 'notion of immanence' developed by Avenarius, Mach, Petzoldt and Schuppe, of which Petzoldt was the last living exponent.[22] Schlick criticized the identification of the actuality of sense data with reality, of being with represented being that characterized the thinking of these writers, showing that identification of this kind makes it impossible to explain how different observers can speak of *one and the same* object even though they have *different* sensory experiences of it.

According to Schlick, given an object O, different observers of it will have different complexes of sensations C_1, C_2, C_3 ....[23] We must then either assume that K_1, K_2, K_3 are the same object because they have some or all sensations in common, or that they are the same object because they all refer to G, or we must abandon the idea that they are the same object. The first option is untenable, because we cannot assert that different people have *the same sensations*, even partially. The second option is rejected by the exponents of the immanence idea themselves, because it would entail assuming the existence of the real object *beyond* sensations, relapsing into the much-maligned Kantian thing-in-itself. The third option leads to solipsism and the abandonment of the task of providing a foundation for knowledge, which no exponent of the philosophy of immanence claims to advocate.

The only way out of this impasse is to stop using sensations as a foundation, shifting the focus to the *connections* between sensations, that is, to the fact that they present themselves in lawful relations.[24] The problem that Schlick sees with this solution is that the object becomes 'something quite shadowy; for it is not anything real, but a concept', so that one ends up 'identifying the validity of abstract propositions with the being of real things'.[25] The philosophical position that had set itself the objective of placing experience, the richness of sensory data, at the centre of the theory of knowledge thus ends up becoming the opposite, identifying reality with abstract concepts such as the laws governing the connections between phenomena.

Schlick also held that the exponents of the philosophy of immanence were unable to explain why *different* observers should *all* register the presence of *the same laws* in the connection of sensations and concepts. Once philosophers give up on the idea of the unitary object that exists beyond multiple individual experiences, they can no longer assume that these laws refer *to the same objects*. There is therefore no explanation for why *the same laws* should hold for everyone, except for some kind of preordained harmony similar to that which binds the monads in Leibniz. Indeed, Schlick maintains that at the end of the day Petzoldt's position is nothing other than a reformulation of Leibnizian monadology, and thus a 'metaphysical system' and not the 'the only natural, metaphysics-free world view' that it claims to be.[26]

Schlick concludes that these contradictory results can only be avoided by abandoning the path taken by the philosophers of immanence and recognizing that lawful connections pertain to real objects, refer to the things of the world, and are not simple connections between sensations or between concepts derived from sensations. Mach and his followers, however, talk in generic terms of 'functional relations', an expression that is equally applicable to relationships between concepts and those between real objects, thus obscuring the fundamental difference between simple logical relations and the actual causal connections that are the subject matter of the natural sciences.[27] Therefore, in order 'to guarantee the univocalness [*Eindeutigkeit*] of causal relations in nature' – an objective clearly held by Petzoldt – one must accept 'the concept of the thing-in-itself', understood as the existence of real objects independently of our representations of them.[28]

As for the theory of relativity, Schlick's position with respect to Petzoldt's interpretation is clear from his first work on the topic, the article entitled 'Die philosophische Bedeutung des Relativitätsprinzips' (The Philosophical Significance of the Principle of Relativity), published in 1915. Here Schlick acknowledges that, apart from the error in his discussion of the Twin Paradox, Petzoldt starts with what is the sine qua non of any interpretation of the theory of relativity, namely, 'a penetrating understanding of the physical aspect of the theory'.[29] Even more importantly, Schlick concedes that 'from the outset, even before the establishment of the relativity principle, there was no set of ideas so closely akin to the notions of the theory as those of positivism', to the extent that 'Einstein could hardly have arrived at his theory if he had not himself already been toying with these ideas'.[30] However, this does not require 'overestimating' positivism to the point of 'thinking that it could so simply have given birth to the physical principle of relativity and thus can have nothing more to learn from it'.[31]

As Schlick notes, Mach's principle of relativity consists of the idea that 'since only relative motions are perceivable, it is they alone that are real'.[32] However much this approach might have caused Einstein to take the path that led him to the theory of relativity, the fact that Mach himself had derived demonstrably erroneous conclusions from his principle shows that the theory of relativity is not a natural predictable consequence of Mach's position.[33] Schlick's conclusion is therefore that:

> The general ideas of the positivist theory of knowledge undoubtedly provide a favourable soil for the principle of relativity and consequences for space and time that are derived from it. The new views can be assimilated by positivism without any difficulty. It is incorrect, however, that positivism has evolved and predicted these views from within itself, as the only correct ones – others would have suited equally well.[34]

This generally positive view of the interpretation of relativity provided by Petzoldt in the footsteps of Mach is partially revised in Schlick's writings in the 1920s. His change of opinion is probably due to Petzoldt's tendency to polarize his own position for the sake of argument, together with his growing emphasis on the sensualistic element. Indeed, Schlick's analyses increasingly tend to distinguish between different degrees of the empiricist-positivistic position. At the start of the 'Kritizistische oder empiristische Deutung der neuen Physik?', written in response to Cassirer's article, Schlick asserts:

> It seemed to me that the principles needed for a philosophical illumination and vindication of that theory could be drawn far more readily from the empiricist than the Kantian theory of knowledge; and even on subsequent occasions I found no reason to abandon this position, more especially since the successful completion of the general theory, which took place soon afterwards, brought victory to an idea that had arisen from the soil of extreme empiricism (namely the positivism of Mach).[35]

Schlick notes that 'there is certainly an empiricism that is distinct from *sensualism*', which is its radical extreme version.[36] Therefore, 'if it is shown (as it is not hard to do) that the theory of relativity cannot be made out upon *purely sensualist* premises', this does not mean that empiricism itself has been rejected.[37] Indeed, whosoever maintains this latter position recognizes 'the necessity of constitutive principles for scientific experience', even if the principles are not interpreted in a transcendental sense, but simply as 'hypotheses' (synthetic but not a priori) or 'conventions' (a priori but not synthetic).[38] A couple of years later, in 'Die Relativitätstheorie in der Philosophie' (The Theory of Relativity in Philosophy), Schlick makes a further distinction between Petzoldt's position and that of Mach, referring to Petzoldt's as '*exaggerated relativistic positivism*', and an '*extreme version of positivism which deems it possible to demolish any idea of an absolute reality with the slogan "everything is relative"*'.[39]

We can thus distinguish three versions of empiricism/positivism in Schlick's writings: his own less radical version; Mach's version, congruent with sensualism; and Petzoldt's version, which takes Mach's sensualism to the extreme, to the point where it merges with relativism.

Like Cassirer, Schlick's arguments against Petzoldt's relativism also often attribute to him positions that he had never held, such as 'everything is relative' in the sceptical sense. For example, Schlick writes that, as *'philosophical relativism'* tends 'to deny a strictly universal subjection to law' in nature, it 'is all the more reason to insist on the totally opposite character of *physical relativity'*, since Einstein's relativity (like all scientific theories) does not aim to 'remove' the laws of nature, but indeed to 'establish' new ones.⁴⁰ However, as we have pointed out, Petzoldt's interpretation never questions the *Eindeutigkeit* of nature, the lawful necessity of the course of events. On the same page, Schlick also speaks of the *'popular opinion'* consisting of the 'incredible misunderstanding that the theory of relativity involves an abolition of strict determinacy, a relaxation of natural laws'.⁴¹ We can therefore surmise that Schlick feared that, even though Petzoldt's relativistic interpretation was distinct from naive relativism, it might foster an incorrect interpretation of Einstein's theories by the general public.

In any case, as can be seen in the criticism elaborated in *Allgemeine Erkenntnislehre*, the fact that Petzoldt's relativism does not deny the lawfulness of nature does not reduce his inability to base this lawfulness on his sensualistic theory of knowledge, as he does not recognize the actual existence of the objects to which such laws pertain. Schlick therefore writes that 'the theory of relativity also – like any scientific theory – is solely concerned to establish objective, universally valid laws, and for this purpose it cannot forego the idea of an objective reality, which all subjects inhabit in common, and all observers must agree about'.⁴² For Schlick, this is the main reason why Petzoldt is unable to interpret the theory of relativity correctly, as his sensualism cannot explain how different observers can refer to *the same* reality, *the same* objects and thus *the same* laws. This also explains why Petzoldt misunderstands the Twin Paradox, incorrectly maintaining that 'the two observers, though both at rest by the clocks, are supposed nevertheless to register quite different states-of-affairs, and thus to read off different pointer positions *from the same clock*'.⁴³ The sensualist position would hold that there is nothing strange in assuming that 'the experiences of different observers' differ, since they 'never require to agree in any fashion'. However, 'Einstein's theory of relativity has never dreamt of advancing such a claim, or of defending a relativism of *that* description', given that if two observers are at rest at the same space-time point and take measurements using the same clock, it is impossible for them to obtain different readings.⁴⁴

Petzoldt's original sin is therefore his sensualistic theory of knowledge, which makes him mistake the relativity of sensations, as discussed by Protagoras, with relativity as a theory of physics, whereas 'man and his sense-qualities do not appear in Einstein's formulae', as his theory 'makes reference only to physical *quantities*, that is, to results of measurement'.⁴⁵

Petzoldt inserts a reply to Schlick, in particular to the criticisms raised in the *Allgemeine Erkenntnislehre*, in a long footnote in the third edition of the *Weltbegriff*. He makes some concessions to Schlick's position, recognizing to some extent the concept of the thing existing independently of sensory perceptions.⁴⁶ In particular, he specifies:

> The *only* thing that can be said of a thing considered independently, conceived as released from its relation to the central nervous system, is that it exists, and exists independently of any perception of it. The question of *how* it is constituted is fundamentally unresolvable and indeed meaningless.⁴⁷

Despite his best efforts, Petzoldt once again ends up drawing closer to Kant. He is well aware of this, as he promptly points out that 'the problem with the concept of thing-in-itself is not the assumption of an existence that is independent' of sensory experience, but the recurring temptation 'to attribute properties to it, to which Kant succumbs when he considers the senses as "affected" by the thing-in-itself'.[48]

Petzoldt therefore sees no problem in assuming that our sensations, being *relative*, refer to an *absolute* substratum that is completely unknowable. Indeed, the error would be to believe that Mach 'had asserted that sensations were components of the world that existed *absolutely*, elements of the world in an *absolute* sense, somewhat like atoms floating freely in space'.[49] In other words, the thing-in-itself in its absolute unconnected nature cannot be known, because all our knowledge is always relative. When we speak of the sensory properties of objects, we must therefore not attribute them to the objects *in themselves* but to the objects in relation to ourselves. However, this does not mean that these properties are mere appearance, as relations are also real and univocally determined.

Therefore either Schlick must concede the completely undetermined absolute existence of the thing-in-itself, and thus agree with Petzoldt (and Kant, one might add), or he must attribute non-empirical properties to it and thus relapse into metaphysics, or he must attribute empirical properties to it, but in that case he is creating a mere duplicate of the sensory object, a 'faded copy' of the 'thing that is visible and tangible', a copy that is 'completely superfluous'.[50] Petzoldt's view is that this third option is in fact the case, as demonstrated by Schlick's assertion that 'transcendent objects are supposed to be the grounds or bases for phenomena; hence to every difference in the phenomena there must also correspond a difference in the objects'.[51]

Turning to the theory of relativity, Petzoldt claims that, in claiming that our intrinsically relative knowledge is projected onto the absolute thing-in-itself, 'Schlick, the most fervent supporter and most excellent exponent of the theory of relativity, does not keep faith with the principle of relativity from which the *erkenntnistheoretisch* core of Einstein's theory originated'.[52] Petzoldt thus rejects Schlick's criticism and reasserts that Einstein's theory of relativity can be subsumed under the more general principle of relativity that holds that 'absolute properties do not exist'.[53]

However, Petzoldt concedes Schlick's point that relativistic positivism replicates Leibniz's position. Indeed, Petzoldt frequently refers to the Leibnizian concept of monad in his last works, showing that it can be adapted to a relativistic conception of the world and above all to Einsteinian relativity:

> Nobody experiences what others experience; this point must be stressed. Each of two observers in relative motion lives within a distinct spatiotemporal system and in this sense can be compared to the monad of Spinoza and Leibniz. [...] As the two systems in reciprocal motion are to be thought of as separate in monadic terms, they cannot be thought of together within a *single* (three-dimensional) space. [...] On this point the comparison with Leibniz's monad fits perfectly.[54]

At the same time, Petzoldt emphasizes that 'our "monads" are to be thought of as interconnected by the same kind of laws', which means that (unlike Leibniz's monads) 'in the end they are not without windows'.[55]

Petzoldt thus replies indirectly to Schlick, while also retracting the negative view of Leibniz's approach expressed in his early works. Putting aside its metaphysical elements, Leibniz's philosophy may provide a valid understanding of reality and of the recent advances in physics, going beyond a simple mechanistic model of the world.

Reichenbach

Although Hans Reichenbach is best known as the foremost representative of the *Berliner Gruppe*, he had not yet moved to Berlin at the time of the debate with Petzoldt on the theory of relativity but was teaching physics at the Technical University of Stuttgart. Like Schlick, Reichenbach had a scientific background which facilitated his understanding of Einstein's theories. The study of physics led him to the philosophy of science, and in 1926 he applied for a permanent position in philosophy at the University of Berlin. However, he succeeded only in securing a post as extraordinary professor (i.e. without a chair) in the narrower field of 'The Epistemological Foundations of Physics'.[56]

From the 1920s onwards, Reichenbach published a large number of works on relativity. These not only explicated the new scientific theory and its philosophical implications but also proposed an axiomatization of the theory, distinguishing between its definitional and its empirical elements.[57] In particular, his first major work on Einstein's theories was *Relativitätstheorie und Erkenntnis apriori* (*The Theory of Relativity and A Priori Knowledge*, 1920), in which he rejected neo-Kantian attempts to claim that relativity could be reconciled with Kant's approach. According to Reichenbach, comparing Einstein's approach with Kant's original doctrine showed that the two theories were contradictory, and that therefore one or the other had to be rejected or modified.[58] Reichenbach obviously came down in favour of maintaining the theory of relativity and revising Kant. He therefore proposed a distinction between two senses of 'a priori' in Kant, that of 'necessarily true or true for all times' and that of 'constituting the concept of object' (*der Gegenstandbegriff konstituierend*). This made it possible to maintain the distinction between the organizing principles of knowledge and cognitive material, that is, between cognitive form and cognitive content, while sacrificing the idea that these principles are eternal and undergo no change consequent upon new experiences.[59] As Reichenbach himself noted in a letter to Schlick, the position expounded in *Relativitätstheorie und Erkenntnis apriori* could equally well be seen as either 'a break with a very profound principle of Kant's' or as 'a newer continuation of the Kantian conception'.[60]

Two years later, Reichenbach published an article in the journal *Logos* in which he provided a critical analysis of the different positions being taken in the debate on the philosophical interpretation of the theory of relativity. In the section on 'Conceptions influenced by Mach', Reichenbach compares Petzoldt's position with those of other followers of Mach who were more critical of Einstein's theories, such as Hugo Dingler. First of all, Reichenbach recognizes the link between Einstein's theory and the ideas developed by Mach. Mach was in fact the first to emphasize not only kinematic relativity (a conception that was implicit already in Leibniz) but also dynamic relativity, by showing that the reference to other bodies is relevant not only to the description of motion but also as a condition for the law of inertia.[61] Given this irrefutable link between the theory of relativity and the ideas of Mach, Reichenbach minimizes the importance of the apocryphal Preface to the *Optik*, attributing it in part to Mach's advanced years and in part to the fact that the theory of relativity had not yet reached its final formulation as the theory of general relativity.[62]

The main idea that Reichenbach attributes to Petzoldt is that Einstein should be credited with basing his theory on coincidences as the only elements of reality that are actually observable, thus contributing to the elimination of metaphysics.[63] Indeed, according to Reichenbach, Petzoldt held that '"real" means "observable"'.[64] However, we have already seen that Petzoldt subscribed to the theory of relativity well before Einstein introduced the concept of 'coincidence'. In fact, the appeal of the new theory of physics to Petzoldt's relativistic positivism was not so much in the emphasis on observability as in the fact that every description of reality was linked to a specific frame of reference, thus rejecting the mechanistic claim that reality could be known in itself and for itself.

Reichenbach's words thus contain the phenomenalistic distortion of Petzoldt's ideas (and Mach's) that represents the heart of their position as the identification of reality with sensations. While it is true that Petzoldt and Mach held that sensory reality *is* reality, and not merely a faded copy of it, this view is completely misrepresented if one fails to precede it with the idea that sensory reality is fundamentally *relative*, and relative in a univocally determined way. Sensory reality is reality *in relation to me*, to my body, to my nervous system. Reality is different for other individuals, but it is no less real, *in relation to them*. The point is that no one can know reality *in itself and for itself*, in the absolute, independently of any relation.

The demonstration that Reichenbach misunderstands Petzoldt's insistence on the *relational* nature of sensory reality is provided by the fact that he attributes to him the idea that 'something can be real for one observer *which is not real for another*'.[65] However, Petzoldt's point is in fact that two different observations, effected by two different observers, *are both real*, each in relation to each of the observers, and both observers can recognize the reality of the observation effected by the other as each observation is univocally determined by its relation to the other observer.

It is therefore necessary to rectify Reichenbach's summary of Petzoldt's position. For Petzoldt, 'real means observable' only in the sense that 'what is observable is real' (perceived sensory reality is actual reality), while the opposite is not necessarily true, namely that 'what is real is what is observable' (in the sense that reality *in itself* is sensory reality, or in the sense that sensory reality exhausts the field of what is real).

In attributing to Petzoldt the idea that reality is *exhausted* by sensory experience, Reichenbach maintains that the value of the theory of relativity to Petzoldt lies in the fact that each individual observer effects observations within their own frame of reference, which, for them, is the only reality. All metaphysics would thus be eliminated and all scientific content based on sensations, as each observer has no need to move outside their own observable reality. Reichenbach therefore sees rejecting Petzoldt's position as an easy task, emphasizing that:

> Even 'the world *of one system*' is a world which goes beyond immediate perception and therefore contains arbitrary elements; it is not possible to eliminate these

elements by reducing them to the *immediate perceptions of one observer*, but only by determining the causal relations between the perceptions which result from a variation of the arbitrary elements. The causal relations are the same for all observers.[66]

Reichenbach adds:

Petzoldt forgets that the knowledge of the transformation formulas provides that element which points *beyond the measurements of a single system*. The transformation formulas are not empty definitions, but empirical discoveries; they express the causal relationships between the observations in different systems. If we are merely given the measurements *in one system* we are not thereby given the transformation formulas and the measurements in other systems.[67]

As we can see, Reichenbach insists on the 'single system', even if Petzoldt never claimed that science could be exhaustive if based on the sensations of a single individual. Petzoldt's relativism was never solipsistic. Although he emphasizes the impossibility of disregarding the relation to the individual, he always maintains that it is only the plurality *of all points of view* that provides the foundation for scientific knowledge. Moreover, we know that Petzoldt held that perceptions in themselves tell us nothing without *Eindeutigkeit*, which is the true foundation of all knowledge and the pillar on which he constructs his conception of reality. It is *Eindeutigkeit* that is the nexus that holds together the sensations of the single individual, that links the sensations of different individuals and more generally that gives substance to a reality that would otherwise appear to be no more than a chaotic stream of perception.

It is also thanks to *Eindeutigkeit* that sensations are objective rather than merely subjective. Reichenbach's misunderstanding of the role of 'univocal determination' in Petzoldt's approach therefore ends up also attributing to him a subjective interpretation of relativity, whereas subjectivism is in fact one of the main targets of Petzoldt's argumentation, as we saw in Chapter 3. Indeed, Reichenbach writes that 'differences in the measurements obtained in the different systems have nothing to do with the *subjectivity of the observer*', with the 'subjectivity of perception'.[68] Therefore 'Petzoldt's positivism deviates from the theory of relativity' in that 'Einstein does not assert the relativity of *truth*' but presents a theory that 'is definitely *objective*'.[69]

Reichenbach and Petzoldt continued their discussion in private correspondence. Reichenbach had already asked Petzoldt to send him his works on relativity as he was preparing *Der gegenwärtige Stand der Relativitätsdiskussion*. They later also agreed to meet in person. We can therefore assume that they had already met when Reichenbach's work was published. On its publication, Reichenbach sent the work to Petzoldt, together with other works of his on relativity. Petzoldt expressed his thanks in a reply, and his appreciation of Reichenbach's work on the axiomatization of Einstein's theory. Petzoldt apologized for not being able to reply in depth to the criticisms of his position, citing health issues that hampered his ability to concentrate on work, and limited himself to rebutting a footnote in which Reichenbach had pointed out the error in his interpretation of the Twin Paradox. Petzoldt also invites Reichenbach to re-read his 'Die Unmöglichkeit mechanischer Modelle zur Veranschaulichung der Relativitätstheorie' (The Impossibility for Mechanistic Models to Illustrate the Theory of Relativity, 1919) and to let him know his impressions, evidently believing that this work might correct his erroneous view of relativistic positivism.[70]

Reichenbach's reply is interesting in that he shifts to a more positive stance with respect to Petzoldt's position. He writes:

> I must say quite openly that I have the feeling that the comments I wrote on your conception in my article in 'Logos' are not my last word on the matter. On the contrary, it appears that I did not do justice to your position. I did not fully understand your considerations. This might also be due to the fact that – as seems to be the case – you encountered some difficulty in understanding the physics of the theory of relativity, and therefore reached philosophical formulations that I did not fully understand. My aim in this work was only to initiate an exchange of views with you, to encourage you to reiterate your views in a different form. It would be even better for me if we could initially conduct this discussion by letter.[71]

It may be the case that Reichenbach's reassessment was also prompted by the words of Einstein, who had sent him a letter with his comments: 'I just find your opposition to Petzoldt-Cassirer lopsided, without intending to concede the point to Petzoldt.'[72] In any case, later in his letter Reichenbach explains the physics aspects of the Twin Paradox to Petzoldt, and how the relativity of simultaneity works from a scientific point of view. With respect to the more clearly philosophical issues, Reichenbach criticizes Petzoldt's approach of wanting to see 'deeper paradoxes in the kinematic assertions of the theory of relativity'.[73] Moreover, Reichenbach reasserts that 'frames [of reference] and observers are not the same thing', as when one speaks of what someone *observes* in one or another frame of reference, one is only dealing with a 'pedagogical tool' whose purpose is to elucidate the theory.[74] Reichenbach closes by asking Petzoldt for a comment on Cassirer's work, to find out whether he thought that he had 'done justice' to the difference between their positions.[75]

Petzoldt's reply conveys the annoyance he must have felt on seeing the distorted account of his position in Reichenbach's article. He writes:

> If we decide to conduct an in-depth discussion by letter, to avoid futile effort, we must at least read the works of others attentively. Otherwise we know nothing of what others may think on many points. But this is very time-consuming, given that our interests have such different starting points. You are concerned with axiomatics, whereas I am concerned with the physical aspects of the theory and its *erkenntnistheoretisch* foundations, by which I do not mean only those that are grounded in physics and mathematics, but also in the physiology of the senses and the psychology of the senses. These are of crucial significance.[76]

Petzoldt clearly feels that he has been misunderstood, and fears embarking on a dialogue of the deaf in which all his philosophical references to relativism and sensations might once again be distorted into phenomenalism, subjectivism and scepticism. It may be that he sees Reichenbach as to some extent a Kantian, in the light of *Relativitätstheorie und Erkenntnis A Priori*, and therefore suggests a route to understanding the perspective of relativistic positivism, the same route that Schuppe took: 'starting with Kant through his doctrine of the empirical reality of phenomena'.[77]

In their further correspondence, Petzoldt and Reichenbach arrange a new meeting, following which Reichenbach sends him his 'La signification philosophique de la théorie de la relativité' (The philosophical meaning of the theory of relativity, 1922), which sparks an exchange of letters in which the argument shifts to the role of philosophy with respect to science. Reichenbach closes the article thus:

> Admittedly, philosophy is sheltered from attacks based upon experience *on the condition that they limit their analysis to the logical structure of scientific systems*. But then it rightly ceases to be in a position to produce any statement about reality, about the world of real things […] The assertion of a predetermined structure of space and time, of causality, etc., to be as general as we wish to consider them, does not constitute reality despite all their special precepts simply because more general precepts than these can be imagined which serve as their foundational concepts. Maximally general concepts do not exist. This is why philosophy cannot consist of the discovery of a more general structure for reality. Philosophy must leave the edification of these structures to the particular sciences and limit itself to explicitly describing the structures that the particular sciences implicitly employ.[78]

Here Petzoldt disagrees with Reichenbach, who he sees as assigning an 'overly narrow' task to philosophy, failing to recognize its 'main task', namely 'to investigate all the preconditions of the prevailing views and to identify the preconceptions, thus clearing the way for research'.[79] Petzoldt's view is that Reichenbach does not grasp the proactive role that philosophy can perform because he is too focused on physics. Thanks to its exchange of ideas with psychology, the physiology of sensations and biology, philosophy had already put behind it any remnants of the 'rationalism' that still lingers in physics.[80] By 'rationalism', Petzoldt means those conceptions that ascribe knowledge to the 'activity of an autonomous "reason" that is in contraposition to things', 'without taking the psychophysical organism into account'.[81]

In his reply, Reichenbach reiterates his view that it is 'premature to want to already construct a fully general theory of knowledge', given that 'much detailed work [*Kleinarbeit*] has yet to be done'.[82] Indeed, the danger of allowing philosophy too much freedom to follow its own path, leaping instantly from the concrete results of the natural sciences to grand generalizations, is evidenced by Petzoldt's own errors. Shortly after, Reichenbach writes that their 'differences of opinion' on Einstein's ideas stem from that fact that Petzoldt 'introduces too many general conceptions about the theory of knowledge into the theory of relativity, in particular in places where they do not belong'.[83]

This exchange on the role of philosophy is particularly interesting in the light of the subsequent passing of the baton of the Berlin *Gesellschaft für empirische Philosophie* from Petzoldt to Reichenbach. Petzoldt had founded the Society with a deep-rooted nineteenth-century mindset, as shown by his goal of an all-embracing philosophical system. He held on to the idea of a necessary evolution of the course of human thinking that was destined to achieve a state of definitive completeness, as well as to the idea of the primary role of psychology, biology and the physiology of the senses, disciplines that had leapt to the fore in nineteenth-century debate. Reichenbach's focus, on the other hand, is directed towards how philosophy will develop in the first half of the twentieth century, at least in the area known as neo-positivism: the centrality of the interface between physics, mathematics and logic, the role of philosophy as the analysis of scientific methods and results, and diffidence with respect to philosophical approaches whose stated aim is to take inspiration from the sciences but which, instead of devoting themselves to the study of science, use it as a springboard to immediately leap to rash generalizations.

Summary

Each of the thinkers that we have considered starts from their own perspective in their confrontation with Petzoldt's position. Nevertheless, the criticisms of Cassirer, Schlick and Reichenbach have a common thread. Petzoldt had always channelled all his energies into elaborating the position of Mach and Avenarius in order to demonstrate that empiriocriticism, and later relativistic positivism, had nothing to do with phenomenalism, subjectivism, scepticism or 'the idealism of sensations'. The words of Cassirer, Schlick and Reichenbach suggest that Petzoldt, following in the footsteps of Mach, wanted to identify reality with sensations, thereby eroding the foundations on which scientific conceptions are based. It is almost as if the notoriety of Mach's assertion that sensations are the elements of the world was to obfuscate all of Petzoldt's attempts to contextualize and elucidate the assertion, while also supplementing it with *Eindeutigkeit*.

As we have noted frequently above, the importance to Petzoldt of the theory of relativity did not derive from the fact that it was based on sensations or on coincidences of sensations but from its recognition that every description of the world is univocally determined by the relation to a specific frame of reference. This is the point of contact between the theory of relativity and the physiology of sensations, the fact that both demonstrate the impossibility of knowing the world in its absolute essence, in itself and for itself, reaffirming, however, the possibility of knowing it *sub specie univocitatis*, in its *Eindeutigkeit*. Just as the *laws* governing the activity of the nervous system univocally determine the sensations experienced by each individual, thereby allowing the experiences of each individual to relate to the experiences of other individuals, so the transformation *laws* of the theory of relativity univocally determine the measurements obtained in each frame of reference, thereby allowing these measurements to interconnect. It is in this sense that Petzoldt speaks of a 'perspectivist physics'.[84]

Petzoldt's interpretation of relativity is clearly a partial one that often distorts Einstein's theories for his own ends, and even includes actual errors, as in the case of the Twin Paradox. However, this does not mean that Petzoldt was not, in turn, the victim of misunderstandings, being attacked for his conception of the theory of relativity for aspects that he had never proposed. Petzoldt's position is that stating that the theory of relativity confirms and rests upon relativistic positivism does not mean maintaining that a unique knowledge of reality *does not exist*, but rather that knowledge of reality is knowledge *of relations*.

Petzoldt's conception cannot be reduced to the notion that Einstein's reference to an 'observer' means that the theory of relativity confirms the phenomenalistic notion that reality is just sensation, sensory experience, thereby making knowledge of the world something that is subjective, mental, psychological. This is especially so when one considers that, even in the field of the physiology of sensations, for Petzoldt the dependence of experience on the brain does not make it subjective, mental or psychological in the sense of dualism or phenomenalism. It must not be overlooked that Petzoldt sees the dependence of experience on the brain, being univocally determined, as part of *physics*, which is knowledge of reality in its *Eindeutigkeit*. In other words, even if one assumes that Petzoldt considers the role of the observer in Einstein's theory as confirmation that all science is based on the sensory experience, this does not imply that physics slides into psychology, as the functional dependence of observations on the observer is itself a matter of physics. He writes:

> Even if empirical psychology, neurology, psychiatry, and the physiology of the brain, all teach the *complete dependence* of psychological processes on physiological processes, philosophy […] still completely lacks any biological thinking […]. Both realists and idealists complain that regarding things as dependent on the human structure necessarily leads to unconstrained subjectivism, and ultimately to the undermining of all of science, whose lifeblood is its generality and objectivity. […] The human cerebral cortex is part of nature and as the physiological process that take place within it *univocally determine* all perception, imagination and thinking, it is essential to pay the utmost attention to these connections when establishing a world view. World views that fail to do this only observe a part of the world, and are fatally flawed.[85]

6

Conclusion

Relativistic positivism, or perspectivism

Our objective in this volume has been to lay out the complex intellectual journey undertaken by Joseph Petzoldt. From his starting point of the search for unified knowledge that characterized the nineteenth century, Petzoldt rejected the solutions offered by the mechanistic materialism of the natural sciences and by the subjectivistic idealism of philosophical circles. He also rejected neo-Kantianism, which tried to reconcile these two positions. Following the lines of Fechner, Avenarius and Mach, Petzoldt aimed to overcome the opposition between the physical and the mental, between subject and object, by constructing a philosophical approach based on the unity of empirical reality, within which human beings – seen as biological organisms – evolve. This reality is not *made of* sensations, as an incorrect reading of Mach might suggest. Sensations, being *reality* in relation to the human organism, *are* reality. Even more importantly, however, sensory reality reveals itself to be regular, governed by laws and univocally determined. Moreover, as human beings are themselves part of reality, our relation to reality is also governed by laws. This means that the different relations that each individual, and, more generally, every being, weaves with reality are not mutually incommensurable, as they are governed by the same laws.

Petzoldt claimed to have revived the relativist message first formulated by Protagoras, namely, that everyone knows reality in relation to themselves. At the same time, this relativism is positivistic in that it does not deny the possibility of a universal and stable knowledge of reality, as the relations that are given in the world are univocally determined. This flavour of relativism thus corresponds to the philosophical position that would today perhaps be called, more specifically, 'perspectivism', given that what it affirms is 'not a relativity of truth but, on the contrary, a truth of the relative' (to use the incisive formulation of Deleuze and Guattari).[1]

In the history of philosophy, the term 'perspectivism' or 'perspectivalism' is usually associated with the works of Leibniz and Nietzsche. However – as noted by James Conant – a result of the spread and the ambiguity of the metaphor of perspective is that several writers have used 'perspectivism' even for philosophical positions that are diametrically opposite.[2] Conant himself concludes by seeing the whole history of the theory of knowledge in the light of this metaphor, identifying different versions of perspectivism, from naive realism to proto-Kantianism.[3] However, Conant claims that, strictly speaking, the only positions that properly come under the heading of 'perspectivism' are those that (1) use the metaphor of 'perspective' in the sense it acquired during the Renaissance as a result of the development of geometrical techniques capable of representing a three-dimensional object from different points of view, and (2) use the metaphor to demonstrate how an individual's point of view is in no way 'subjectivist' (in the sense of being a mere representation, an illusory event with no actual reality), as 'every perspective forms a part of a rational system of determinable relationships between itself and other possible perspectives'.[4] This is indeed the sense found in Leibniz,[5] Nietzsche[6] and in Petzoldt, as can be seen in his interpretation of the philosophy of Protagoras cited above, where the metaphor of perspective is used precisely in this sense:

> All perceptions of the world, all our intuitions about the world, as they are equally actual so they are equally true, *in the same way that perspectives of an object seen from different points of view are absolutely equally valid and equally correct.*[7]

As Petzoldt's use of 'perspectivism' is close to that of Leibniz and Nietzsche, it is worth devoting some space to the relationship between these three thinkers. With respect to Leibniz, Petzoldt initially does not appear to grasp the affinity between his own position and monadology. The first edition of the *Weltproblem* affirms that Leibniz placed 'minds (monads) in total reciprocal isolation from the world' and that the result of this 'logical inadequacy' is that its conception 'should not concern us further, being tangential to the straight line of continuous development'.[8] Perhaps somebody pointed out to Petzoldt that his interpretation of the Leibniz doctrine was unfair, so in subsequent editions of the *Weltproblem* the section on Leibniz is revised almost in full. Petzoldt changes his overall view, stating that Leibniz's doctrine, even though it is a 'rocky wall that humanity's most able souls are trying to climb', 'it is nevertheless an advance', mainly because it 'overcomes the mechanistic conception of nature'.[9] Nevertheless, Petzoldt stands firm on his condemnation of monadology for its 'metaphysical', 'theological' and even 'neo-Platonic' and 'medieval' aspects (ibid.). Petzoldt's criticisms of Leibniz suggest that the similarities between their approaches are the result of an at best indirect influence of Leibniz on Petzoldt. In fact, Leibniz was a constant reference point in the science and philosophy on the nineteenth century. In particular, we can surmise that Petzoldt took on board the Leibnizian aspects of Fechner's approach.[10] In any case, Petzoldt draws closer to Leibniz when he addresses Einstein's theory of relativity, underlining that the relationship between different reference systems resembles that between different monads in Leibniz's position.[11] It may be the case that Petzoldt's reassessment of Leibniz was prompted by the general reappraisal of Leibniz's views in relevant circles in the light of the new scientific theories.[12]

Nietzsche is mentioned in positive terms several times by Petzoldt in the *Einführung in die Philosophie der reinen Erfahrung*, but in reference to his theories of aesthetics and morals, rather than to his gnoseological position. A nod to the latter is, however, found in one of Petzoldt's later works, where he underlines the affinity between the 'History of an Error' in *Twilight of the Idols* and Mach's position: 'even Nietzsche –

perhaps influenced by Mach? – wrote "The true world – we have abolished. What world has remained? The apparent one perhaps? But no! With the true world we also have abolished the apparent one."[13]

In view of the above, we can say that Petzoldt sympathized with other philosophers who belong to the perspectivist tradition. More importantly, he can be regarded as a representative of a wider trend that characterized the nineteenth century, when the psychophysical study of sense perception (thanks to thinkers such as Weber, Fechner, Helmholtz and Mach himself) led to a flowering of perspectivism, based on the idea that 'representations are always physiologically, psychologically, and perspectively relative to the observer', but 'precisely for that reason, they can be used as scientific evidence, provided that relativity to the observer itself can be analyzed scientifically'.[14] In other words, it was the possibility of a scientific investigation of the relationship between the subject and the world (first provided by the recent advances in the physiology of sense perception and experimental psychology) that paved the way for a relativism that was not sceptical but realist.

It should be noted that perspectivism has recently gained increasing attention in the world of philosophy and especially in the field of philosophy of science.[15] However, precisely because of this, the term has become even more ambiguous, making it difficult to present a universally valid definition. Michela Massimi defines it as the idea that 'scientific knowledge is historically' or 'culturally situated'.[16] Ronald N. Giere considers it a middle position between objective realism and social constructivism.[17] We do not think it would be useful to locate Petzoldt's perspectivism in this contemporary debate, since this strand of contemporary thinkers engaged in perspectivism has interests and goals that differ considerably from those that occupied Petzoldt in his time. Nevertheless, we think that the concept of perspectivism is the name that best fits his philosophy, in order to avoid possible misunderstandings that might arise from adopting other concepts, such as relativism and phenomenalism. Furthermore, calling Petzoldt's philosophy a form of perspectivism can help highlight his connections to other philosophies of his time with whom he did indeed share common ground, such as the aforementioned Nietzsche and Helmholtz.

Petzoldt's role in the history of philosophy

In building his system of philosophy, Petzoldt picked up elements from the work of his masters Avenarius and Mach. Avenarius focused on how human knowledge is entirely conditioned by the cerebral substratum, and in particular by the evolution of the brain. This evolution, proceeding as it does towards increasing stability, leads to the elimination of metaphysical and unempirical content from our knowledge of the world. Mach, on the other hand, attacked the claim of mechanistic physics that it was possible to achieve absolute knowledge, reminding us that human beings, with their experiences and needs, are a component from which one might be able to abstract, but which cannot be eliminated from scientific knowledge.

Convinced of the historical inevitability of this philosophical prospect, one that was destined to prevail and to eradicate the endless debates between materialists, idealists and Kantians, Petzoldt poured all his energies into spreading and developing his relativistic positivism. When Einstein published his theory of relativity, Petzoldt saw it as formidable support of his cause, containing as it did the notion of a reality that presents itself differently in the different frames of reference of different observers, but remains a reality that can be known scientifically because these differences are univocally determined precisely by the relation between the different frames of reference. Petzoldt thus believed that he could enlist Einstein into the service of relativistic positivism, bolstered by the acknowledged influence of Mach's criticism of Newtonian physics on the development of the theory of relativity.

However, Petzoldt's great hopes would crumble over the years. Mach's apocryphal Preface to the *Optik* undermined his quite legitimate arguments in support of an intrinsic link between Mach's position and the theory of relativity. The attacks of renowned thinkers such as Cassirer, Schlick and Reichenbach demonstrated that relativistic positivism continued to be misunderstood in the same ways as Mach's thinking. Moreover, the very idea of producing a grand synthesis of all of the sciences, guided by philosophy, became increasingly anachronistic due to the fragmentation of psychology into a myriad of schools, the withdrawal of physiology into specializations, and physics achieving such levels of complexity as to make it ever more difficult for an ordinary thinker to be able to say anything significant as a result of insufficient fundamental knowledge of mathematics.

Petzoldt persisted to the bitter end in his endeavours to get philosophers and scientists from different fields to talk to one another and to promote relativistic positivism through the *Gesellschaft für empirische Philosophie*. However, on his death his main legacy was the society itself rather than his philosophical works, a society that had now passed into the hands of Reichenbach and the other logical positivists of Berlin. Even though Petzoldt had been the main exponent of Mach's approach in the early twentieth century, the Ernst Mach that will continue to be a beacon for twentieth-century philosophy is completely different from the one depicted by Petzoldt. In fact, Petzoldt tried to transform Mach's *critique of scientific knowledge* into a *philosophical system* that was tainted with some dogmatism. It will be Mach's critical approach, with his emphasis on the provisional and unstable nature of all knowledge, that will be taken up by the exponents of neo-positivism.

Nevertheless, Petzoldt deserves to be remembered and studied.

Firstly, he represents a case study of the complex dialogue between philosophy and the world of science on the cusp of the twentieth century, with his aim of achieving a synthesis that could overcome the traditional antinomies between different philosophical approaches and at the same time incorporate the results of the natural sciences (biology, psychophysiology and physics).

Secondly, he was the most vocal proponent of the philosophy of Avenarius and Mach and the main person responsible for shaping 'empiriocriticism' as a philosophical school based on the convergence of the ideas of these two thinkers. Thus, it would be impossible to fully understand empiriocriticism or the legacy of Avenarius and Mach without taking into account the role that Petzoldt played in the dissemination, elaboration and even distortion of their ideas.

Thirdly, our study of Petzoldt gives us a fuller picture of the debate over the early philosophical interpretations of the theory of relativity, in which he was one of the main participants. When speaking of Einstein's connections to the Machian world view, one should always distinguish two aspects: on the one hand, the influence that Mach exerted on Einstein, and on the other hand, the quasi-Machian interpretation of relativity that Petzoldt provided and that was (partially and initially) supported by Einstein himself. In particular, Petzoldt's notion of *Eindeutigkeit* represented a departure from Mach's indeterministic tendency which appealed to Einstein, who always resisted the idea that phenomena could happen without being fully determined.

Finally, even though Petzoldt transformed Mach's theory of knowledge into a philosophical system far removed from Mach's original intentions, his continued efforts to combat the sensualist and phenomenalist interpretations of Mach's philosophy are nevertheless of value. Since the 'received view of Mach' still misrepresents him as the 'champion of an "anti-metaphysical" phenomenalism, having sense-data (sensations) as its proper ontology',[18] Petzoldt's antithetical reading of Mach's philosophy as a form of realist and anti-sceptical relativism, or perspectivism, can help counterbalance these widely held trivializing interpretations of Mach's thought.[19]

Notes

Chapter 1

1. The biographical information presented here is taken from 'Autobiographie bis zum Beginn der Lehrtätigkeit an der Technischen Hochschule Berlin' (1904), in PetzoldtNachlass, PE1, and from the commemorative works published on Petzoldt's death: Lily Herzberg, 'Joseph Petzoldt Tot', *Monistische Monatshefte*, 14 (1929), 223–4; Walter Dubislav, 'Joseph Petzoldt in Memoriam', *Annalen Der Philosophie Und Philosophischen Kritik*, 8 (1929), 289–95; Christian Herrmann, 'Joseph Petzoldt', *Kant-Studien*, 34 (1929), 508–10. Further details on Petzoldt's life can be found in Hermann Wagner, 'Professor Joseph Petzoldt Zum Gedächtnis. Von Einem Ehemaligen Schüler', *Kai Nein Kai Grammata. 1853–1953* (1953), 27–9; Horst Müller, 'Joseph Petzoldt', *Humanismus Und Technik. Technische Universität Berlin*, 11 (1966), 33–6. See further references in notes 7 and 10.
2. J. Petzoldt, 'Autobiographie', 2.
3. J. Petzoldt, 'Autobiographie', 4. Cited from Ernst Mach, *Die Mechanik in ihrer Entwickelung; historisch-kritisch dargestellt*, 1st edn (Leipzig: Brockhaus, 1883), iv; *The Science of Mechanics: A Critical and Historical Account of Its Development*, trans. by Thomas J. McCormack (Chicago, London: Open Court, 1919), x.
4. J. Petzoldt, 'Autobiographie', 7.
5. J. Petzoldt, 'Autobiographie', 8.
6. Joseph Petzoldt, 'Anzeige von R. Avenarius "Kritik der reinen Erfahrung"', *Das Magazin für die Litteratur des In- und Auslandes*, 58/8 (1889), 120–36.
7. Cf. 'Briefentwurf an das Ministerium betr. Bitte um einen Lehrauftrag' (1920), in PetzoldtNachlass, PE04, 3–4. Cf. also Harmut Hecht and Dieter Hoffmann, 'The Berlin "Society for Scientific Philosophy" as Organizational Form of Philosophizing in the Medium of Natural Science', in *World Views and Scientific Discipline Formation*, ed. by William R. Woodward and Robert S. Cohen (Dordrecht et al.: Kluwer, 1991), 78. For a perspective on the antipositivist and apolitical climate at the University of Berlin in the second half of the nineteenth century, in particular after the failed uprising of 1848, see the Introduction to Klaus Christian Köhnke, *The Rise of Neo-Kantianism: German Academic Philosophy between Idealism and Positivism* (Cambridge: Cambridge University Press, 1991).
8. A *Privatdozent* was an academic licenced to teach but with no contractual relationship with the university. The position did not attract a salary, and compensation depended on the number of students attending lectures.
9. For further details on Petzoldt and, more generally, on the role of technical universities in Germany at the start of the twentieth century in the teaching of philosophy as an alternative occupation for thinkers rejected by more traditionalist universities as a result of their positivist views, see Christian Tilitzki, *Die deutsche Universitätsphilosophie in der Weimarer Republik und im Dritten Reich*, 2 vols

(Berlin: De Gruyter, 2002), II, 186 ff., cf. in particular the chapter 'TH Berlin: Joseph Petzoldt, Walter Dubislav und Hans Reichenbach' (ibid. 230 ff.).

10 For more on this topic, see the relevant paragraph on *Gesellschaft für positivistische Philosophie* in Klaus Hentschel, *Die Korrespondenz Petzoldt-Reichenbach: Zur Entwicklung der „wissenschaftliche Philosophie" in Berlin* (Berlin: Sigma, 1990), 16 ff. It should be noted that creating societies to further particular philosophical directions was not unusual in Germany in those times. For example, three of the most important ones were: the *Jakob-Friedrich-Fries-Gesellschaft* founded by Leonard Nelson (active from 1913 to 1921 but built on the *Neufriesische Schule* founded in 1904); the *Deutsche Gesellschaft für ethische Kultur*, founded in 1892, originally to oppose religious teaching in schools, but later broadened to include offices even in Vienna and Zurich; and above all the *Deutsche Monistenbund* founded by Ernst Haeckel (1906–33), which had more than 6,000 members. The latter was a society with which Petzoldt collaborated on several occasions, and on his death its journal published a memorial piece that opened with the words 'A close friend of our society, Joseph Petzoldt, has died' (Herzberg, 'Joseph Petzoldt Tot', 223).

11 'Aufruf', *Archiv Für Geschichte Der Philosophie*, 25 (1912), 502. See Gerald J. Holton, *Science and Anti-Science* (Cambridge: Harvard University Press, 1993), 14.

12 Cf. Hentschel, *Die Korrespondenz Petzoldt-Reichenbach*, 22 ff.

13 Cf. 'Gesellschäftliche Mitteilungen', *Zeitschrift Für Positivistische Philosophie*, 1 (1913), 1.

14 See the society's promotional leaflet, included in PetzoldtNachlass, PE32. Cf. also Hentschel, *Die Korrespondenz Petzoldt-Reichenbach*, 24 ff., 30. Translation from Dieter Hoffmann, 'The Society for Empirical/Scientific Philosophy', in *The Cambridge Companion to Logical Empiricism*, ed. by Alan Richardson and Thomas Uebel (Cambridge: Cambridge University Press, 2007), 43–4.

15 Cf. 'Mitgliederliste der Ortsgruppe Berlin der Internationale Gesellschaft für empirische Philosophie', in PetzoldtNachlass, PE32.16.

16 In particular, Vaihinger proposed a 'fictionalist' interpretation of Kantian philosophy, based on a reinterpretation of Kant's notion of *als-ob* (as if). This view suggested that the understanding as the law-giver applies its own schemata and categories to things. Even if these schemata and categories are 'fictions', they must be treated *as if* they were true. Cf. Hans Vaihinger, *Die Philosophie des Als Ob. System der theoretischen, praktischen und religiösen Fiktionen der Menschheit auf Grund eines idealistischen Positivismus* (Berlin: Reuter und Richard, 1911). See also Klaus Hentschel, *Interpretationen und Fehlinterpretationen der speziellen und der allgemeinen Relativitätstheorie durch Zeitgenossen Albert Einsteins*, Science Networks. Historical Studies (Basel: Birkhäuser, 1990), 168–77 and 276–92 as well as Klaus Hentschel, 'Zur Rezeption von Vaihingers Philosophie des Als Ob in der Physik', in *Fiktion und Fiktionalismus. Beiträge zu Hans Vaihingers 'Philosophie des Als-Ob'*, ed. by Matthias Neuber (Würzburg: Königshausen & Neumann, 2014), 161–86.

17 Cf. Reiner Hegselmann and Geo Siegwart, 'Zur Geschichte der "Erkenntnis"', *Erkenntnis*, 35 (1991), 461–2.

18 On the Berliner Gruppe see Lutz Danneberg, Andreas Kamlah and Lothar Schäfer (eds.), *Hans Reichenbach und die Berliner Gruppe* (Braunschweig: Vieweg, 1994); Nikolay Milkov (ed.), *Die Berliner Gruppe. Texte zum Logischen Empirismus* (Hamburg: Meiner, 2015); Nikolay Milkov and Volker Peckhaus (eds.), *The Berlin Group and the Philosophy of Logical Empiricism*, Boston Studies in the Philosophy and

History of Science (Cham: Springer, 2013); Harmut Hecht and Dieter Hoffmann, 'The Berlin "Society for Scientific Philosophy"'; Nicolas Rescher, 'The Berlin School of Logical Empiricism and Its Legacy', *Erkenntnis* 64 (2006), 281–304.

19 H. Reichenbach, 'Über die philosophischen Grundlagen der Mathematik' (1927), 'Kausalität oder Wahrscheinlichkeit' (1928), 'Raum und Zeit' (1929); W. Dubislav, 'Konventionelle und moderne Logik' (1927), 'Zur sogenannte Strukturtheorie der Wahrheit' (1929), 'Das Problem der Induktion' (1929). Cf. 'Vollständige Liste der Vorträge für die Gesellschaft für empirische Philosophie, Berlin, 1927–1933', in Hentschel, *Die Korrespondenz Petzoldt-Reichenbach*, 75–7.

20 Hans Reichenbach, 'Lichtgeschwindigkeit und Gleichzeitigkeit', *Annalen der Philosophie und philosophische Kritik*, 6 (1927), 128–44.

21 The wording 'Published by the *Gesellschaft für wissenschaftliche Philosophie* in Berlin and the *Ernst Mach Verein* in Vienna' appeared below the main title of the journal from 1930 to 1933, when the two societies were dissolved by the newly installed fascist regimes in Germany and Austria.

22 The initial popularity of this term, now in disuse, is certainly due to the fact that it was used by Lenin in his critical examination of the ideas of Avenarius and Mach in *Materialism and Empirio-Criticism. Critical Notes Concerning a Reactionary Philosophy*, Collected Works of V. I. Lenin (Moscow: Progress Publisher, 1970), XIII.

23 Cf. Herbert Feigl's 'The critical positivism of Mach and Avenarius' in 'Positivism and Logical Empiricism', in *The New Encyclopaedia Britannica* (Chicago et al.: Encyclopaedia Britannica, 1978), 878.

24 Leopold Stubenberg, 'Neutral Monism', in *The Stanford Encyclopedia of Philosophy*, ed. by Edward N. Zalta, Fall 2018 (Metaphysics Research Lab, Stanford University, 2018), https://plato.stanford.edu/archives/fall2018/entries/neutral-monism/ [accessed 28 October 2021].

25 Cf. Erik C. Banks, *The Realistic Empiricism of Mach, James, and Russell. Neutral Monism Reconceived* (Cambridge: Cambridge University Press, 2014), 79.

26 In his famous lecture of 1872, Du Bois Reymond claimed that the ideal of a mechanistic explanation of the entire universe comes up against the impossibility of understanding the relationship between material reality and consciousness, requiring researchers to conclude that – on this subject – we do not know and will not know, '*ignoramus et ignorabimus*' (Emil Du Bois-Reymond, *Über Die Grenzen Des Naturerkennens* (Leipzig, 1872); 'The Limits of Our Knowledge of Nature', *The Popular Science Monthly*, 5 (1874), 17–32).

27 Also relevant here is the emergence of quantum physics and Heisenberg's uncertainty principle, which had an extraordinary impact on the philosophical and scientific debate of the early twentieth century, a little later than Einsteinian relativity and therefore only marginally overlapping with the period in which Petzoldt was active.

28 Cf. Klaus Hentschel, 'Philosophical interpretations of Einstein's theories of relativity', *PSA (Philosophy of Science Association)*, 2 (1990), 169–79 on the 'belt of defenders' (*Verteidigergürtel*) around Einstein and his theory of relativity.

29 See especially Hentschel, *Interpretationen und Fehlinterpretationen*, and Don Howard, 'Einstein and Eindeutigkeit: A Neglected Theme in the Philosophical Background to General Relativity', in *Studies in the History of General Relativity*, ed. by Jean Eisenstaedt and A. J. Kox (Boston, Basel, Berlin: Birkhäuser, 1992), 154–243.

Chapter 2

1. There is a vast body of literature on this subject, but two very recent works provide an excellent historical and philosophical overview of the debate on the life sciences in Germany at the cusp of the nineteenth century: J. H. Zammito, *The Gestation of German Biology. Philosophy and Physiology from Stahl to Schelling* (Chicago-London: The University of Chicago Press, 2018); A. Gambarotto, *Vital Forces, Teleology and Organization. Philosophy of Nature and the Rise of Biology in Germany* (Cham: Springer, 2018). For a more general account of the life sciences between the late eighteenth and the early nineteenth centuries, see S. Poggi, and M. Bossi, *Romanticism in Science. Science in Europe, 1790–1840* (Springer, Boston et al., 1994), in particular the Introduction from XI ff.
2. For more on *Materialismusstreit*, see the recent work of the *Zentrum für interdisziplinäre Forschung* of the University of Bielefeld, published in K. Bayertz, W. Jaeschke, M. Gerhard (eds.), Kurt Bayertz, Myriam Gerhard and Walter Jaeschke, *Der Materialismus-Streit*, Weltanschauung, Philosophie und Naturwissenschaft im 19. Jahrhundert (Hamburg, 2007), I; and the anthology Kurt Bayertz, Myriam Gerhard and Walter Jaeschke (eds.), *Der Materialismus-Streit: Texte von L. Büchner, H. Czolbe, L. Feuerbach, I. H. Fichte, J. Frauenstädt, J. Froschammer, J. Henle, J. Moleschott, M. J. Schleiden, C. Vogt und R. Wagner* (Hamburg: Meiner, 2012).
3. See the other volume in the series edited by the *Zentrum für interdisziplinäre Forschung* of the University of Bielefeld: Kurt Bayertz, Myriam Gerhard and Walter Jaeschke, *Der Darwinismus-Streit: Texte von L. Büchner, B. von Carneri, F. Fabri. G. von Gyzicki, E. Haeckel, E. von Hartmann, F. A. Lange, R. Stoeckl und K. Zittel* (Hamburg, 2012), in particular the Introduction, where the links with *Materialismusstreit* are highlighted.
4. This is the fundamental thesis of Friedrich Albert Lange, *History of Materialism and Criticism of Its Present Importance* [1873–1875], trans. by Ernest Chester Thomas, 3 vols (London, 1879), which exerted an influence that cannot be underestimated in the second half of the nineteenth century.
5. On neo-Kantianism from the 1850s to the 1870s, see Scott Edgar, 'The Physiology of the Sense Organs and Early Neo-Kantian Conceptions of Objectivity: Helmholtz, Lange, Liebmann', in *Objectivity in Science: New Perspectives from Science and Technology Studies*, ed. by Flavia Padovani, Alan Richardson, and Jonathan Y. Tsou, Boston Studies in the Philosophy and History of Science (Cham, 2015), 101–22; Frederick C. Beiser, *The Genesis of Neo-Kantianism, 1796–1880* (Oxford, 2014), in particular the final chapter of the first part ('The Interim Years, 1840–1860') and the whole of the second part ('The Coming of Age'); and the book by Köhnke, *The Rise of Neo-Kantianism*, even though this work focuses more on the sociopolitical backdrop to the rise of Neo-Kantianism than on the issues raised by the scientific advances of those years.
6. Köhnke, *The Rise of Neo-Kantianism*, 95–6.
7. Avenarius1887, 177. My italics.
8. For a more in-depth exposition of these themes in Fechner's work, and of how his pioneering steps foreshadowed modern theories of 'self-organization' that take the same direction of overcoming the antinomy between causality and teleology, see the chapter 'Self-Organization and Irreversibility. Order Originating from Chaos'

in Michael Heidelberger, *Nature from Within: Gustav Theodor Fechner and His Psychophysical Worldview* (Pittsburgh: University of Pittsburgh Press, 2004), 248 ff.
9   Gustav Theodor Fechner, *Einige Ideen zur Schöpfungs- und Entwickelungsgeschichte der Organismen* (Leipzig: Breitkopf und Härtel, 1873), 1–2.
10 Fechner, *Einige Ideen zur Schöpfungs- und Entwickelungsgeschichte der Organismen*, 3.
11 Cf. Johann Karl Friedrich Zöllner, *Über die Natur der Cometen. Beiträge zur Geschichte und Theorie der Erkenntnis* (Leipzig: Engelmann, 1872). Combining his interest in astrophysics and spiritism, Zöllner puts forward two fundamental ideas in this work: the existence of the fourth dimension, which – being also the locus of the spiritual world – meant that the entire universe was animate; and the fact that variations in the universe occur so as to reduce the sum total of displeasure to a minimum. According to Zöllner, the fundamental laws of mechanics – such as the Gaussian principle of least constraint – were contained in his principle. Therefore, the fact that the number of collisions in any system of atoms in motion tends to decrease to a minimum was due to the fact that this would also diminish associated feelings of displeasure.
12 Fechner, *Einige Ideen zur Schöpfungs- und Entwickelungsgeschichte der Organismen*, 26.
13 Although Fechner did not refer to them explicitly, in citing the solar system as an example of approximate stability he was harking back – either directly or via Zöllner – to the eighteenth-century research of mathematicians such as Euler and Lagrange on celestial mechanics and in particular on the problem of gravitational perturbation.
14 Fechner, *Einige Ideen zur Schöpfungs- und Entwickelungsgeschichte der Organismen*, 30. My italics.
15 Ibid. Fechner excludes the possibility of achieving states that are 'absolutely stable' as the law of the conservation of energy prevents energy being lost and therefore reaching a state of total stasis. At the final stage of development, what remains is therefore the thermal oscillation of the particles, which is why Fechner is careful to state 'if we exclude infinitesimal motion'.
16 Fechner, *Einige Ideen zur Schöpfungs- und Entwickelungsgeschichte der Organismen*, 43.
17 Fechner, *Einige Ideen zur Schöpfungs- und Entwickelungsgeschichte der Organismen*, 34–5.
18 Fechner, *Einige Ideen zur Schöpfungs- und Entwickelungsgeschichte der Organismen*, iv.
19 On why the tendency towards stability is a necessary consequence of the relationships between the internal forces of a system (or on external forces that are constant), see Fechner, *Einige Ideen zur Schöpfungs- und Entwickelungsgeschichte der Organismen*, 27–8, where Fechner insists that the principle of the tendency towards stability must be considered a priori valid.
20 Fechner, *Einige Ideen zur Schöpfungs- und Entwickelungsgeschichte der Organismen*, 91.
21 For further details on the elaboration of the second law of thermodynamics, see Ingo Müller, *A History of Thermodynamics: The Doctrine of Energy and Entropy* (Berlin: Springer Nature, 2010), 47 ff.
22 See S. J. Holmes, 'The Principle of Stability as a Cause of Evolution. A Review of Some Theories', *The Quarterly Review of Biology*, 23 (1948), 324–32. The most notable examples of this debate include the chapter on 'Darwinism and teleology' in Lange, *History of Materialism*. On attempts to unify mechanism and teleology, in particular in biology, see Timothy Lenoir, *The Strategy of Life: Teleology and Mechanics in Nineteenth Century German Biology* (Dordrecht: Reidel, 2011). It is worth emphasizing that while it is clear that the relationship between teleology and

Darwinism, or more generally between teleology and biology, has been studied extensively, the same cannot be said of the relationship between teleology and the second law of thermodynamics, and – even more so – the complex interconnections between entropy and Darwinian evolution in the teleological debate of the late nineteenth century.

23   On the adoption of Darwinian theory by German materialists, see Alfred Kelly, *The Descent of Darwin: The Popularization of Darwinism in Germany, 1860–1914* (Chapel Hill: The University of North Carolina Press, 2012), 18 ff.

24   Ludwig Büchner, *Sechs Vorlesungen über die Darwin'sche Theorie von der Verwandlung der Arten und die erste Entstehung der Organismenwelt*, 3rd edn (Leipzig: Thomas, 1872), 166.

25   See Peter J. Bowler, *The Non-Darwinian Revolution: Reinterpreting a Historical Myth* (Baltimore: Johns Hopkins University Press, 1988), 198 ff. Bowler demonstrates how the evolutionism that established itself in Germany after the publication of *On the Origin of Species* was in fact pseudo-Darwinism, based not on the theory of natural selection but on laws of development internal to the organism itself. This found favour among many scientists – who preferred the existence of *laws* to the role of chance in Darwin's theory – and of many who had an interest in maintaining a teleological element in the development of the world.

26   Fechner, *Einige Ideen zur Schöpfungs- und Entwickelungsgeschichte der Organismen*, iv.

27   Fechner, *Einige Ideen zur Schöpfungs- und Entwickelungsgeschichte der Organismen*, v.

28   Fechner, *Einige Ideen zur Schöpfungs- und Entwickelungsgeschichte der Organismen*, 63. Herbert Spencer formulated a view of evolution that was similar to Fechner's, defining three universal laws in his *First Principles*: the persistence of force, the multiplicity of effects and the instability of the homogeneous. The third law specified that the universe tended to move towards states of increased stability and heterogeneity (Herbert Spencer, *First Principles* (London: Williams & Norgate, 1862), 337 ff.).

29   Johannes E. Kuntze, *Gustav Theodor Fechner (Dr. Mises). Ein deutsches Gelehrtenleben* (Leipzig: Breitkopf und Härtel, 1892), 39 f., cited in Heidelberger, *Nature from Within*, 22. On the influence of Oken and Schelling's *Naturphilosophie* on Fechner, see section 1.2 of Heidelberger's book, ibid. 21 ff.

30   Fechner, *Einige Ideen zur Schöpfungs- und Entwickelungsgeschichte der Organismen*, 32.

31   Avenarius1887, 183. My italics.

32   Ibid.

33   On Herbart in general, and his metaphysics and psychology in particular, see 'Johann Friedrich Herbart, Neo-Kantian Metaphysician' in Beiser, *The Genesis of Neo-Kantianism*, 89–141.

34   On Herbart's critique of idealist teleology, see Beiser, *The Genesis of Neo-Kantianism*, 91–2, 125. See also Frederick Beiser, 'Herbart's Monadology', *British Journal for the History of Philosophy*, 23/6 (2015), 1060; and Frederick Beiser, 'Two Traditions of Idealism', in *From Hegel to Windelband: Historiography of Philosophy in the 19th Century*, ed. by Gerald Hartung and Valentin Pluder (Berlin, 2015), 81–98, in particular the section 'The clash over teleology', 87 ff.

35   This is the title of the first section of Heymann Steinthal, *Abriss der Sprachwissenschaften. Einleitung in die Psychologie und Sprachwissenschaft*, 2 vols (Berlin: Dümmler, 1871), I, 91 ff.

36   Steinthal, *Abriss der Sprachwissenschaften*, I, 93.

37   Steinthal, *Abriss der Sprachwissenschaften*, I, 92.

38 In fact, Steinthal only cites Fechner once in his works, in the fourth edition of *Der Ursprung der Sprache im Zusammenhange mit den letzten Fragen alles Wissens*, 4th edn (Berlin: Dümmler, 1888). However, the important point is that there is no doubt that Steinthal's work addresses the psychophysics of the time, irrespective of whether his sources were Fechner, Helmholtz or others.
39 Steinthal, *Abriss der Sprachwissenschaften*, I, 183–4. See also 277.
40 Steinthal, *Abriss der Sprachwissenschaften*, I, 133.
41 Steinthal, *Abriss der Sprachwissenschaften*, I, 132–3.
42 Steinthal, *Abriss der Sprachwissenschaften*, I, 136–7.
43 Steinthal, *Abriss der Sprachwissenschaften*, I, 163.
44 Steinthal, *Abriss der Sprachwissenschaften*, I, 134 ff.
45 Steinthal, *Abriss der Sprachwissenschaften*, I, 133.
46 Steinthal, *Abriss der Sprachwissenschaften*, I, 293, 292.
47 See Avenarius's biography in Ludwig Avenarius, *Avenarianische Chronik : Blätter aus drei Jahrhunderten einer deutschen Bürgerfamilie* (Leipzig: Reisland, 1912), 126, 131–2.
48 Richard Avenarius, *Philosophie als Denken der Welt gemäss dem Princip des kleinsten Kraftmasses: Prolegomena zu einer Kritik der reinen Erfahrung* (Leipzig: Fues, 1876), 3.
49 Avenarius, *Prolegomena zu einer Kritik der reinen Erfahrung*, 1.
50 Avenarius, *Prolegomena zu einer Kritik der reinen Erfahrung*, iii-iv. My italics.
51 Avenarius, *Prolegomena zu einer Kritik der reinen Erfahrung*, iv. My italics.
52 It should be added that Lamarck's theories had nevertheless inflicted an initial blow on the finalism that characterized 'natural theology'. In fact, Lamarck overturned the relationship between organs and the environment by asserting that it was not divine knowledge that provided animals with the organs they needed to survive in a certain environment but that living in a certain environment determined the development of adequate organs in animals.
53 Cf. Gustav Theodor Fechner, 'Über das Lustprincip des Handeln', *Zeitschrift für Philosophie und philosophische Kritik*, 19 (1848), 1–30, 163–94.
54 Avenarius, *Prolegomena zu einer Kritik der reinen Erfahrung*, 66.
55 J. Petzoldt, 'Meine Begegnung mit Avenarius', in PetzoldtNachlass, PE03, 4.
56 Avenarius1887, 177–8.
57 Cf. Avenarius1887, 184.
58 We do not know whether Avenarius had read Fechner's *Einige Ideen zur Schöpfungs- und Entwickelungsgeschichte der Organismen*. His works contain only one reference to *Elemente der Psychophysik*, in the second volume of the *Kritik der reinen Erfahrung* (1888–90). Avenarius' *Nachlass*, however, contains notes he made during Fechner's lectures, some sixty pages for *Über die Grundbeziehung des materiellen und geistigen Princips* (summer semester, 1864) and a similar amount for *Allgemeine Aesthetik* (winter semester, 1864).
59 Cf. Avenarius1887, 189 ff.
60 Avenarius1887, 186–7.
61 Petzoldt's position could also be termed 'teleomatic', to use one of the terms proposed by Ernst Mayr to distinguish different forms of teleology. Mayr saw as 'teleomatic' 'any process which persist toward an endpoint under varying conditions or in which the endstate of the process is determined by its properties at the beginning', thus encompassing 'all processes in inorganic nature that have an endpoint'. Processes of this kind 'are end-directed only in a passive, automatic way, regulated by external

forces or conditions, that is by natural laws'. Mayr exemplifies this by citing 'the law of gravity and the second law of thermodynamics' (Cf. Ernst Mayr, 'The Idea of Teleology', *Journal of the History of Ideas*, 53/1 (1992), 125).

62 The Austrian physicist Anton Lampa, also influenced by Mach, wrote an article in which he argued against Petzoldt on this point, pointing out that the concept of states with a lower degree of stability moving to ones with a higher degree is improbable from a statistical point of view, and comes into conflict with a statistical interpretation of the law of entropy as the tendency for the emergence of states with greater probability. Cf. Anton Lampa, 'Über die Tendenz zur Stabilität', in *Festschrift für Wilhelm Jerusalem zu seinem 60. Geburtstag* (Vienna: Braumüller, 1915), 147–53. Petzoldt responded to Lampa, underlining that the principle of stability was more general than that of entropy, which only represents a special case of it (cf. Joseph Petzoldt, *Das allgemeinste Entwicklungsgesetz* (München: Rösl, 1923), 215 ff.).

63 Addy Pross, 'Paradoxes of Stability: How Life Began and Why It Can't Rest', *Aeon*, https://aeon.co/essays/paradoxes-of-stability-how-life-began-and-why-it-can-t-rest [accessed 28 October 2021]. Probably unaware of his nineteenth-century predecessors, Addy Pross has recently attempted to develop ideas similar to those of Fechner and Petzoldt, using the concept of stability to arrive at the unification of the laws of physics and those of biology, and at a possible explanation of the emergence of living organisms. Cf. Addy Pross, *What Is Life?: How Chemistry Becomes Biology* (Oxford: Oxford University Press, 2012), 58 ff.

64 J. Petzoldt, *Autobiographie*, 3. My italics.

65 This move to produce a hybrid concept encompassing both evolutionary and biological thermodynamics may seem strange, especially as the principle of entropy, according to which the universe moves to states of increasing disorder, is generally considered to be antithetical to the biological view according to which living beings exhibit increasingly sophisticated degrees of organization (cf. Joseph Needham, 'Evolution and Thermodynamics: A Paradox with Social Significance', *Science & Society*, 6/4 (1942), 352–75). However, it is important to remember that the concept of organization was only introduced relatively late into the elaboration of the second law of thermodynamics. The early definitions of the law could therefore in fact favour a teleological interpretation. For example, before elaborating the concept of entropy, Clausius had successfully launched the concept of 'disgregation': the degree of distribution of the molecules of a body (cf. Emilio Marco Pellegrino, Elena Ghibaudi and Luigi Cerruti, 'Clausius' Disgregation: A Conceptual Relic That Sheds Light on the Second Law', *Entropy*, 17/7 (2015), 4500–18). Clausius stated that he had introduced the concept of disgregation because he needed 'a quantity that was valid for every change in a system and whose value could change in only one direction' (Rudolf Clausius, 'Zur Geschichte Der Mechanischen Wärmetheorie', *Annalen Der Physik Und Chemie*, 145 (1872), 145–6). In consequence, when the second law of thermodynamics was still being expressed using similar concepts, it would have been much easier to reconcile it, for example, with the widespread idea that the cosmos developed from the simple to the complex, from the homogeneous to the heterogeneous, as promoted by Herbert Spencer in England and Karl Ernst von Baer in Germany. In any case, as we have noted, there is a need for deeper research into the links between teleology and the second law of thermodynamics in the nineteenth century.

66 Avenarius1887, 188.

67 Avenarius1887, 189–90. My italics.

68 Einführung1904, 129.
69 This states that the motion of a system of material points, constrained in an arbitrary manner, and subjected to arbitrary forces at any moment of time, takes place in a manner which is as similar as possible to the motion that would be performed by these points if they were free, i.e. with least-possible forcing – the measure of forcing during the time dt being defined as the sum of the products of the mass of each point and the square of the distance of the point from the position which it would occupy if it were free. See Carl Friedrich Gauss, 'Ueber ein allgemeines Grundgesetz der Mechanik', *Journal für die reine und angewandte Mathematik*, 4 (1829), 232–5.
70 Avenarius1887, 191.
71 Avenarius1887, 192–3.
72 Avenarius1887, 193–4.
73 Avenarius1887, 194. My italics.
74 Ibid. My italics.
75 Ibid. My italics.
76 Mach, *Die Mechanik* (1883), 338; *The Science of Mechanics*, 362–3. Cited in Avenarius1887, 192.
77 Cf. Richard Avenarius, *Kritik der reinen Erfahrung*, 2 vols (Leipzig: Fues, 1888–1890), I, 59 ff.
78 Maxima1890, 354–66, 417–42, 418.
79 Cf. Richard Avenarius, *Nachlass*, Staatsbibliothek zu Berlin, Kasten 14.
80 Maxima1890, 206.
81 Ibid.
82 This information is taken from Morris Kline, *Mathematical Thought from Ancient to Modern Times* (Oxford: Oxford University Press, 1990), II. See also Michael Stöltzner, 'Action Principles and Teleology', in *Inside Versus Outside. Endo- and Exo-Concepts of Observation and Knowledge in Physics, Philosophy and Cognitive Science*, ed. by Harald Atmanspacher and Gerhard J. Dalenoort (Berlin, Heidelberg: Springer, 1994), 36 ff.
83 Quoted in Kline, *Mathematical Thought from Ancient to Modern Times*, II, 740.
84 Maxima1890, 207.
85 Ibid.
86 Maxima1890, 209.
87 Maxima1890, 210.
88 Maxima1890, 215–16.
89 Don Howard, 'Relativity, Eindeutigkeit, and Monomorphism: Rudolf Carnap and the Development of the Categoricity Concept in Formal Semantics', in *Origins of Logical Empiricism*, ed. by Ronald N. Giere and Alan W. Richardson (Minneapolis and London: University of Minnesota Press, 1996), 158–9.
90 Maxima1890, 211.
91 Maxima1890, 215–16.
92 Maxima1890, 216.
93 Maxima1890, 216–17.
94 Maxima1890, 234.
95 Maxima1890, 222.
96 Cf. Maxima1890, 229.
97 Maxima1890, 231.
98 Maxima1890, 231.

99 Cf. Maxima1890, 355–6: 'When we speak of "force" or "internal" "causes", of "predispositions", "tendencies", etc., any metaphysical representations must be avoided […]. We must only describe *facts*; we use these exclusively to facilitate understanding and reject any anthropomorphic interpretation. Nothing is assumed apart from the dependence of phenomena on preceding or simultaneous phenomena.'
100 Maxima1890, 226.
101 Maxima1890, 355.
102 Maxima1890, 354–5. Cf. also 361.
103 Maxima1890, 363–4.
104 Maxima1890, 365.
105 Maxima1890, 363. My italics. Petzoldt also discusses his criticism of the concept of economy in his correspondence with Mach. In reply to a letter of Mach's that is not available, Petzoldt specifies that it is not his intention the get rid of the concept of economy in general but only to show that it is insufficient 'for an *objective* description of processes, whether these are physical or mental'. As the aim is to formulate an 'account of what happens without subjective additions', he considers it necessary to dispense with the 'subjective concept of finality, replacing it with that of the outcome: the outcome of a development is a stationary state'. Stability is thus the objective outcome of the process of evolution, whereas the concept of economy is a subjective interpretation of that outcome, and therefore an 'indirect description', 'an image we can use to visualize the outcome of any evolutionary process'. Therefore, 'precisely because it is an image, it has all the advantages and all the disadvantages of images' (letter from Petzoldt to Mach dated 12 April 1897, MachNachlass, NL 174/2429). We do not know how Mach replied to Petzoldt, but the subject reappears several years later in a letter Mach wrote to the physicist Friedrich Adler: 'Petzoldt argues against the economy of thought and only wants to talk of stability [*Beständigkeit*]. I doubt that we can gain much from that. But I don't want to argue about terminology' (letter from Mach to Friedrich Adler dated 20 August 1909, in John T. Blackmore and Klaus Hentschel, *Ernst Mach als Aussenseiter: Machs Briefwechsel über Philosophie und Relativitätstheorie mit Persönlichkeiten seiner Zeit* (Wien, 1985), 61).
106 Reference to the incipit of Ernst Mach, *Die Ökonomische Natur Der Physikalischen Forschung. Vortrag Gehalten in Der Feierlichen Sitzung Der Kaiserlichen Akademie Der Wissenschaften Am 25 Mai 1882* (Wien, 1882); 'The Economical Nature of Physical Inquiry', in *Popular Scientific Lectures*, by Ernst Mach, trans. by Thomas J. McCormack (Cambridge, 2014), 186–213: 'When the human mind, with its limited powers, attempts to mirror in itself the rich life of the world, of which it is itself only a small part, and which it can never hope to exhaust, it has every reason for proceeding economically'. In the letter cited in the previous note, Petzoldt specifies that also in the case of the 'mirroring of facts' his criticism is not to be understood in an absolute sense, given that the concept has value as an 'image' that captures at least one aspect of the situation. However, compared to the image of thought-economy, the metaphor of thought as a 'mirror' of reality is a more dangerous image, because it leads directly to the subject-object dichotomy, and to the redoubling of reality into things and their representations (letter from Petzoldt to Mach dated 12 April 1897, MachNachlass, NL 174/2429).
107 Maxima1890, 428.
108 Maxima1890, 429.

109　Maxima1890, 430.
110　Maxima1890, 442.
111　Petzoldt's criticism of Wundt was also a reply to his attacks on Avenarius in a series of articles entitled 'Über naiven und kritischen Realismus. II. Der Empiriokritizismus', *Philosophische Studien*, 13 (1898), 1–105, 323–433. After Avenarius's death in 1896, Petzoldt took on the role of defending his ideas, elevating himself to a spokesman for empiriocriticism. On the exchanges between Avenarius and Wundt, see Chiara Russo Krauss, *Wundt, Avenarius, and Scientific Psychology: A Debate at the Turn of the Twentieth Century* (New York: Palgrave Macmillan, 2019).
112　Sittenlehre1894, 197.
113　On Comte's position on causality, see the chapter entitled 'The Question of Causal Laws' in Robert C. Scharff, *Comte after Positivism*, Modern European Philosophy (Cambridge: Cambridge University Press, 1995), 63 ff.
114　Julius Robert Mayer, 'Bemerkungen über die Kräfte der unbelebten Natur', *Annalen der Chemie und Pharmazie*, 47 (1842), 1.
115　Hermann von Helmholtz, 'Ueber die Erhaltung der Kraft. Zusätze', in *Wissenschaftliche Abhandlungen*, by Hermann von Helmholtz (Leipzig: Barth), I, 68.
116　Ibid.
117　Emil Du Bois-Reymond, *Untersuchungen über thierische Elektricität* (Berlin: Reimer, 1848), xl.
118　Du Bois-Reymond, *Untersuchungen über thierische Elektricität*, xlii. My italics.
119　See Michael Heidelberger, 'Functional Relations and Causality in Fechner and Mach', *Philosophical Psychology*, 23/2 (2010), 163–72.
120　Cf. Eindeutigkeit1895, 146. The physicist Kirchhoff had criticized the concept of cause in the Preface to his *Vorlesungen über mathematische Physik* (Leipzig: Teubner, 1874). For more on Kirchhoff, see the chapter entitled 'Helmholtz, Kirchhoff, and Physics at Berlin University', in Christa Jungnickel and Russell McCormmach, *Intellectual Mastery of Nature. Theoretical Physics from Ohm to Einstein, Volume 2: The Now Mighty Theoretical Physics, 1870 to 1925* (Chicago: University Of Chicago Press, 1990), 17 ff.
121　Cf. Eindeutigkeit1895, 147.
122　Ernst Mach, *Die Geschichte und die Wurzel des Satzes von der Erhaltung der Arbeit* (Prag, 1872), 31; Ernst Mach, *History and Root of the Principle of the Conservation of Energy*, trans. by Philip E. B. Jourdain (Chicago: Open Court, 1911), 56. Cited in Eindeutigkeit1895, 161. My italics.
123　Mach, *Die Geschichte und die Wurzel des Satzes von der Erhaltung der Arbeit*, 35–6; *History and Root of the Principle of the Conservation of Energy*, 61.
124　Eindeutigkeit1895, 160.
125　Eindeutigkeit1895, 168.
126　Eindeutigkeit1895, 168.
127　Cf. Eindeutigkeit1895, 162–3.
128　Eindeutigkeit1895, 184–5. Petzoldt is referring to Wilhelm Ostwald, 'Ueber das Prinzip des ausgezeichneten Falles', *Königlich sächsische Gesellschaft der Wissenschaften. Mathematisch-physikalische Classe. Berichte über die Verhandlungen*, 1893, 599–603.
129　Gustav Theodor Fechner, *Elemente der Psychophysik*, 2 vols (Leipzig: Breitkopf und Härtel, 1860), I, 35. My italics.

130 On Fechner's position on indeterminism, see Heidelberger, *Nature from Within*, 273 ff.
131 In the debates of the day, the principle of the conservation of energy consisted of two principles, a 'principle of equivalence', which states that 'every charge of energy causes the discharge of a specific quantum of energy, and vice versa', and a 'principle of constancy', which states that 'the amount of actual and potential energy in the process of the transformation of energy remains unchanged' (Cf. Rudolf Eisler, 'Energie', *Handwörterbuch Der Philosophie* (Berlin, 1922), 169).
132 Wilhelm Wundt, *Grundzüge der physiologischen Psychologie*, 3rd edn, 2 vols (Leipzig: Engelmann, 1887), II, 483–4. My italics.
133 Sittenlehre1894, 51.
134 Sittenlehre1894, 52. On Wundt's notion of mental causality in the light of the principle of the conservation of energy, cf. Saulo De Freitas Araujo, *Wundt and the Philosophical Foundations of Psychology: A Reappraisal* (Cham: Springer, 2016), 198 ff.
135 Sittenlehre1894, 47.
136 Eindeutigkeit1895, 164–5.
137 Mach, *Die Geschichte und die Wurzel des Satzes von der Erhaltung der Arbeit*, 36; *History and Root of the Principle of the Conservation of Energy*, 61–2. My italics.
138 Mach, *Die Geschichte und die Wurzel des Satzes von der Erhaltung der Arbeit*, 37; *History and Root of the Principle of the Conservation of Energy*, 62. Cited in Eindeutigkeit1895, 165.
139 Mach, *Die Geschichte und die Wurzel des Satzes von der Erhaltung der Arbeit*, 35; *History and Root of the Principle of the Conservation of Energy*, 61. Cited in Eindeutigkeit1895, 162.
140 Eindeutigkeit1895, 160–1.
141 Mach, *Die Geschichte und die Wurzel des Satzes von der Erhaltung der Arbeit*, 56; *History and Root of the Principle of the Conservation of Energy*, 89. Cited in Eindeutigkeit1895, 161.
142 Mach, *Die Geschichte und die Wurzel des Satzes von der Erhaltung der Arbeit*, 37; *History and Root of the Principle of the Conservation of Energy*, 63. Cited in Eindeutigkeit1895, 165.
143 Mach, *Die Geschichte und die Wurzel des Satzes von der Erhaltung der Arbeit*, 57; *History and Root of the Principle of the Conservation of Energy*, 91.
144 Eindeutigkeit1895, 164.
145 Eindeutigkeit1895, 177.
146 Eindeutigkeit1895, 156, see also 180 ff.
147 Parallelismus1902, 297. The first use of italics is mine.
148 Cf. Eindeutigkeit1895, 172.
149 Eindeutigkeit1895, 173.
150 Eindeutigkeit1895, 177.
151 Eindeutigkeit1895, 178.
152 Eindeutigkeit1895, 179.
153 Eindeutigkeit1895, 181.
154 Eindeutigkeit1895, 167.
155 Eindeutigkeit1895, 168.
156 Eindeutigkeit1895, 169.
157 Eindeutigkeit1895, 202–3.
158 Eindeutigkeit1895, 203.

159 Einführung1900, 40.
160 Ibid.
161 Einführung1904, 293.
162 Pierre Simon Laplace, *A Philosophical Essay on Probabilities* [1814], trans. by Frederick Wilson Truscott and Frederick Lincoln Emory (New York, 1951), 4.
163 Eindeutigkeit1895, 192.
164 Eindeutigkeit1895, 192 footnote.
165 Einführung1900, 32.
166 Ibid.
167 Einführung1900, 32, 34.
168 Einführung1900, 44.
169 Einführung1900, 39.
170 Einführung1900, 44–5.
171 Einführung1900, 35.
172 Einführung1900, 36. My italics.
173 Cf. Einführung1900, 39.
174 Einführung1900, 37.
175 Einführung1900, 38.
176 Einführung1900, 36.
177 For an analysis of the inconsistencies in Petzoldt's concept of *Eindeutigkeit*, see also Heinrich Grünbaum, *Zur Kritik der modernen Causalanschauungen* (Würzburg: Stürtz, 1899), 382 ff.
178 Sittenlehre1894, 197.
179 Parallelismus1902, 287–8.
180 Einführung1904, 311.
181 Ibid.
182 Ibid.
183 Cf. Fechner, *Elemente der Psychophysik*, I, 3–4.
184 Fechner, *Elemente der Psychophysik*, I, 4.
185 Ernst Mach, *Beiträge zur Analyse der Empfindungen* (Jena: Gustav Fischer, 1886), 12; *Contributions to the Analysis of the Sensations*, trans. by C. M. Williams (Chicago: Open Court, 1897), 14.
186 Mach, *Beiträge zur Analyse der Empfindungen*, 10; *Contributions to the Analysis of the Sensations*, 11.
187 Mach, *Beiträge zur Analyse der Empfindungen*, 13; *Contributions to the Analysis of the Sensations*, 14–15.
188 Cf. Richard Avenarius, 'Bemerkungen zum Begriff des Gegenstandes der Psychologie', *Vierteljahrsschrift für wissenschaftliche Philosophie*, 18–19 (1894–1895), I, 137 ff.
189 Richard Avenarius, *Der menschliche Weltbegriff* [1891], 2nd edn (Leipzig, 1905), 82.
190 Cf. Avenarius, 'Bemerkungen zum Begriff des Gegenstandes der Psychologie', I, 414.
191 Avenarius, *Der menschliche Weltbegriff*, 15.
192 Ibid.
193 Einführung1904, 311–12.
194 This is why Petzoldt specifies that his 'distinction between psychological perspective and physical perspective must not be confused with Avenarius's 'absolute' and 'relative' perspectives' (Einführung1904, 312).
195 Ibid.

196 Cf. Einführung1900, 51, 84.
197 Parallelismus1902, 289. See also Einführung1900, 84.
198 Cf. Einführung1900, 84. On the debate on the measurability of sensations, and more generally of mental events, see David J. Murray, *The Creation of Scientific Psychology* (Routledge, 2020). One of the first to reject the applicability of the quantitative concept of intensity to sensations was the physiologist Ewald Hering, whose works Petzoldt studied (cf. Einführung1904, 130, 136, 138).
199 Einführung1900, 59, 60.
200 Einführung1900, 61.
201 Eindeutigkeit1895, 198.
202 Parallelismus1902, 317.
203 Ibid.
204 Parallelismus1902, 317.
205 Parallelismus1902, 318.
206 Parallelismus1902, 319. For a critique of Wundt's mental causality, cf. Sittenlehre1894, 41 ff., in particular 44.
207 This was, for example, the conclusion reached by Kant in the Preface to the *Metaphysical Foundations of Natural Science*. In this work, a cornerstone of the nineteenth-century debate on psychology, Kant claimed that psychology ('the empirical doctrine of the soul') could never be a proper science, that is, based on necessary natural laws, but at best an 'historical doctrine of nature', driven by classificatory goals and elaborated using empirical generalizations (see Immanuel Kant, *Metaphysical Foundations of Natural Science* [1786], trans. by Michael Friedman (Cambridge, 2004), 7).
208 Eindeutigkeit1895, 199.
209 Ibid.
210 Eindeutigkeit1895, 201.
211 Parallelismus1902, 283.
212 Parallelismus1902, 284. My italics.
213 Sittenlehre1894, 52, footnote.
214 Einführung1900, 92.
215 Einführung1900, 83.
216 Einführung1900, 83, see also 351 ff.
217 Sittenlehre1894, 38. My italics.
218 Parallelismus1902, 332.
219 Ibid.
220 Einführung1900, 90.
221 Einführung1900, 88.
222 Ibid.
223 Ibid.
224 Eindeutigkeit1895, 201.
225 Einführung1900, 10.
226 Ibid.
227 Ibid.
228 Einführung1900, 11.
229 Ibid. However, it should be emphasized that appealing to the principle of the conservation of energy to demonstrate that mental events depend on the brain is not an entirely convincing argument. The fact that the entirety of physical events must be seen as a closed process, as this principle requires, could itself also be read

as a counterargument to psychophysical parallelism, that is, as an argument for the *independence* of mental events. An interpretation along these lines can be found, for example, in Wundt, who holds that psychophysical parallelism states that physical causality (based on the principle of the conservation of energy) and mental causality coexist without interfering with each other (cf. Wilhelm Wundt, 'Über psychische Causalität und das Princip des psychophysischen Parallelismus', *Philosophische Studien*, 10 (1894), 1–124). This may also be why Petzoldt returns to the issue of the relationship between the conservation of energy and psychophysical parallelism in *Die Notwendigkeit und Allgemeinheit des psychophysischen Parallelismus*, in reply to an article by the American philosopher Edward Gleason Spaulding. Disputing some of Petzoldt's terminology, such as indeed the term 'psychophysical parallelism', Spaulding had written that '[Petzoldt], *quite correctly* and in line with current scientific thinking, considers the principle [of parallelism] to be a direct consequence of the law of energy, as mental phenomena cannot be measured physically, and are therefore not a form of energy' (Edward Gleason Spaulding, *Beiträge Zur Kritik Des Psychophysischen Parallelismus Vom Standpunkte Der Energetik* [1900] (Hildesheim, 1985), 31. My italics). Instead of making the most of Spaulding's support, Petzoldt replies – somewhat surprisingly – that he had been 'misunderstood', as his intention had never been to 'base the demonstration of psychophysical parallelism on the law of energy'. Indeed, he states that he 'considers the fact that the law of energy is in fact undisturbed to be one of the main advantages of [his own] line of argumentation' (Parallelismus1902, 315). Even assuming that Petzoldt is speaking in good faith, and that this is not a feigned change of view but simply an unfortunate formulation, the fact remains that at the start of *Einführung in die Philosophie der reinen Erfahrung* he explicitly presents the principle of the conservation of energy as an argument that 'supports' (*unterstüzt*) the two facts on which the assumption of the correlation between cerebral processes and mental activity is based, that of the disappearance of mental events if the link between the core and the periphery of the nervous system is severed, and that of the joint development of the brain and mental capacities (Einführung1900, 11).

230 Friedrich Kirchner, 'Energy', *Wörterbuch Der Philosophischen Grundbegriffe* (Leipzig: Dürr, 1907), 177–8. Cf. Einführung1900, 14.

231 The first psychophysiological studies into the mechanisms controlling reactions to stimuli were unsuccessful in including the cerebral cortex. This led to the view that, as it was not exposed to *external* stimuli, it was exposed to purely *internal* causes; that is, it was the locus of consciousness in the brain. However, this position came undone with the discovery by Gustav Fritsch and Eduard Hitzig that the cerebral cortex could be stimulated (1870), which laid the foundations for a further extension of the physiological approach. See Stanley Finger, *Minds behind the Brain: A History of the Pioneers and Their Discoveries* (Oxford: Oxford University Press, 1999), 159 ff.

232 Einführung1900, 16. My italics.

233 Petzoldt later extends this criticism also to Johannes Rehmke, who, in *Die Seele des Menschen* (Leipzig: Teubner, 1905) had adopted a position similar to that of Wundt (cf. Parallelismus1902, 315–16, footnote).

234 Wilhelm Wundt, *Grundzüge der physiologischen Psychologie*, 1st edn (Leipzig: Engelmann, 1874), 727.

235 Einführung1900, 17.

236 Ibid.

237   Ibid.
238   Ibid.
239   Einführung1900, 20. See also Parallelismus1902, 315–16, footnote.
240   Einführung1900, 22.
241   Parallelismus1902, 336.
242   Einführung1900, 15.
243   Avenarius, *Kritik der reinen Erfahrung*, I, 202, endnote 7.
244   Joseph Petzoldt, 'Vorwort des Herausgebers', in *Kritik der reinen Erfahrung*, by Richard Avenarius, 2nd edn, 2 vols (Leipzig: Reisland, 1907), I, vi.
245   Petzoldt, 'Vorwort des Herausgebers', I, vii.
246   Einführung1900, 350.
247   Wundt, 'Über naiven und kritischen Realismus. II. Der Empiriokritizismus', 47–8.
248   Einführung1900, 349.
249   Einführung1900, 93. Cf. Avenarius, *Kritik der reinen Erfahrung*, I, 16.
250   Ibid.
251   Einführung1900, 352.
252   Lange, *History of Materialism*, III, 112–13. Cited in Einführung1900, 354.
253   Cf. Sittenlehre1894, 218, footnote.
254   Einführung1900, 119 ff.
255   Sittenlehre1894, 240, 242.
256   Einführung1900, 111.
257   Einführung1904, 180. In a letter to Petzoldt, Mach reveals that when he first read that 'the evolution of the brain is the goal of all evolution' the notion seemed 'repugnant'. However, he later acknowledged that 'it emerges at least as an apparent *goal*, to the extent that [the evolution of the brain] encompasses every other kind of evolution' (letter from Mach to Petzoldt dated 10 December 1914, in Blackmore and Hentschel, *Ernst Mach als Aussenseiter*, 147).
258   Sittenlehre1894, 208.
259   Wilhelm Roux held that the struggle for survival and Darwinian selection occurs not only between the individuals of a species but also between the parts of the organism itself, which are competing with one another for nutrients. These intraorganic selective processes thus contribute to the morphology of living beings (cf. Wilhelm Roux, *Der Kampf der Teile im Organismus. Ein Beitrag zur Vervollständigung der mechanischen Zweckmäßigkeitslehre* (Leipzig: Engelmann, 1881)).
260   Einführung1904, 188–9.
261   Petzoldt replied by letter to Mach on the similarities between this position and the theories of the biologist August Weismann, who had developed Darwin's theories and had promoted them in Germany, highlighting the unusual aspects of Roux's theories: 'I agree that here, too, there is a process of selection. However, there is a significant difference with respect to Weismann, who holds that every variation occurs in a *specific* individual and, unless it receives reinforcement from the environment, destroys not only itself but also that individual; for Roux, it is the individual that exhibits a range of variations or of possible variations, with the actual stimuli of the environment deciding between them. […] I think it would be better if Roux, rather than talking of the "functional self-organization of purposeful structures" [*funktionelle Selbstgestaltung der Zweckmässigen*] talked of "the organization of tissue through the environment (through the relatively constant environment)"' (letter from Petzoldt to Mach dated 15 September 1904, in MachNachlass, NL 174/2455).

262 Einführung1904, 19.
263 Ibid. Cf. Weltproblem1906, 50.
264 Ibid.
265 Einführung1904, 287.
266 Einführung1904, 100.
267 Cf. Einführung1900, 330, 345.
268 Einführung1904, 287.
269 Sittenlehre1894, 209.
270 Einführung1900, 322.
271 Einführung1900, 339.
272 Cf. Einführung1904, 101.
273 Cf. Einführung1904, 80.
274 Einführung1904, 99.
275 Sittenlehre1894, 212, 229.
276 Einführung1904, 101.
277 Cf. Einführung1904, 15 ff.
278 Joseph Petzoldt, 'Sonderschulen für hervorragend Befähigte', *Neue Jahrbücher für Pädagogik*, 7 (1904), 425–56.
279 Einführung1904, 17.
280 Weltproblem1906, 10–11.
281 See Charles Darwin, *On the Origin of Species* [1959], ed. by G. Beer (Oxford: Oxford University Press, 2008), 44: 'It is the most flourishing, or, as they may be called, the dominant species, – those which range widely over the world, are the most diffused in their own country, and are the most numerous in individuals, – which oftenest produce well-marked varieties, or, as I consider them, incipient species.'
282 Weltproblem1906, 11.

Chapter 3

1 Avenarius1887, 179–80. My italics. Extract from Richard Avenarius, 'Über die Stellung der Psychologie zur Philosophie', *Vierteljahrsschrift für wissenschaftliche Philosophie*, 1 (1877), 480.
2 Avenarius1887, 179.
3 Avenarius, *Der menschliche Weltbegriff*, ix.
4 Avenarius, *Der menschliche Weltbegriff*, 54.
5 Avenarius, *Der menschliche Weltbegriff*, ix–x.
6 Avenarius, *Der menschliche Weltbegriff*, xi.
7 For further information, see Russo Krauss, *Wundt, Avenarius, and Scientific Psychology*, 13 ff.
8 Avenarius, *Der menschliche Weltbegriff*, xii.
9 See Edgar, 'The Physiology of the Sense Organs and Early Neo-Kantian Conceptions of Objectivity', 101–22.
10 Avenarius, *Der menschliche Weltbegriff*, 108–9. My italics.
11 Avenarius, *Der menschliche Weltbegriff*, xii.
12 Avenarius, *Der menschliche Weltbegriff*, 4–5.
13 Avenarius, *Der menschliche Weltbegriff*, 6 ff.

14   For further details on introjection in the work of Avenarius cf. Russo Krauss, *Wundt, Avenarius, and Scientific Psychology*, 18 ff.
15   J. Petzoldt, 'Autobiographie', 8.
16   Letter from Avenarius to Petzoldt dated 27 February 1889 (PetzoldtNachlass, PE44-13).
17   Petzoldt, 'Anzeige von R. Avenarius "Kritik der reinen Erfahrung"', 120.
18   Petzoldt, 'Anzeige von R. Avenarius "Kritik der reinen Erfahrung"', 121.
19   Mach noted something similar with respect to the relationship between naive realism and philosophical conceptualization in 'Introductory Anti-Metaphysical Remarks': 'The philosophical point of view of the average man [...] is a product of *nature*, and is *preserved* and sustained by nature. Everything that philosophy has accomplished – the biological value of every advance, nay, of every error, admitted – is, as compared with it, but an insignificant and *ephemeral* product of art. (Mach, *Beiträge zur Analyse der Empfindungen*, 23-4; *Contributions to the Analysis of the Sensations*, 26. My italics.).
20   Sittenlehre1894, 33.
21   Sittenlehre1894, 33-4.
22   Sittenlehre1894, 35.
23   Sittenlehre1894, 37.
24   The most prominent criticisms of Mach's ideas by his fellow scientists can be found in Max Planck, 'Die Einheit Des Physikalischen Weltbildes', *Physikalische Zeitschrift*, 10 (1909), 62-75; English translation in 'The Mach-Planck Polemics', in *Ernst Mach. A Deeper Look: Documents and New Perspectives*, ed. by John Blackmore, Boston Studies in the Philosophy of Science (Dordrecht: Springer, 1992), 127-50. Also worthy of mention is Lenin's condemnation of Mach's position in *Materialism and Empirio-Criticism*. However, the interpretation of Mach as an exponent of subjective idealism persisted over the years, with Popper writing an article on the affinity between Berkeley and Mach (Karl R. Popper, 'A Note on Berkeley as Precursor of Mach', *The British Journal for the Philosophy of Science*, 4/13 (1953), 26-36).
25   Sittenlehre1894, 38.
26   Sittenlehre1894, 35.
27   Letter from Avenarius to Petzoldt dated 25 December 1892 (PetzoldtNachlass, PE44-31).
28   Sittenlehre1894, 53. My italics.
29   Sittenlehre1894, 35.
30   Einführung1900, 1.
31   Ibid.
32   Maxima1890, 428.
33   Cf. Einführung1900, 1.
34   Einführung1900, 2-3.
35   Einführung1900, 3.
36   Ibid.
37   In the closing sections of the *Einführung in die Philosophie der reinen Erfahrung*, Petzoldt rejects the phenomenalist position of Hans Cornelius and Theodor Ziehen as inaccurate subjectivist interpretations of the positions of Mach and Avenarius (cf. Einführung1904, 298-310).
38   Joseph Petzoldt, 'Solipsismus auf praktischem Gebiet', *Vierteljahrsschrift für wissenschaftliche Philosophie*, 25 (1901), 340.
39   Petzoldt, 'Solipsismus auf praktischem Gebiet', 340-1.

40 Petzoldt, 'Solipsismus auf praktischem Gebiet', 341.
41 Einführung1904, 304.
42 Einführung1904, 305.
43 Ibid.
44 Einführung1904, 306.
45 Ibid.
46 Einführung1904, 307–8.
47 Einführung1904, 310.
48 Further proof that Petzoldt continued to support the philosophical system he developed in the early years of the twentieth century is that he published two short monographs in 1923 and 1927 that were in fact just a re-elaboration of parts of the second volume of the 1904 *Einführung*, namely: *Das allgemeinste Entwicklungsgesetz*; *Das natürliche Höhenziel der menschheitlichen Entwicklung* (Berlin, Leipzig: Paetel, 1927).
49 The expression 'empiriocriticism' was initially in adjectival form, associated with some key concepts in Avenarius's work. In the *Kritik der reinen Erfahrung* (1888–1890) and the *Menschliche Weltbegriff* (1891) there are references to an 'empiriocritical point of view', to 'empiriocritical axioms', etc. Avenarius's student Friedrich Carstanjen describes how 'the term "empiriocriticism" was first used only some time after the publication of *Kritik der reinen Erfahrung* and in particular in informal exchanges, around 1893, by the circle of [Avenarius's] friends and students, after other proposed terms had all been rejected' (Friedrich Carstanjen, 'Der Empiriokritizismus: Zugleich eine Erwiderung auf W. Wundts Aufsätze "Über naiven und kritischen Realismus"', *Vierteljahrsschrift für wissenschaftliche Philosophie*, 22 (1898), 54). The first time that Avenarius uses the term 'empiriocriticism' to define his own approach is in 1894, in a note written in reply to the open letter addressed to him by Wilhelm Schuppe (Richard Avenarius, 'Anmerkung zu der vorstehenden Abhandlung [R. Willy, Das Erkenntnisstheoretische Ich und der natürliche Weltbegriff]', *Vierteljahrsschrift für wissenschaftliche Philosophie*, 18 (1894), 31).
50 Einführung1904, 295 ff.
51 Letter from Petzoldt to Mach dated 21 May 1901, MachNachlass, NL 174/2434.
52 See, for example, the letter from Petzoldt to Mach dated 2 May 1903, MachNachlass, NL 174/2444. More generally, on Petzoldt's professional qualification (*Habilitation*), see the correspondence with Mach from the letter dated 21 May 1901, where it is first mentioned, to that dated 8 May 1904, in which Petzoldt informs Mach that the matter has been concluded satisfactorily (NL 174/2452).
53 See the letter from Petzoldt to Mach dated 22 April 1903, MachNachlass, NL 174/2443.
54 Ernst Mach, 'Über Das Princip Der Vergleichung in Der Physik', *Verhandlungen Deutscher Naturforscher*, 1894, 44–56.
55 Letter from Petzoldt to Mach dated 14 October 1894, MachNachlass, NL 174/2426.
56 Einführung1900, 352.
57 Weltproblem1906, I.
58 Weltproblem1906, 72. Petzoldt does not use Avenarius's original term *Introjektion* here but a different one that is also used in his work to define introjection: *Einlegung*. We have therefore translated it as 'introduction' here.
59 Weltproblem1906, 72.
60 Weltproblem1906, 71.
61 Weltproblem1906, 7.

62 Weltproblem1906, 6, 8. It should be noted that ideas of this kind were not proposed solely by Avenarius but were common in nineteenth-century positivism, starting with August Comte and his law of three stages, according to which teleology and metaphysics, despite being flawed, are nevertheless necessary phases in the development of humanity (cf. Auguste Comte, *Discours Sur l'esprit Positif* (Paris: Vrin, 1844)).
63 Cf. Weltproblem1906, 10 ff.
64 Weltproblem1906, 21.
65 Ibid.
66 Weltproblem1906, 23, 24–5.
67 Weltproblem1906, 43 ff.
68 Weltproblem1906, 55 ff.
69 Weltproblem1906, 60.
70 Mach, *Beiträge zur Analyse der Empfindungen*, 5; *Contributions to the Analysis of the Sensations*, 6.
71 Weltproblem1906, 60. My italics.
72 Weltproblem1906, 63.
73 Weltproblem1906, 65.
74 Petzoldt expresses this view even more clearly in the letter that accompanies the copy of the *Weltproblem* that he sends to Mach. He emphasizes that his own interpretation of Protagoras is independent of that of Theodor Gomperz, *Griechische Denker: eine Geschichte der antiken Philosophie*, 3 vols (Leipzig: Veit, 1896) and summarizes his interpretation thus: '1. He eliminates the concept of substance, thus establishing his opposition to metaphysics (positivism), 2. He eliminates the distinction between existence and appearance (relativism). He is therefore a relativistic positivist' (letter from Petzoldt to Mach dated 5 December 1906, MachNachlass, NL 174/2478).
75 Weltproblem1906, 62–3.
76 Weltproblem1906, 64.
77 Weltproblem1906, 66.
78 Weltproblem1906, 67.
79 Weltproblem1906, 63–4. My italics.
80 Weltproblem1906, 66.
81 Ibid.
82 Ibid.
83 Ibid.
84 Weltproblem1906, 65.
85 Weltproblem1906, 67.
86 Weltproblem1906, 74.
87 Ibid.
88 Joseph Petzoldt, 'Zur positivistische Philosophie', *Zeitschrift für positivistische Philosophie*, 1 (1913), 10.
89 Cf. Weltproblem1906, 84, 88.
90 Weltproblem1906, 90.
91 Weltproblem1906, 94.
92 Weltproblem1906, 106.
93 Weltproblem1906, 62.
94 See Weltproblem1906, 113–14.
95 Weltproblem1906, 114.
96 Weltproblem1906, 116.

97   Weltproblem1906, 117.
98   Weltproblem1906, 119.
99   Weltproblem1906, 120.
100 David Hume, *A Treatise of Human Nature* [1739], ed. by David Norton and Mary Norton (Oxford: Clarendon, 2007), 148.
101 Weltproblem1906, 116.
102 Weltproblem1906, 121.
103 Weltproblem1906, 118–19. Mach refers to these elements as 'undifferentiated' or 'neutral' with respect to the physical and the mental in his correspondence (letter from Mach to Gabrielle Rabel dated 11 November 1906, in Blackmore and Hentschel, *Ernst Mach als Aussenseiter*, 47–8). We have not found other instances of elements being described in these terms in Mach's work, in spite of the fact that he is considered to be one of the prime exponents of what is known as 'neutral monism'.
104 Weltproblem1906, 124–5.
105 Weltproblem1906, 122–3.
106 Weltproblem1906, 125.
107 Ibid.
108 Otto Liebmann, *Kant und die Epigonen. Eine kritische Abhandlung* (Stuttgart: Schober, 1865); Beiser, *The Genesis of Neo-Kantianism*, 283 ff. We can find further details of Petzoldt's view of Liebmann's neo-Kantianism in his correspondence with Mach. In a 1902 letter, Petzoldt writes favourably of having heard him speak in person in his youth, and of having read some of his works (Otto Liebmann, *Zur Analysis der Wirklichkeit: philosophische Untersuchungen* (Strasburg: Trübner, 1876); *Gedanken und Tatsachen. Philosophische Abhandlungen, Aphorismen und Studien* (Strasburg: Trübner, 1882). Moreover, as we have seen, Petzoldt initially held a Kantian position. However, Petzoldt also criticizes the 'almost Chinese deference to Kant' that characterizes Liebmann (due also to his philological training) and 'most of our philosophers', who analyse Kant like a 'very rigorous scholastic might have analysed Aristotle' (letter from Petzoldt to Mach dated 31 January 1902, MachNachlass, NL 174/2440).
109 Weltproblem1906, 125.
110 In his correspondence with Mach, Petzoldt refers several times to his *Philosophische Gesellschaft* activity to promote positivist ideas. In a 1903 letter, he reports that 'in our often lively conflicts at the *Philosophische Gesellschaft* it appears that the drop is beginning to wear away the stone. Even those of a Hegelian persuasion, like [Adolf] Lasson, have at least begun to appreciate modern positivism' (letter from Petzoldt to Mach dated 11 August 1903, MachNachlass, NL 174/2445). Then, in 1904, Petzoldt rejoices over having attended a lecture by a certain Georg Ulrich (later published Georg Ulrich, 'Bewusstsein und Ichheit', *Zeitschrift für Philosophie und philosophische Kritik*, 124 (1905), 58–79), which expressed ideas similar to those of Avenarius. Petzoldt writes that now 'the new [philosophical] direction is no longer the *quantité négligeable* that it once was', and that therefore 'these gentlemen, even the youngest members, have seen that there exists an alternative philosophy to apriorism'. Petzoldt therefore also states that he has confirmed his 'intention to continue to be a member of the Society' (letter from Petzoldt to Mach dated 28 February 1904, MachNachlass, NL 174/2451). This letter shows that Petzoldt, frustrated after years of being one of the few that had rejected Kantianism, had contemplated leaving the Berlin *Philosophische Gesellschaft*.

111 Weltproblem1906, 126.
112 See Lange, *History of Materialism*, II, 153 ff. and especially 193 ff.
113 Weltproblem1906, 126–7.
114 Weltproblem1906, 129.
115 See the passage in which, discussing the need to eliminate not only the remaining substantialist concept of the thing-in-itself but also the remaining substantialist Kantian concept of the subject, Petzoldt writes, 'Where is it that experience shows such an organization of the understanding? It only ever shows an organized brain, and never a mental organization. All concepts that belong to pure reason, to mental faculties, capacities, predispositions, are mere abstractions' (Weltproblem1906, 136).
116 Cf. Weltproblem1906, 131–2.
117 Cf. Weltproblem1906, 132.
118 Cf. ibid.
119 Cf. Weltproblem1906, 146.
120 Cf. Weltproblem1906, 142.
121 Letter from Petzoldt to Mach dated 25 January 1905, MachNachlass, NL 174/2460.
122 Einführung1904, 313.
123 Einführung1904, 316.
124 Ibid.
125 Einführung1904, 317.
126 Ibid.
127 Ibid.
128 Ibid.
129 Kant introduces the concept of *Bewusstsein überhaupt* in the first *Critique*, in the section on the 'Transcendental deduction of the pure concepts of the understanding', where he writes 'Therefore all manifold, insofar as it is given in one empirical intuition, is determined in regard to one of the logical functions for judgment, by means of which, namely, it is brought to a consciousness in general [*zu einem Bewusstsein überhaupt*]' (Immanuel Kant, *Critique of Pure Reason* [1787], ed. by Allen Wood and Paul Guyer (Cambridge: Cambridge University Press, 1999), B143). However, a more precise definition can be found in the posthumous manuscripts on *Logic*, where he writes 'The representation of the manner in which various concepts (as such) belong to a consciousness in general [*Bewusstsein überhaupt*], not only my own, is judgment. They belong to a consciousness in part according to the laws of the imagination, and hence subjectively, or according to the laws of the understanding, that is, in a manner which is universally valid for any being possessing an understanding' (Immanuel Kant, *Handschriftlicher Nachlass: Logik*, Gesammelten Werken, 23 vols (Akademie Ausgabe), XVI, 633, footnote 3051.). For an analysis of this passage see Beatrice Longuenesse, *Kant and the Capacity to Judge: Sensibility and Discursivity in the Transcendental Analytic of the 'Critique of Pure Reason'* (Princeton: Princeton University Press, 1998), 88 ff.
130 An overview of the different positions on *Bewusstsein überhaupt*, made necessary by the success that this concept enjoyed in the nineteenth and twentieth centuries, can be found in Hans Amrhein, *Kants Lehre vom 'Bewusstsein überhaupt' und ihre Weiterbildung bis auf die Gegenwart* (Berlin: Reuther und Reichard, 1909). In the preface of this book, Hans Vaihinger affirms that the concept of '*Bewusstsein überhaupt*', even though 'it is mentioned just nine times in the works of Kant', could be used to write 'the entire history of the last forty years of German philosophy'

(Hans Vaihinger, 'Geleitwort', in *Kants Lehre vom 'Bewusstsein überhaupt' und ihre Weiterbildung bis auf die Gegenwart*, ed. by Hans Amrhein (Berlin: Reuther und Reichard, 1909), iv).

131 In the preface cited above, Vaihinger notes that the first person to address *Bewusstsein überhaupt* was in fact Laas 'about thirty years ago' (Vaihinger, 'Geleitwort', iv). Vaihinger is probably referring to Ernst Laas, *Kants Analogien der Erfahrung. Eine kritische Studie über die Gründlagen der theoretisehen Philosophie* (Berlin: Weidmann, 1876), in which this concept plays a central role (see sections 22 ff.). The concept of *Bewusstsein überhaupt* also appears in Laas's main work, *Idealismus und positivismus. Eine kritische Auseinandersetzung*, 3 vols (Berlin: Weidmann, 1879), III, 47ff.

132 See the section 'Das urteilende Bewusstsein überhaupt' in Heinrich Rickert, *Der Gegenstand der Erkenntnis. Einführung in die Transcendentalphilosophie* (Tübingen: Mohr, 1904), 142ff.

133 The word 'immanence' in fact hardly appears at all in Schuppe's works, but can be found in those of the other main exponents of this philosophical approach, such as Richard von Schubert-Soldern, *Ueber Transcendenz des Objects und Subjects* (Leipzig: Fues, 1882). However, they coined the term 'philosophy of immanence' together when they founded the *Zeitschrift für immanente Philosophie*, which appeared over four years, from 1896 to 1899.

134 Wilhelm Schuppe, 'Die Bestätigung des naiven Realismus. Offener Brief an Herrn Prof. Dr. Richard Avenarius', *Vierteljahrsschrift für wissenschaftliche Philosophie*, 17 (1893), 365–88.

135 Avenarius, *Der menschliche Weltbegriff*, 83 ff.

136 Schuppe, 'Die Bestätigung des naiven Realismus', 386.

137 Schuppe, 'Die Bestätigung des naiven Realismus', 386–7.

138 Schuppe, 'Die Bestätigung des naiven Realismus', 387, see also 372.

139 Rudolf Willy, 'Das erkenntnistheoretische Ich und der natürliche Weltbegriff', *Vierteljahrsschrift für wissenschaftliche Philosophie*, 18 (1894), 1–29.

140 Cf. Avenarius, 'Anmerkung zu der vorstehenden Abhandlung [R. Willy, Das Erkenntnisstheoretische Ich und der natürliche Weltbegriff]', 30.

141 Ibid.

142 Mach himself provides an account of this episode in a letter to Schuppe dated 9 November 1902 ('Briefe von Richard Avenarius und Ernst Mach an Wilhelm Schuppe', *Erkenntnis*, 6 (1936), 75–6). We know of Avenarius' requests from a letter sent to Mach in 1895 to extend his compliments for having been offered a professorship in philosophy at Vienna: 'Now you have joined "your fellow philosophers" and – to the joy of your admirers – you have decided to devote yourself to the work of philosophy in one way or another, which should satisfy the conditions on which your collaboration once depended' (Joachim Thiele, 'Briefe deutscher Philosophen an Ernst Mach', *Synthese*, 18/2/3 (1968), 285–301).

143 This was probably a sideswipe at the abstruse lexicon favoured by Avenarius, which had induced Mach to say that 'it is asking rather much of an elderly man that to the labour of learning the languages of the nations he should add that of learning the language of an individual' (Ernst Mach, *Die Analyse der Empfindungen und das Verhältniss des Physischen zum Psychischen*, Zweite vermehrte Auflage der Beiträge zur Analyse der Empfindungen (Jena: Fischer, 1900), 36; *The Analysis of Sensations, and the Relation of the Physical to the Psychical*, trans. by C. M. Williams (Chicago, London: Open Court, 1914), 48).

144 Letter from Mach to Schuppe dated 29 March 1902, in 'Briefe von Richard Avenarius und Ernst Mach an Wilhelm Schuppe', 74–5. On the subject of solipsism, it is worth noting what Mach writes in the later editions of *The Analysis of Sensations*: 'Anyone who has at some time or another been influenced by Kant, anyone who has adopted an idealistic standpoint, and has been unable to get rid of the last traces of the notion of the "thing in itself," retains a certain inclination towards solipsism, which will appear more or less clearly. Having been through it in my early youth, I know this condition of mind well, and can easily understand it' (Ernst Mach, *Die Analyse der Empfindungen und das Verhältniss des Physischen zum Psychischen*, Vierte vermehrte Auflage (Jena: Fischer, 1903), 280; *Die Analysis of Sensations*, 357–8).

145 Mach writes that he has sent 'one work for a readership of physicists and another for a broader readership' (letter from Mach to Schuppe dated 29 March 1902, in 'Briefe von Richard Avenarius und Ernst Mach an Wilhelm Schuppe', 75). As the year is 1902, the most significant works of Mach that match this description are *The Science of Mechanics*, whose fourth edition appeared in 1901, and *The Analysis of Sensations*, published in 1900 as an elaboration of the *Beiträge zur Analyse der Empfindungen* of 1886.

146 Mach, *Die Analyse der Empfindungen* (1903), 38; *Die Analysis of Sensations*, 46.

147 For the sections in which he addresses Petzoldt, see Wilhelm Schuppe, *Der Zusammenhang von Leib und Seele und das Grundproblem der Psychologie* (Wiesbaden: Bergmann, 1902), 17–22 and 63–5.

148 Petzoldt tells Mach that he met Schuppe in his letter dated 21 April 1905, noting his pleasure in now having met in person all three of the founders of positivism (NachlassMach, NL 174/2463).

149 Schuppe, *Der Zusammenhang von Leib und Seele*, 65.

150 Schuppe, *Der Zusammenhang von Leib und Seele*, 19.

151 Schuppe, *Der Zusammenhang von Leib und Seele*, 65. My italics.

152 Schuppe, *Der Zusammenhang von Leib und Seele*, 18.

153 Letter from Petzoldt to Mach dated 9 May 1902, MachNachlass, NL 174/2441.

154 Cf. Parallelismus1902, 306 ff.

155 Parallelismus1902, 282.

156 Einführung1904, 295.

157 See Weltproblem1906, I.

158 Wilhelm Schuppe, *Grundriss der Erkenntnistheorie und Logik* (Berlin: Gaertners, 1894), 13–16.

159 Schuppe, *Grundriss der Erkenntnistheorie und Logik*, 10.

160 Schuppe, *Grundriss der Erkenntnistheorie und Logik*, 30.

161 Letter from Petzoldt to Mach dated 31 July 1906, MachNachlass, NL 174/2475.

162 Ibid.

163 Einführung1904, 319–20.

164 Einführung1904, 320.

165 Parallelismus1902, 335.

166 Ibid.

167 Ibid.

168 Parallelismus1902, 334.

169 Einführung1904, 321–2.

170 Einführung1904, 305.

171 Cf. Weltproblem1906, 139. My italics.

172 It is worth noting that Schuppe had himself initially misunderstood Mach's position to be a subjectivist one. In a letter to Mach about his visit to Schuppe, Petzoldt states that he 'had assured him that you [Mach] are not an idealist, not a subjectivist, nor a psychomonist, or anything similar, but a true positivist, which he had not been certain of, as in your *Analysis of Sensations* you sometimes describe the elements of the world as "sensations", which is also often used in an exclusively subjectivist sense' (letter from Petzoldt to Mach dated 21 April 1905, MachNachlass, NL 174/2463).

173 Cf. Weltproblem1906, I.

174 These words reveal the influence of Fechner, who had contrasted the 'night view' of materialism and more generally of all philosophical approaches that assume a supposed true reality beyond the facade of the senses, with the 'day view' that believes in the actual reality of sensations (Gustav Theodor Fechner, *Die Tagesansicht gegenüber der Nachtansicht* (Leipzig: Breitkopf und Härtel, 1879)).

175 Einführung1904, 329–30.

176 Cf. Weltproblem1906, 141–2.

177 Einführung1904, 324.

178 Avenarius, *Der menschliche Weltbegriff*, 83. My italics.

179 Avenarius, 'Bemerkungen zum Begriff des Gegenstandes der Psychologie', II, 144.

180 Ibid.

181 Ibid.

182 Einführung1904, 325. My italics.

183 Ibid.

184 Einführung1904, 326.

185 Mach, *Die Analyse der Empfindungen* (1900), 16–17; *Die Analysis of Sensations*, 24–5. Cited in Einführung1904, 326–7.

186 For an anti-Kantian and anti-Cartesian perspective on Mach's 'unsavable I', see Manfred Sommer, 'Das unrettbare Ich und die heitere Passivität des Ernst Mach', in *La metafisica del positivismo*, ed. by Luca Guidetti, Discipline Filosofiche (Macerata: Quodlibet), 149–59.

187 Einführung1904, 329.

188 Einführung1904, 328.

189 Einführung1904, 327–8.

190 Einführung1904, 317.

191 Einführung1904, 318.

192 Letter from Petzoldt to Mach dated 6 May 1906, MachNachlass, NL 174/2472.

193 Petzoldt defines phenomenalism as a form of 'positivist idealism', including in it the new 'psychomonistic approaches (of, for example, Cornelius, Kleinpeter, Ziehen, Veroworn) that result from a misinterpretation of Mach and Avenarius' (Weltproblem1906, 137).

194 Weltproblem1906, 142–3.

195 Weltproblem1906, 143–4.

196 Weltproblem1906, 144. We note here the emergence of the topic of the relativity of systems of coordinates as a consequence of the absence of an absolutely stationary system of coordinates. Clearly, even though Petzoldt's work is from 1906, it was still too early for him to have incorporated Einstein's 1905 theory of special relativity. Rather, Petzoldt is referring to Mach's criticism of the Newtonian idea of absolute

space. In Mach's conception of physics there is no absolute system of coordinates, even though we usually assume implicitly that the fixed stars are a system of coordinates. See the next chapter for further elaboration of this issue.
197 Weltproblem1906, 144.
198 Weltproblem1906, 145.
199 E. Mach, *Erkenntnis und Irrtum*, Barth, Leipzig, 1905, VII footnote; En. tr. *Knowledge and Error* (Dordrecht-Boston: Reidel Publishing Company, 1976), XXXIII footnote.

Chapter 4

1 Let us mention as example the following works by Petzoldt: 'Metaphysikfreie Naturwissenschaft', *Naturwissenschaftliche Wochenschrift*, 17 (1902), 31, 361–4; 'Die mechanische Naturansicht und das Weltproblem. Vorlesungen bei den Hochschulkursen in Salzburg. 1906', *Das Wissen für alle*, 1907; 'Die Relativitätstheorie im erkenntnistheoretischen Zusammenhange des relativistischen Positivismus', *Berichte der Deutschen physikalischen Gesellschaft*, 23 (1912), 1055–64; 'Die Existenz der Atome', *Chemiker-Zeitung*, 120 (1916), 846–7; 'Philosophie auf naturwissenschaftlicher Grundlage', *Literarische Ratgeber und Jahresbericht des Dürerbundes*, 1916, 123–8; 'Verbietet die Relativitätstheorie Raum und Zeit als etwas wirkliches zu denken?', *Verhandlungen der Deutsche Physikalische Gesellschaft*, 20 (1918), 189–201; 'Die Unmöglichkeit mechanischer Modelle zur Veranschaulichung der Relativitätstheorie', *Berichte der Deutsche Physikalische Gesellschaft*, 13–14 (1919), 495–503; 'Mach und die Atomistik', *Die Wissenschaft*, 10 (1922), 230–1; 'Zur Krisis des Kausalitätsbegriffes', *Die Wissenschaft*, 32 (1922), 693–5; 'An der Pforte der Relativitätstheorie', *Velhagen und Klasings Monatshefte*, 36 (1922–1923), 468–72; 'Dämmerung der mechanische Naturansicht', *Monatsschrift für höhere Schulen*, 23 (1924), 83–7; 'Philosophie der Technik', *Monatsblätter des Berliner Bezirksverein Deutscher Ingenieure*, 6–7 (1926).
2 For further details on the discussion of absolute space presented below, see Max Jammer, *Concepts of Space. The History of Theories of Space in Physics*, 3rd enlarged edn (New York: Dover, 1993), in particular chapters 4 and 5.
3 Isaac Newton, *The Principia: Mathematical Principles of Natural Philosophy: the Authoritative Translation*, trans. by I. Bernard Cohen, Anne Whitman, and Julia Budenz (Berkeley; Los Angeles; London: University of California Press, 1999), 54.
4 Newton, *The Principia*, 69.
5 Newton, *The Principia*, 50. Definition 3.
6 Although the fundamental link between absolute space and inertia was implicit in Newton's reasoning, it was Euler who made it explicit, regarding the truth of the principle of inertia as a demonstration of the otherwise indemonstrable existence of absolute space (cf. Leonhard Euler, 'Réflexions sur l'espace et le temps', *Histoire de l'Académie des Sciences et Belles Lettres de Prusse*, 4 (1748), 324–33, §§ 4–5).
7 Newton, *The Principia*, 57.
8 Newton, *The Principia*, 58.
9 Cf. Newton, *The Principia*, 58 ff.

10 Centrifugal force is in fact nothing other than a specific manifestation of the force of inertia, which is why it is also called a 'pseudo-force' or a 'fictitious force'. The link between the two forces was pointed out as early as by Descartes in the *Principia Philosophiae*. However, other interpretations of centrifugal force had also been proposed over the years. For example, Newton himself favoured its interpretation as a force that was opposite to the force of gravity, basing this on his third law, that is, on the principle of action and reaction. It was only in the eighteenth century, with Euler and Bernoulli, that the identification of centrifugal force as the force of inertia was confirmed (cf. Domenico Bertoloni Meli, 'The Relativization of Centrifugal Force', *Isis*, 81/1 (1990), 23–43).

11 For further details on Mach's criticism of Newton, see J. Bradley, *Mach's Philosophy of Science* (London: Bloomsbury, 2014); Erwin N. Hiebert, 'The Influence of Mach's Thought on Science', *Philosophia Naturalis*, 21 (1984), 598–615; Erwin N. Hiebert, 'Mach's Philosophical Use of the History of Science', in *Historical and Philosophical Perspectives of Science*, ed. by Roger H. Stuewer, Minnesota Studies in the Philosophy of Science, 5 (Minneapolis: University of Minnesota Press), 184–203; Richard Staley, 'Mother's Milk and More: On the Role of Ernst Mach's Relational Physics in the Development of Einstein's Theory of Relativity', in *Interpreting Mach: Critical Essays*, ed. by John Preston (Cambridge: Cambridge University Press, 2021), 28–47.

12 Mach, *Die Mechanik* (1883), 213; *The Science of Mechanics*, 229.
13 Cf. Mach, *Die Mechanik* (1883), 216; *The Science of Mechanics*, 232.
14 Ibid.
15 Mach, *Die Mechanik* (1883), 219; *The Science of Mechanics*, 235.
16 Mach, *Die Mechanik* (1883), 217; *The Science of Mechanics*, 233.
17 Mach, *Die Mechanik* (1883), 216; *The Science of Mechanics*, 232.
18 The passage cited first appears in the fifth edition: Ernst Mach, *Die Mechanik in ihrer Entwickelung; historisch-kritisch dargestellt, Fünfte vermehrte Auflage* (Leipzig: Brockhaus, 1904), 261. On Mach's 'research programme', see Gereon Wolters, Mach I, Mach II, *Einstein und die Relativitätstheorie* (Berlin, New York: De Gruyter, 1987), 49 ff.

19 Cf. Albert Einstein, 'Prinzipielles zur allgemeinen Relativitätstheorie', *Annalen der Physik*, 55 (1918), 241; 'On the Foundations of the General Theory of Relativity', in *The Collected Papers of Albert Einstein. Volume 7. The Berlin Years: Writings, 1918-1921 (English translation supplement)*, trans. by Alfred Engel (Princeton: Princeton University Press, 2002), 33. The most exhaustive resource on Einstein's use of Mach's principle is certainly the volume edited by Julian B. Barbour and Herbert Pfister, *Mach's Principle: From Newton's Bucket to Quantum Gravity* (Boston: Birkhäuser, 1995), in particular the first section, which is of a historical nature. See also Julian B. Barbour, 'Einstein and Mach's Principle', in *The Genesis of General Relativity*, ed. by Michel Janssen and others, Boston Studies in the Philosophy of Science (Dordrecht: Springer Netherlands, 2007), 1492–527, https://doi.org/10.1007/978-1-4020-4000-9_32 [accessed 29 October 2021]; John T. Blackmore, Ryōichi Itagaki and S. Tanaka (eds.), *Ernst Mach's Science: Its Character and Influence on Einstein and Others* (Minamiyana, Hadano-shi, Kanagawa: Tokai University Press, 2006), 193 ff. For a further relevant work, albeit overpolemical on Mach, see also Howard Stein, 'Some Philosophical Prehistory of General Relativity', in *Foundations of Space-Time Theories*, Minnesota Studies in the Philosophy of Science, 8 (Minneapolis: University of Minnesota Press), in particular 14 ff.

20 See note 10 of this chapter.
21 Mach, *Die Mechanik* (1883), 216–17; *The Science of Mechanics*, 232. My italics.
22 Mach, *Die Mechanik* (1904), 253; *The Science of Mechanics*, 543.
23 On the fixed stars as a frame of reference in Mach, see Wolters, *Mach I, Mach II*, 37 ff., in particular 43.
24 On Neumann, see Robert DiSalle, 'Carl Gottfried Neumann', *Science in Context*, 6/1 (1993), 345–53.
25 On Lange, an interesting figure who was also a psychophysicist, and studied and worked under Wundt, see Max von Laue, 'Dr. Ludwig Lange (1863–1936). Ein zu Unrecht Vergessener', *Die Naturwissenschaften*, 35/7 (1948), 193–6; Herbert Pfister, 'Ludwig Lange on the Law of Inertia', *The European Physical Journal H*, 39/2 (2014), 245–50; David K. Robinson, 'Reaction-Time Experiments in Wundt's Institute and Beyond', in *Wilhelm Wundt in History*, ed. by Robert W. Rieber and David K. Robinson (Springer, Boston, MA, 2001), in particular 175 ff.
26 For an overview of the debate between Mach, Neumann, Lange and others on Newtonian physics in the second half of the nineteenth century, see Julian B. Barbour, *The Discovery of Dynamics: A Study from a Machian Point of View of the Discovery and the Structure of Dynamical theories* (Oxford: Oxford University Press, 2001), 645ff.
27 Carl Gottfried Neumann, *Ueber die Prinzipien der Galilei-Newtonschen Theorie* (Leipzig: Teubner, 1870), 14; 'On the Principles of the Galilean-Newtonian Theory', trans. by Robert DiSalle, *Science in Context*, 6/1 (1993), 361.
28 Neumann, *Ueber die Prinzipien der Galilei-Newtonschen Theorie*, 15; 'On the Principles of the Galilean-Newtonian Theory', 362.
29 Neumann, *Ueber die Prinzipien der Galilei-Newtonschen Theorie*, 15; 'On the Principles of the Galilean-Newtonian Theory', 362. Neumann returns to this topic, responding to criticisms and emphasising the eminently mathematical nature of the concept of the Body Alpha, in Carl Neumann, 'Über den Körper Alpha', *Berichte über die Verhandlungen der Königlich Sächsischen Gesellschaft der Wissenschaften. Physische Klasse*, 62 (1910), 69–86.
30 Neumann, *Ueber die Prinzipien der Galilei-Newtonschen Theorie*, 15, 21–2; 'On the Principles of the Galilean-Newtonian Theory', 362, 366.
31 Neumann, *Ueber die Prinzipien der Galilei-Newtonschen Theorie*, 20; 'On the Principles of the Galilean-Newtonian Theory', 365.
32 DiSalle, 'Carl Gottfried Neumann', 349.
33 Neumann, *Ueber die Prinzipien der Galilei-Newtonschen Theorie*, 18; 'On the Principles of the Galilean-Newtonian Theory', 364.
34 Neumann, *Ueber die Prinzipien der Galilei-Newtonschen Theorie*, 27; 'On the Principles of the Galilean-Newtonian Theory', 366.
35 Ibid.
36 Mach, *Die Mechanik* (1904), 290–1; *The Science of Mechanics*, 572.
37 Ludwig Lange, 'Über die wissenschaftliche Fassung des Galilei'schen Beharrungsgesetzes', *Philosophische Studien*, 2 (1885), 266–97; 'Nochmals über das Beharrungsgesetz', *Philosophische Studien*, 2 (1885), 539–45; 'Über das Beharrungsgesetz', *Berichte über Verhandlungen der Königliche Sächsischen Gesellschaft der Wissenschaften. Mathematische-physikalische Klasse*, 1885, 333–51; translated in 'On the Law of Inertia', trans. by Herbert Pfister, *European Physical Journal H*, 39 (2014), 251–62; *Die geschichtliche Entwickelung des Bewegungsbegriffes und ihr voraussichtliches Endergebniss. Ein Beitrag zur historischen Kritik der mechanischen Principien* (Leipzig: Engelmann, 1886).

38 Lange, 'Über das Beharrungsgesetz', 334–5; 'On the Law of Inertia', 252.
39 Lange, 'Über das Beharrungsgesetz', 336; 'On the Law of Inertia', 253. My italics.
40 Lange, 'Über das Beharrungsgesetz', 336; 'On the Law of Inertia', 253.
41 Lange, 'Über das Beharrungsgesetz', 336–7; 'On the Law of Inertia', 253.
42 Lange, 'Über das Beharrungsgesetz', 337–8; 'On the Law of Inertia', 253. See also Robert DiSalle, 'Space and Time: Inertial Frames', in *The Stanford Encyclopedia of Philosophy*, ed. by Edward N. Zalta, Winter 2020 (Metaphysics Research Lab, Stanford University, 2020), https://plato.stanford.edu/archives/win2020/entries/spacetime-iframes/ [accessed 29 October 2021].
43 Lange, 'Über das Beharrungsgesetz', 338; 'On the Law of Inertia', 254.
44 Ernst Mach, *Die Mechanik in ihrer Entwickelung; historisch-kritisch dargestellt*, Dritte verbesserte und vermehrte Auflage (Leipzig: Brockhaus, 1897), 234; *The Science of Mechanics*, 544.
45 Mach, *Die Mechanik* (1897), 235; *The Science of Mechanics*, 545–6.
46 Mach, *Die Mechanik* (1897), 235; *The Science of Mechanics*, 546.
47 Mach, *Die Mechanik* (1897), 236; *The Science of Mechanics*, 546. My italics.
48 Leonhard Euler, *Mechanica sive motus scientia analytice exposita* (Petropoli: Typographia Acadamiae Scientiarum, 1736), §56. See also Eric Watkins, 'The Laws of Motion from Newton to Kant', *Perspectives on Science*, 5 (1997), 330 ff.
49 Eindeutigkeit1895, 189.
50 Eindeutigkeit1895, 189.
51 Ibid.
52 Cf. Eindeutigkeit1895, 190.
53 Eindeutigkeit1895, 190.
54 Ibid.
55 Ernst Mach, *Die Mechanik in ihrer Entwickelung; historisch-kritisch dargestellt*, Vierte verbesserte und vermehrte Auflage (Leipzig: Brockhaus, 1901), 282–3; *The Science of Mechanics*, 562–3.
56 Eindeutigkeit1895, 191, footnote, quoting from Ernst Mach, *Die Mechanik in ihrer Entwickelung; historisch-kritisch dargestellt*, Zweite verbesserte Auflage (Leipzig: Brockhaus, 1889), 486; *The Science of Mechanics*, 586.
57 Eindeutigkeit1895, 192.
58 Ibid.
59 Ibid.
60 Eindeutigkeit1895, 193, footnote.
61 Ibid.
62 Cf. Emil Wiechert, 'The Relativity Concept in Physics', in *Ernst Mach - A Deeper Look: Documents and New Perspectives*, ed. by John Blackmore (Dordrecht, Boston: Kluwer, 1992), 165–6.
63 Mach, *Die Mechanik* (1901), 290; *The Science of Mechanics*, 571. My italics. Unlike the other passages where Mach cites Petzoldt, this passage appears only in the 1901 and 1908 editions.
64 Mach, *Die Mechanik* (1901), 290; *The Science of Mechanics*, 571–2. Translation slightly modified.
65 Mach suffered a stroke in July 1898 which left him paralysed on one side and almost completely unable to speak. This condition was accompanied by urinary tract problems that required the use of a catheter, sleep problems, neuralgia and frequent falls that confined him to bed for weeks or months. Following his stroke,

Mach became convinced that he would not live much longer, even though in fact he died only in 1916. He therefore delegated to his followers the task of preparing new editions of his works, to ensure that they were in safe hands in the event of his death. On this point, see his comment in the Preface to the 1900 edition of *The Analysis of Sensations*: 'I was unwilling to let slip this last opportunity without once again saying something on a subject which I have so much at heart' (Mach, *Die Analyse der Empfindungen* (1900), vii; *Die Analysis of Sensations*, ix). On Mach's poor health, see Wolters, *Mach I, Mach II*, 276 ff. The letters in which Petzoldt refers to the tasks assigned to him by Mach are those from 21 May 1901 onwards, MachNachlass, NL 174/2434 ff. In particular, in the letter of 14 June 1901 Petzoldt expresses the hope that 'benevolent destiny will give us not only the next edition but also the one after that, by your own hand' (letter from Petzoldt to Mach dated 14 June 1901, MachNachlass, NL 174/2436).

66 Letter from Petzoldt to Mach dated 14 June 1901, MachNachlass, NL 174/2436.
67 Einführung1904, 311–12.
68 Letter from Petzoldt to Mach dated 26 May 1901, MachNachlass, NL 174/2435. Reference to 'Experimental proposition a': 'Bodies set opposite each other induce in each other, under certain circumstances to be specified by experimental physics, contrary accelerations in the direction of their line of junction. (The principle of inertia is included in this.)' (Mach, *Die Mechanik* (1901), 258; *The Science of Mechanics*, 243).
69 Letter from Petzoldt to Mach dated 26 May 1901, MachNachlass, NL 174/2435. My italics.
70 Ibid.
71 Ibid.
72 Ibid.
73 Ibid.
74 Ibid.
75 Ibid.
76 Ibid.
77 Ibid.
78 Mach, *Die Mechanik* (1883), 217; *The Science of Mechanics*, 232.
79 Letter from Petzoldt to Mach dated 26 May 1901, MachNachlass, NL 174/2435. Citation from Mach, *Die Mechanik* (1901), 291; *The Science of Mechanics*, 572.
80 Letter from Petzoldt to Mach dated 26 May 1901, MachNachlass, NL 174/2435. My italics.
81 Ibid.
82 Ibid.
83 Mach, *Die Mechanik* (1901), 245; *The Science of Mechanics*, 235. Cited in Petzoldt's letter to Mach dated 26 May 1901, MachNachlass, NL 174/2435.
84 Mach, *Die Analyse der Empfindungen* (1900), 11; *Die Analysis of Sensations*, 17. Cited in Petzoldt's letter to Mach dated 26 May 1901, MachNachlass, NL 174/2435.
85 Letter from Petzoldt to Mach, dated 26 May 1901, MachNachlass, NL 174/2435.
86 Eindeutigkeit1895, 194, footnote.
87 Letter from Petzoldt to Mach dated 14 June 1901, MachNachlass, NL 174/2436.
88 Letter from Petzoldt to Mach dated 20 September 1901, MachNachlass, NL 174/2437.

89 Letter from Petzoldt to Mach dated 1 September 1902, MachNachlass, NL 174/2442. Petzoldt is probably referring to James Clerk Maxwell, *A Treatise on Electricity and Magnetism* (Oxford: Clarendon, 1873); *Lehrbuch der Electricität und des Magnetismus*, trans. by Max Bernhard Weinstein (Berlin: Julius Springer, 1883), in which – as the title indicates – Maxwell presents a comprehensive overview of electromagnetism. Petzoldt's letter to Mach comes almost forty years after the publication of Maxwell's treatise, but the work continued to be at the centre of scientific debate as a result of a number of 'asymmetries', as Einstein called them in the incipit of his 'Zur Elektrodynamic bewegter Körper', *Annalen der Physik*, 17 (1905), 891–921; 'On the Electrodynamics of Moving Bodies', in *The Collected Papers of Albert Einstein. Volume 2: The Swiss Years: Writings, 1900–1909 (English translation supplement)* (Princeton: Princeton University Press, 1989), 140–71. The most important of these questions was whether the equations presented could be considered invariant with respect to all inertial reference systems, given that some experiments appeared to indicate the contrary, thus contradicting the Galilean principle of relativity.
90 Letter from Petzoldt to Mach dated 9 August 1904, MachNachlass, NL 174/2453.
91 See the letter from Petzoldt to Mach dated 11 July 1906, MachNachlass, NL 174/2473.
92 Letter from Petzoldt to Mach dated 18 July 1906, MachNachlass, NL 174/2474.
93 Ibid. This is a reference to Arwed Fuhrmann, *Aufgaben aus der analytischen Mechanik: ein Übungsbuch für Studierende der Mathematik, Physik, Technik* (Leipzig: Teubner, 1867).
94 Letter from Petzoldt to Mach dated 8 August 1907, MachNachlass, NL 174/2482.
95 Bewegung1908, 30.
96 Cf. Bewegung1908, 30–1.
97 Bewegung1908, 31.
98 Bewegung1908, 33. My italics.
99 Ibid.
100 Ibid.
101 Cf. Letter from Petzoldt to Mach dated 4 September 1907, MachNachlass, NL 174/2483.
102 Bewegung1908, 33–4.
103 Bewegung1908, 34 and 56.
104 Ernst Mach, *Erkenntnis und Irrtum. Skizzen zur Psychologie der Forschung* (Leipzig: Barth, 1905), 331; *Knowledge and Error: Sketches on the Psychology of Enquiry*, trans. by Thomas J. McCormack, Vienna Circle Collection (Dordrecht, Boston: D. Reidel, 1976), 251.
105 Letter from Petzoldt to Mach dated 11 October 1905, MachNachlass, NL 174/2467.
106 Bewegung1908, 34.
107 Ibid.
108 Bewegung1908, 35.
109 Ibid.
110 Cf. Bewegung1908, 36.
111 Bewegung1908, 37.
112 Ibid.
113 Cf. ibid., reference to Mach, *Die Geschichte und die Wurzel des Satzes von der Erhaltung der Arbeit*, 162 ff.; *Knowledge and Error*, 120 ff.

114	Cf. Bewegung1908, 37.
115	Cf. Bewegung1908, 38–9.
116	Bewegung1908, 39.
117	Ibid.
118	Bewegung1908, 44.
119	Lange, 'Nochmals über das Beharrungsgesetz', 542. Cited in Bewegung1908, 41.
120	Lange, 'Über das Beharrungsgesetz', 337–8; 'On the Law of Inertia', 253. My italics.
121	Bewegung1908, 42.
122	Bewegung1908, 45. Cf. also 48.
123	Bewegung1908, 46.
124	Cf. Bewegung1908, 45 ff., in particular 49.
125	Bewegung1908, 50.
126	Cf. Bewegung1908, 52.
127	Bewegung1908, 54.
128	Bewegung1908, 55.
129	Ibid.
130	Bewegung1908, 56.
131	Bewegung1908, 59, footnote.
132	Bewegung1908, 57.
133	Ibid.
134	Ibid.
135	Ibid. My italics.
136	Cf. Bewegung1908, 59.
137	Bewegung1908, 60.
138	Bewegung1908, 61.
139	Bewegung1908, 55.
140	Cf. John D. Norton, 'Einstein's Special Theory of Relativity and the Problems in the Electrodynamics of Moving Bodies That Led Him to It', in *The Cambridge Companion to Einstein*, ed. by Christoph Lehner and Michel Janssen (Cambridge: Cambridge University Press, 2014), 72–102; John D. Norton, 'Einstein's Investigations of Galilean Covariant Electrodynamics Prior to 1905', *Archive for History of Exact Sciences*, 59/1 (2004), 45–105.
141	Our task here is clearly not to provide an exhaustive introduction to the complex field of Einstein's theory of relativity, and we will therefore limit ourselves to those aspects that are indispensable to an appreciation of Petzoldt's approach. At the same time, it is impossible to provide the reader with a sample bibliography of the immense number of publications on the theory of relativity. We have found very useful Vladimir A. Ugarov, *Special Theory of Relativity* (Moscow: Mir, 1979) but can suggest as more standard reference Arthur I. Miller, *Albert Einstein's Special Theory of Relativity: Emergence (1905) and Early Interpretation (1905–1911)* (Reading: Addison-Wesley, 1981).
142	For more on Fizeau's experiment, see Alejandro Cassini and Marcelo Leonardo Levinas, 'Einstein's Reinterpretation of the Fizeau Experiment: How It Turned Out to Be Crucial for Special Relativity', *Studies in History and Philosophy of Science Part B: Studies in History and Philosophy of Modern Physics*, 65 (2019), 55–72.
143	On the historical relationship between the theories of Lorentz and those of Einstein, leaving to one side the vast debate on whether and how the validity of one theory over the other might be established, see the first chapters of Arthur I. Miller, *Albert Einstein's Special Theory of Relativity*.

144 For an analysis of how Einstein's theory of special relativity addressed the issues raised, see Robert Rynasiewicz, 'The Construction of the Special Theory: Some Queries and Considerations', in *Einstein: The Formative Years. 1879–1909*, ed. by Don Howard and John Stachel (Cham: Springer, 2000), 159–201.

145 Klaus Hentschel, *Interpretationen und Fehlinterpretationen der speziellen und der allgemeinen Relativitätstheorie durch Zeitgenossen Albert Einsteins*, Science Networks. Historical Studies (Basel: Birkhäuser, 1990), xi, see also 55 ff. On the reception of Einstein's theory of relativity from a historical rather than a philosophical perspective, see also Thomas F. Glick, ed., *The Comparative Reception of Relativity*, Boston Studies in the Philosophy of Science, 103 (Dordrecht: Reidel, 1987); and in particular, on Germany, the paper by Lewis Pyenson, 'The Relativity Revolution in Germany', in *The Comparative Reception of Relativity*, ed. by Thomas F. Glick, Boston Studies in the Philosophy of Science, 103 (Dordrecht: Reidel, 1987), 59–111; see also Hubert F. M. Goenner, 'The Reception of the Theory of Relativity in Germany as Reflected by Books Published between 1908 and 1945', in *Studies in the History of General Relativity*, ed. by Jean Eisenstaedt and A. J. Kox (Boston, Basel, Berlin: Birkhäuser, 1992), 15–38.

146 Petzoldt is referring to the elucidatory piece by Johannes Classen, 'Über das Relativitätsprinzip in der modernen Physik', *Zeitschrift für Physikalischen und Chemischen Unterricht*, 23 (1910), 257–67; which replicates in large measure the outline published by Max Planck, 'Das Prinzip der Relativität und die Grundgleichungen der Mechanik', *Verhandlungen der Deutschen Physikalischen Gesellschaft*, 8 (1906), 136–41.

147 Letter from Petzoldt to Mach dated 22 September 1910, published in Blackmore and Hentschel, *Ernst Mach als Aussenseiter*, 84–5.

148 Hermann Minkowski, *Raum und Zeit: Vortrag, gehalten auf der 80. Naturforscher-Versammlung zu Köln am 21. September 1908*, 1909th edn (Leipzig: Teubner, 1909); 'Space and Time', in The Principle of Relativity. A Collection of Original Memoirs on the Special and General Theory of Relativity, ed. by Wilfrid Perrett and George B. Jeffrey, Fundamental Theories of Physics (New York: Dover, 1923), 73–91.

149 Frank himself refers to this in a letter sent in 1959 to the historian of science Friedrich Herneck (cf. Friedrich Herneck, 'Ernst Mach Und Albert Einstein', in *Symposium Aus Anlass Des 50. Todestages von Ernst Mach*, ed. by Wolfgang Merzkirch (Freiburg: Ernst-Mach-Institut, 1966), 49). The paper he sent to Mach is Philipp Frank, 'Das Relativitätsprinzip und die Darstellung der physikalischen Erscheinungen im vierdimensionalen Raum', *Zeitschrift für physikalische Chemie, Stöchiometrie und Verwandtschaftslehre*, 74 (1910), 466–95.

150 Ernst Mach, *Die Geschichte und die Wurzel des Satzes von der Erhaltung der Arbeit*, 2nd edn (Leipzig: Barth, 1909), 60; *History and Root of the Principle of the Conservation of Energy*, 95.

151 Cf. Planck, 'Die Einheit Des Physikalischen Weltbildes', 62–75; English translation in 'The Mach-Planck Polemics', 127–50.

152 Cf. Ernst Mach, 'Die Leitgedanken meiner wissenschaftlichen Erkenntnislehre und ihre Aufnahme durch die Zeitgenossen', *Scientia*, 7 (1910), 237.

153 Letter from Einstein to Mach dated 9 August 1909, in Albert Einstein, *The Collected Papers of Albert Einstein. Volume 5. The Swiss Years: Correspondence, 1902–1914 (English translation supplement)* (Princeton: Princeton University Press, 1995), 130, also published in Blackmore and Hentschel, *Ernst Mach als Aussenseiter*, 59. For an

analysis of the brief correspondence between Einstein and Mach, see also Wolters, *Mach I, Mach II*, 148 ff.

154   Otto Berg, 'Das Relativitätsprinzip Der Elektrodynamik', *Abhandlungen Der Frieschen Schule*, 3 (1910), 333–82.

155   Paul Gruner, 'Elementare Darlegung Der Relativitätstheorie', *Mitteilungen Der Naturforschenden Gesellschaft in Bern*, 1910.

156   According to Wolters, Mach probably pointed out that the theory of special relativity only applied to frames of reference in uniform rectilinear motion. Cf. Wolters, *Mach I, Mach II*, 177.

157   Letter from Petzoldt to Mach dated 1 June 1911, in Blackmore and Hentschel, *Ernst Mach als Aussenseiter*, 91. In a later document, Petzoldt will even suggest that Protagoras was 'the first precursor to the theory of relativity' (Stellung1923, 3).

158   Klaus Hentschel, '(Mis-)Interpretations of the Theory of Relativity – Considerations on how they arise and how to analyze them', in Chiara Russo Krauss and Luigi Laino (eds.), *Philosophers and Einstein's Relativity* (Cham: Springer, forthcoming).

159   Weltproblem1911, 201.

160   See the letter from Petzoldt to Mach dated 10 January 1912, in MachNachlass, NL 174/2488.

161   The rejection of the being/appearing dichotomy is a recurring theme in Petzoldt's writings. The clearest expression of the identification of this theme as the cornerstone of relativism is Petzoldt's letter to Mach in which he summarizes Protagoras as follows: '1. he eliminates the concept of substance, thereby becoming an opponent of all of metaphysics (positivism), 2. he eliminates the opposition between being and appearing (relativism). He is therefore a relativistic positivist' (letter from Petzoldt to Mach dated 5 December 1906, MachNachlass, NL 174/2478).

162   Joseph Petzoldt, 'Die Relativitätstheorie im erkenntnistheoretischen Zusammenhange des relativistischen Positivismus', *Berichte der Deutschen physikalischen Gesellschaft*, 23 (1912), 1056.

163   See in particular sections 11 and 12 of Chapter 7, 'The principle of relativity according to Lorentz, Einstein and Minkowski' and 'Critical elucidation of the principle of relativity and the confirmation of idealism' (Paul Natorp, *Die logischen Grundlagen der exakten Wissenschaften* (Leipzig: Teubner, 1910), 392 ff.). On Natorp's interpretation of Einstein's theory of relativity, see Hentschel, *Interpretationen und Fehlinterpretationen*, 212 ff.; Don Howard, 'Einstein and Eindeutigkeit: A Neglected Theme in the Philosophical Background to General Relativity', in *Studies in the History of General Relativity*, ed. by Jean Eisenstaedt and A. J. Kox (Boston, Basel, Berlin: Birkhäuser, 1992), 154–243; Don Howard, 'Einstein, Kant and the Origins of Logical Positivism', in *Logic, Language, and the Structure of Scientific Theories: Proceedings of the Carnap-Reichenbach Centennial*, ed. by Wesley C. Salmon and Gereon Wolters (Pittsburgh: University of Pittsburgh Press, 1994), 45–105. Although even Petzoldt had supported the *logical* validity of the concepts of absolute space, time and motion in 1908, in contrast to their *empirical* relativity, it must be remembered that he considered these to be the result of a mathematical/geometrical conceptual construction, whereas Natorp saw them as transcendental, as conditions for the possibility of experience.

164   Petzoldt, 'Die Relativitätstheorie im erkenntnistheoretischen Zusammenhange des relativistischen Positivismus', 1059.

165 Letter from Petzoldt to Mach dated 15 June 1913, in Blackmore and Hentschel, *Ernst Mach als Aussenseiter*, 120.

166 Letter from Einstein to Mach dated 25 June 1913, in *The Collected Papers of Albert Einstein. Volume 5. The Swiss Years: Correspondence, 1902–1914 (English translation supplement)*, 340; also published in Blackmore and Hentschel, *Ernst Mach als Aussenseiter*, 121.

167 Even though the theory of general relativity made use of 'Mach's principle' – thus adding weight to the argument of the connections between Mach's ideas and Einstein's relativity – Petzoldt never devoted as much attention to general relativity as he had to special relativity, possibly as a result of struggling to assimilate the details of the new theory. Where he does address it, he adopts the same interpretation as he had given to special relativity, seeing general relativity as a further argument against the concepts of absolute space and time. As he writes: 'The theory of general relativity is even closer to Mach's ideas than special relativity. Absolute space and time are put completely to one side here. […] The spatiotemporal properties of events, or rather, their four-dimensional properties […] depend entirely on the distribution of gravitational masses, so one cannot speak of the space or the time in which nature takes its course. Space and time lose all of their mechanistic properties as containers or "shapes"' (Weltproblem1921, 206). In other words, Petzoldt claims that the theory of general relativity makes it impossible to speak of space-time independently of the bodies that exist within it, as the masses of these bodies determine changes in the shape of space-time itself. It is therefore not space-time that is real but the bodies and the dependency relations between them. Petzoldt concludes that 'a physics that was complete [*vollendete*], one that recognized this fundamental dependency, would have no further need to consider time or space, given that these would already have been exhaustively accounted for by it' (Weltproblem1921, 207).

168 Letter from Petzoldt to Mach dated 5 March 1914, in MachNachlass, NL 174/2494 and in Blackmore and Hentschel, *Ernst Mach als Aussenseiter*, 135. The latter volume maintains that the date written on the back of the letter is an error, and that the letter is in fact from May 1914, being the reply to the letter dated 1 May in which Mach mentions an attached letter from Einstein (ibid., 134).

169 Letter from Einstein to Petzoldt dated 14 April 1914, in *The Collected Papers of Albert Einstein. Volume 8. The Berlin Years: Correspondence, 1914–1918 (English translation supplement)* (Princeton: Princeton University Press, 1998), 12. In this letter, Einstein also points out some of Petzoldt's errors in his exposition of the clock paradox, later known widely as the Twin Paradox thanks to Paul Langevin's 1911 reformulation in 'L'Évolution de l'espace et du temps', *Scientia*, 10 (1911), 31–54. The problem is as follows. In two frames of reference in moving reciprocally at relativistic speed, each of the two observers can claim to be at rest while the other is in motion, and therefore each can say that their own time is running normally and that the other's clock is running more slowly. What would happen to a pair of twins if one remained on Earth while the other took a journey in space? Would it be legitimate for each of them to say that the other twin was ageing more slowly? What would happen on the return to Earth of the twin that had taken the space flight? Petzoldt rejected the standard reply that the travelling twin was indeed younger than his earthbound brother, maintaining that even when they meet again each of the twins continues to consider the other to be younger. In Petzoldt's view, to accept that one of the twins is right and the other is wrong,

that one of them had indeed aged more slowly, meant 'relapsing into absolutist thinking' (Relativitätstheorie1914, 50). There are in fact various ways of resolving the paradox. One is to point out that this is not a case of the uniform rectilinear motion assumed by special relativity, given that the journey into space and back to Earth would require accelerating, turning and decelerating. For more on this topic, see Mendel Sachs, 'On Einstein's Later View of the Twin Paradox', *Foundations of Physics Volume*, 15 (1985), 977–80; Yakov P. Terletskii, *Paradoxes in the Theory of Relativity* (New York: Springer, 1978), 38 ff.; Galina Weinstein, 'Einstein's Clocks and Langevin's Twins', https://arxiv.org/abs/1205.0922.

170 Letter from Einstein to Petzoldt dated 14 April 1914, cit.
171 Alice Calaprice, *The Einstein Almanac* (Baltimore: Johns Hopkins University Press, 2004), 37 ff.
172 Emil Cohn, *Physikalisches über Raum und Zeit* (Leipzig: Teubner, 1913).
173 Albert Einstein, 'Vom Relativitäts-Prinzip', *Vossische Zeitung*, 209 (1914), 34; 'On the Principle of Relativity', in *The Collected Papers of Albert Einstein. Volume 6. The Berlin Years: Writings, 1914–1917 (English translation supplement)* (Princeton: Princeton University Press, 1997), 4.
174 Letter from Einstein to Petzoldt dated 11 June 1914, in *The Collected Papers of Albert Einstein. Volume 8*, 24. As mentioned in the notes to the letter, Ehrenfest writes about the visit in his personal diary on 3 June 1914.
175 Cf. Joachim Thiele, 'Briefe Albert Einsteins an Joseph Petzoldt', *NTM. Schriftenreihe Für Geschichte Der Naturwissenschaften, Technik Und Medizin*, 8 (1971), 70–4.
176 Letter from Einstein to Elsa Einstein dated 3 August 1914, in *The Collected Papers of Albert Einstein. Volume 8*, 40.
177 Letter from G. Helm to Einstein dated 22 March 1918, in *The Collected Papers of Albert Einstein. Volume 8*, 511.
178 Letter from Einstein to Petzoldt dated 19 August 1919, in *The Collected Papers of Albert Einstein. Volume 9. The Berlin Years: Correspondence, January 1919-April 1920 (English translation supplement)* (Princeton: Princeton University Press, 2004), 74–5.
179 Albert Einstein, 'Ernst Mach' [1916], in *The Collected Papers of Albert Einstein. Volume 6. The Berlin Years: Writings, 1914–1917 (English translation supplement)* (Princeton: Princeton University Press, 1997), 141–5.
180 Einstein, 'Ernst Mach', 142. By placing the absolute and the a priori on the same level, Einstein appears to be siding with Petzoldt here against neo-Kantian interpretations of the theory of relativity, such as that propounded by Natorp.
181 Einstein, 'Ernst Mach', 143.
182 Einstein, 'Ernst Mach', 144.
183 Einstein, 'Ernst Mach', 145.
184 Cf. letter from Petzoldt to Einstein dated 26 July 1919, in *The Collected Papers of Albert Einstein. Volume 9*, 65–6. On the problem of the rotation of the rigid disc, including an analysis of the correspondence between Einstein and Petzoldt on the issue, see John Stachel, *Einstein from 'B' to 'Z'* (Boston: Birkhäuser, 2001), 245–60.
185 Cf. letters from Einstein to Petzoldt dated 19 and 23 August 1919, in *The Collected Papers of Albert Einstein. Volume 9*, 74–5, 77.
186 Letter from Petzoldt to Einstein dated 6 July 1920, in *The Collected Papers of Albert Einstein. Volume 10. The Berlin Years: Correspondence, May-December 1920, and Supplementary Correspondence, 1909–1920 (English translation supplement)* (Princeton: Princeton University Press, 2006), 205–6.

187 For a detailed reconstruction of the events of the conference, of Einstein's non-participation and of the role played by Petzoldt, see Hentschel, *Interpretationen und Fehlinterpretationen*, 168–77.
188 Moritz Schlick, 'Die Philosophische Bedeutung Des Relativitätsprinzips', *Zeitschrift Für Philosophie Und Philosophische Kritik*, 159 (1915), 129–75; 'The Philosophical Significance of the Principle of Relativity', in *Philosophical Papers, Volume I (1909–1922)*, ed. by Henk L. Mulder and Barbara F. B. van de Velde-Schlick, Vienna Circle Collection (Dordrecht: Reidel, 1979), 153–89.
189 Letter from Petzoldt to Einstein dated 6 July 2020, in *The Collected Papers of Albert Einstein. Volume 10*, 205–6.
190 Letter from Einstein to Petzoldt dated 21 July 2020, in *The Collected Papers of Albert Einstein. Volume 10*, 212.
191 The extent of Einstein's commitments from 1920 onwards can be seen from the trips he undertook to present his theories in The Netherlands, Norway and Denmark (1920), the Unites States and the United Kingdom (1921), Japan, Palestine and Spain (1922), France and Sweden, to accept his Nobel Prize (1923). Cf. Calaprice, *The Einstein Almanac*, 58 ff.
192 Ernst Mach, *Die Prinzipien der physikalischen Optik. Historisch und erkenntnispsychologisch entwickelt* (Leipzig: Barth, 1921), vii–ix; *The Principles Of Physical Optics*, trans. by John S. Anderson and A. F. A. Young (London: Methuen, 1926), vii–viii.
193 Cf. Wolters, *Mach I, Mach II*, in particular 274 ff. and 328 ff. Given the quantity of philological evidence presented by Wolters, the spurious nature of the Preface to the *Optik* can be considered an unconfutable fact. Or rather, it could be confuted by someone who provided equally or more convincing evidence of its veracity. However, those who still consider the Preface to the *Optik* to be the original words of Mach tend to simply ignore the work of Wolters, or counter the *philological* evidence adduced by Wolters using *philosophical* arguments intended to show that Mach's rejection of the theory of relativity is based on his position on the theory of knowledge. See also the later work of Wolters: 'Mach and Einstein, or, Clearing Troubled Waters in the History of Science', in *Einstein and the Changing World Views of Physics*, ed. by Christoph Lehner, Jürgen Renn, and Matthias Schemmel (New York: Springer, 2012), 39–57; 'Globalized Parochialism: Consequences of English As Lingua Franca in Philosophy of Science', *International Studies in the Philosophy of Science*, 29 (2015), 189–200.
194 Joseph Petzoldt, 'Das Verhältnis der Machschen Gedankenwelt zur Relativitätstheorie', in *Die Mechanik in ihrer Entwicklung historisch-kritisch dargestellt*, by Ernst Mach, 8th edn (Leipzig: Brockhaus, 1921), 490–517.
195 See the reviews of the *Optik* in John Blackmore, ed., *Ernst Mach - A Deeper Look: Documents and New Perspectives* (Dordrecht; Boston: Kluwer, 1992), 65 ff.
196 Cf. Hubert Goenner, 'The Reaction to Relativity Theory I: The Anti-Einstein Campaign in Germany in 1920', *Science in Context*, 6/1 (1993), 107–33; Jeroen van Dongen, 'Reactionaries and Einstein's Fame: "German Scientists for the Preservation of Pure Science," Relativity, and the Bad Nauheim Meeting', *Physics in Perspective*, 9/2 (2007), 212–30; Milena Wazeck, *Einstein's Opponents: The Public Controversy about the Theory of Relativity in the 1920s*, trans. by Geoffrey S. Koby (Cambridge; New York: Cambridge University Press, 2014). Ludwig Mach also drew closer to several exponents of the anti-Einstein movement (cf. Wolters, *Mach I, Mach II*, 367 ff.).

197   Société française de Philosophie, 'Comptes rendus des séances. Séance du 6 avril 1922. La théorie de la relativité', *Bulletin de la Société française de Philosophie*, 22 (1922), 91–113; 'The Theory of Relativity. Discussion Remarks at a Meeting of the Société française de Philosophie', in *The Collected Papers of Albert Einstein. Volume 13: The Berlin Years: Writings & Correspondence January 1922-March 1923 (English translation supplement)*, ed. by Diana Kormos Buchwald and others (Princeton: Princeton University Press, 2012); on the topic see also Wolters, *Mach I, Mach II*, 108 ff.
198   Letter from Petzoldt to Ludwig Mach dated 18 May 1923, cited in Wolters, *Mach I, Mach II*, 366–7 ff.
199   Joseph Petzoldt, 'Postulat der absoluten und relativen Welt', *Zeitschrift für Physikalischen und Chemischen Unterricht*, 21 (1924), 143, footnote.
200   Petzoldt was engaged in a number of long-distance arguments. One started with his 'Verbietet die Relativitätstheorie Raum und Zeit als etwas wirkliches zu denken?', *Verhandlungen der Deutsche Physikalische Gesellschaft*, 20 (1918), 189–201, which took on Emil Cohn's *Physikalisches über Raum und Zeit*, which, as we noted above, was the other work recommended by Einstein in the 'Vossische Zeitung'. The physicist Max Jakob replied to Petzoldt's criticisms in 'Bemerkung zu dem Aufsatz von J. Petzoldt: 'Verbietet die Relativitätstheorie Raum und Zeit als etwas Wirkliches zu denken?', *Verhandlungen der physikalischen Gesellschaft zu Berlin*, 21 (1919), 159–61. This was followed by a reply by Petzoldt ('Die Unmöglichkeit mechanischer Modelle zur Veranschaulichung der Relativitätstheorie', *Berichte der Deutsche Physikalische Gesellschaft*, 13–14 (1919), 495–503) and a counter-reply by Jakob ('Bemerkungen zu dem Aufsatz von J. Petzoldt: 'Über die Unmöglichkeit mechanischer Modelle zur Veranschaulichung der Relativitätstheorie', *Verhandlungen der physikalischen Gesellschaft zu Berlin*, 21 (1919), 501–3). The contentious topic was Petzoldt's assertion that relativistic 'variations' in the measurements of clocks and distances 'can never find an intuitive [*anschauliche*] representation in a single spatiotemporal system, but only in a multiplicity of systems' (Petzoldt, 'Verbietet die Relativitätstheorie Raum und Zeit als etwas wirkliches zu denken?', 193). It should be noted that Einstein sided with Cohn and Jakob here, albeit not publicly. In a letter to the physicist Karl Scheel, who had sent him Jakob's work, Einstein writes that Jakob's criticism of Petzoldt is 'appropriate', and that the reflections contained in his work were 'rather confused and pretentious', as Cohn 'is well aware that a model is not a copy, but something that by its very nature reflects only some of the properties of the original'. Einstein says that he is 'disappointed' in Petzoldt, pointing out that this latest work of his on the theory of relativity does not match up to his 1914 work (letter from Einstein to Karl Scheel, 17 March 1919, in *The Collected Papers of Albert Einstein. Volume 9*, page 15 of the German edition, letter not selected for translation into English). Another of Petzoldt's targets (see 'Kausalität und Relativitätstheorie', *Zeitschrift für Physik*, 1/5 (1920), 467–74; 'Mechanistische Naturauffassung und Relativitätstheorie', *Annalen der Philosophie*, 2/3 (1921), 447–62) was the Danish physicist Helge Holst, who questioned the equivalence of different frames of reference, considering the fixed stars to be a special frame of reference allowing a *causal* explanation of Lorentzian contraction that was valid for all ('Die kausale Relativitätsforderung und Einsteins Relativitätstheorie', *Det Kgl. Danske Videnskabernes Selskab, Matematisk-fysiske Meddelelser*, 2 (1919); 'Wirft die Relativitätstheorie den Ursachsbegriff über Bord?', *Zeitschrift für Physik*, 1 (1920), 32–9). In 'Postulat der absoluten und relativen

Welt', Petzoldt takes on Aloys Müller over his attempt to maintain the concepts of absolute space and time ('Probleme der speziellen Relativitätstheorie', *Zeitschrift für Physik*, 17 (1923), 409–20; *Die philosophischen Probleme der Einsteinschen Relativitätstheorie* (Braunschweig: Vieweg, 1922)).

201 Relativitätstheorie1914, 2–3.
202 Relativitätstheorie1914, 3–4. It should be noted that the expression 'temporal order of events' is somewhat inaccurate here, as relativity is about the different measurement of temporal intervals, modifying the simultaneity of events between one frame of reference and another, but it does not change the order, that is, the sequence, of events (which would have compromised the principle of causality).
203 Relativitätstheorie1914, 4.
204 Mach, *Die Geschichte und die Wurzel des Satzes von der Erhaltung der Arbeit*, 38; *History and Root of the Principle of the Conservation of Energy*, 56, cited in Relativitätstheorie1914, 5.
205 Relativitätstheorie1914, 4–5 and 6.
206 Relativitätstheorie1914, 6.
207 Ibid. Petzoldt clearly now attributes to Mach his own perspective, one that is based less on the concept of 'economy' and more on that of 'stability'.
208 Relativitätstheorie1914, 7–8.
209 Relativitätstheorie1914, 9.
210 Relativitätstheorie1914, 9–10.
211 Relativitätstheorie1914, 11.
212 Ibid.
213 Cf. Relativitätstheorie1914, 12.
214 Cf. ibid.
215 Cf. Relativitätstheorie1914, 33.
216 Relativitätstheorie1914, 16.
217 Minkowski, *Raum und Zeit*, 6; 'Space and Time', 81, my italics. Cited in Relativitätstheorie1914, 16.
218 See the chapter entitled 'The Opposition between Mechanical and Phenomenological Physics', in Ernst Mach, *Die Prinzipien der Wärmelehre historisch-kritisch entwickelt* (Leipzig: Barth, 1896), 362 ff.; *Principles of the Theory of Heat. Historically and Critically Elucidated*, ed. by Brian McGuinness (Dordrecht: Reidel, 1986), 333 ff.
219 Cf. Relativitätstheorie1914, 39.
220 Relativitätstheorie1914, 17.
221 Ibid. See also the passage where Petzoldt writes that 'the facts of perspective and the theory of relativity of space coincide in their key points like two species of the same genus' (Relativitätstheorie1914, 46).
222 Relativitätstheorie1914, 21.
223 Cf. Relativitätstheorie1914, 31.
224 Cf. Relativitätstheorie1914, 52 ff.
225 Relativitätstheorie1914, 42.
226 Cf. Relativitätstheorie1914, 42 ff.
227 Cf. Relativitätstheorie1914, 45.
228 Relativitätstheorie1914, 44.
229 Relativitätstheorie1914, 36.
230 Weltproblem1921, 216.
231 Cf. Hentschel, *Interpretationen und Fehlinterpretationen*, xii–xiii.

232 Hentschel, '(Mis-)Interpretations of the Theory of Relativity'.
233 Relativitätstheorie1914, 27.
234 Cf. Relativitätstheorie1914, 27–8 and 32–3.
235 Relativitätstheorie1914, 48–9.
236 Mach, *Die Prinzipien der Wärmelehre*, 52 ff.; *Principles of the Theory of Heat*, 59 ff.
237 Relativitätstheorie1914, 30.
238 Relativitätstheorie1914, 40.
239 Relativitätstheorie1914, 41.
240 Cf. Relativitätstheorie1914, 40–1.
241 Relativitätstheorie1914, 41.
242 Relativitätstheorie1914, 21–2.
243 Relativitätstheorie1914, 21.
244 Relativitätstheorie1914, 36–7.
245 Relativitätstheorie1914, 37.
246 Minkowski, *Raum und Zeit*, 2; 'Space and Time', 76. My italics.
247 Minkowski, *Raum und Zeit*, 4; 'Space and Time', 79.
248 Relativitätstheorie1914, 45.
249 Minkowski, *Raum und Zeit*, 2; 'Space and Time', 76.
250 Minkowski, *Raum und Zeit*, 7; 'Space and Time', 83.
251 Ibid. Cited in Relativitätstheorie1914, 37. Petzoldt's italics.
252 Relativitätstheorie1914, 37.
253 Minkowski, *Raum und Zeit*, 14; 'Space and Time', 91. On the established harmony between physics and mathematics in those times, see Lewis Pyenson, *The Young Einstein. The Advent of Relativity* (Bristol; Boston: CRC Press, 1985), 80–100.
254 It is worth recalling here Einstein's famous expression in *Geometrie und Erfahrung* 'As far as the propositions of mathematics refer to reality, they are not certain; and as far as they are certain, they do not refer to reality'. (Albert Einstein, *Geometrie und Erfahrung. Erweiterte Fassung des Festvortrages gehalten an der Preußischen Akademie der Wissenschaften zu Berlin am 27. Januar 1921* (Berlin: Springer, 1921), 1–2); 'Geometry and Experience', in *The Collected Papers of Albert Einstein. Volume 7: The Berlin Years: Writings, 1918-1921 (English translation supplement)* (Princeton: Princeton University Press, 2002), 209.
255 Stellung1923, 37. My italics. Compare Petzoldt's words with the following passage from Mach's *The Analysis of the Sensations*: 'In the investigation of purely physical processes we generally employ concepts of so abstract a character that as a rule we think only cursorily, or not at all, of the sensations (elements) that lie at their base. For example, when I ascertain the fact that an electric current having the intensity of 1 ampere develops 10½ cubic centimetres of oxyhydrogen gas at 0 °C and 760 mm. mercury-pressure in a minute, I am readily disposed to attribute to the objects defined a reality wholly independent of my sensations. But I am obliged, in order to arrive at what I have defined, to conduct the current, for the existence of which my sensations are my only warrant, through a circular wire having a definite radius, so that the current, the intensity of terrestrial magnetism being given, shall turn the magnetic needle a certain angular distance out of the meridian. The determination of the magnetic intensity, of the volume of the oxyhydrogen gas, etc., is no less intricate. *The whole statement is based upon an almost unending series of sensations*, particularly if we take into consideration the adjustment of the apparatus, which must precede the actual experiment. [...] Now I maintain that every physical concept means nothing but a certain definite kind of connexion

of the sensory elements' (Mach, *Die Analyse der Empfindungen* (1900), 30–1; *Die Analysis of Sensations*, 41–1).

256   The concept of 'coincidence' actually makes its first appearance in a letter from Einstein to Ehrenfest dated 26 December 1915 (in Einstein, *The Collected Papers of Albert Einstein. Volume 8*, 167), with 1916 being the date of its first published use.

257   Albert Einstein, 'Die Grundlage der allgemeinen Relativitätstheorie', *Annalen der Physik*, 49 (769), 771; 'The Foundation of the General Theory of Relativity', in *The Collected Papers of Albert Einstein. Volume 6. The Berlin Years: Writings, 1914–1917 (English translation supplement)* (Princeton: Princeton University Press, 1997), 153.

258   Einstein, 'Die Grundlage der allgemeinen Relativitätstheorie', 771; 'The Foundation of the General Theory of Relativity', 153–4. My italics. For a detailed analysis of Einstein's concept of 'coincidence' in the context of the evolution of his position on the meaning of frames of reference, see Marco Giovanelli, 'Nothing but Coincidences: The Point-Coincidence and Einstein's Struggle with the Meaning of Coordinates in Physics', *European Journal for Philosophy of Science*, 11/2 (2021), 45.

259   Stellung1923, 74–5.
260   Stellung1923, 73–4. My italics.
261   Stellung1923, 3–4.
262   Stellung1923, 37. See also 63.
263   Stellung1923, 63.
264   Stellung1923, 67.
265   Ibid.
266   Stellung1923, 66.

Chapter 5

1   On the debate between these writers on the theory of relativity, see the opening chapters of Thomas Ryckman, *The Reign of Relativity: Philosophy in Physics 1915–1925* (Oxford: Oxford University Press, 2004).

2   On the criticisms levelled at Petzoldt by Cassirer, Schlick and Reichenbach, see also the brief outline in Hentschel, *Interpretationen und Fehlinterpretationen*, 416 ff.

3   Cf. Ernst Cassirer, *Zur Einsteinschen Relativitätstheorie. Erkenntnistheoretische Betrachtungen* (Berlin: Bruno Cassirer, 1921); 'Einstein's Theory of Relativity', in *Substance and Function and Einstein's Theory of Relativity*, trans. by William Curtis Swabey and Marie Taylor Swabey (Chicago, London: Open Court, 1923), 351–460. See also the Vorlesungen delivered in those years: Ernst Cassirer, 'Die Philosophischen Probleme Der Relativitätstheorie', in *Vorlesungen Und Vorträge Zu Philosophischen Problemen Der Wissenschaften. 1907–1945*, ed. by Jörg Fingerhut, Gerald Hartung, and Rüdiger Kramme, Nachgelassene Manuskripte Und Texte, 8 (Hamburg: Meiner, 2010), 29–116. On Cassirer's interpretation of relativity, see also Andreas Bartels, 'Die Auflösung der Dinge. Schlick und Cassirer über wissenschaftliche Erkenntnis und Relativitätstheorie', in *Philosophie und Wissenschaften. Formen und Prozesse ihrer Interaktion*, ed. by Hans Jörg Sandkühler (Frankfurt am Main: Peter Lang, 1997), 193–210; Luigi Laino, 'The Conditions of Possibility of Scientific Experience: Cassirer's Interpretation of the Theory of Relativity', in *The Changing Faces of Space*, ed. by Maria Teresa Catena and Felice

Masi, Studies in Applied Philosophy, Epistemology and Rational Ethics, 39 (Cham: Springer, 2017), 235–54; Maja Lovrenov, 'The Role of Invariance in Cassirer's Interpretation of the Theory of Relativity', *Synthesis Philosophica*, 21 (2006); Matthias Neuber, *Die Grenzen des Revisionismus. Schlick, Cassirer und das Raumproblem* (Vienna, New York: Springer, 2012); Enno Rudolph, *Ernst Cassirer Im Kontext: Kulturphilosophie Zwischen Metaphysik Und Historismus* (Tübingen: Mohr Siebeck, 2003), in particular 16–55; Thomas Ryckman, 'Einstein, Cassirer, and General Covariance. Then and Now', *Science in Context*, 12 (1999), 585–619; Thomas Ryckman, 'Conditio Sine qua Non? Zuordnung in the Early Epistemologies of Cassirer and Schlick', *Synthese*, 88 (1991), 57–95.
4   Cassirer, 'Einstein's Theory of Relativity', 367.
5   Ernst Cassirer, 'Substance and Function', in *Substance and Function and Einstein's Theory of Relativity*, trans. by William Curtis Swabey and Marie Taylor Swabey (Chicago, London: Open Court, 1923), 261.
6   Ibid.
7   Cassirer, 'Einstein's Theory of Relativity', 392–3. My italics.
8   Cassirer, 'Die Philosophischen Probleme Der Relativitätstheorie', 107–8.
9   Cassirer addresses Petzoldt's principle of *Eindeutigkeit* a decade later, when he discusses the philosophical interpretation of the principles of maximum and minimum (cf. Ernst Cassirer, *Determinismus und Indeterminismus in der modernen Physik. Historische und systematische Studien zum Kausalproblem* (Göteborg: Wettergren & Kerber, 1936), 71; *Determinism and Indeterminism in Modern Physics. Historical and Systematic Studies of the Problem of Causality*, trans. by Theodor Benfey (New Haven: Yale University Press, 1956), 56).
10 Stellung1923, 95.
11 Ibid.
12 Letter from Petzoldt to Reichenbach dated 24 May 1922, Hentschel, *Die Korrespondenz Petzoldt-Reichenbach*, 49.
13 Weltproblem1921, 208, footnote. My italics.
14 Ernst Cassirer, 'Erkenntnistheorie nebst den Grenzfragen der Logik' [1913], in *Aufsätze und kleine Schriften. 1902–1921*, ed. by Birgit Recki, Gesammelte Werke, 9 (Hamburg: Meiner, 2001), 190–1.
15 Cassirer, 'Einstein's Theory of Relativity', 393. My italics.
16 Cassirer, 'Einstein's Theory of Relativity', 428.
17 Ibid.
18 Cassirer, 'Einstein's Theory of Relativity', 429.
19 Ibid.
20 On Schlick's interpretation of the theory of relativity, see Neuber, *Die Grenzen des Revisionismus*; Fynn Ole Engler, 'Über das erkenntnistheoretische Raumproblem bei Moritz Schlick, Wilhelm Wundt und Albert Einstein', in *Stationen. Dem Philosophen Und Physiker Moritz Schlick Zum 125. Geburtstag*, ed. by Friedrich Stadler and others (Wien: Springer Nature, 2008), 107–45; Edwin Glassner, 'Was heißt Koinzidenz bei Schlick?', in *Stationen. Dem Philosophen Und Physiker Moritz Schlick Zum 125. Geburtstag*, ed. by Friedrich Stadler and others (Wien: Springer Nature, 2008), 146–66; Thomas Oberdan, 'Geometry, Convention, and the Relativized Apriori: The Schlick–Reichenbach Correspondence', in *Stationen. Dem Philosophen Und Physiker Moritz Schlick Zum 125. Geburtstag*, ed. by Friedrich Stadler and others (Wien: Springer Nature, 2008), 186–211.

21 Moritz Schlick, 'Kritizistische oder empiristische Deutung der neuen Physik? Bemerkungen zu Ernst Cassirers Buch "Zur Einsteinschen Relativitätstheorie"' [1921], in *Rostock, Kiel, Wien: Aufsatze, Beitrage, Rezensionen 1919–1925: Aufsätze, Beiträge, Rezensionen 1919–1925*, ed. by Edwin Glassner, Heidi Konig-Porstner, and Karsten Boger (Wien; New York: Springer Nature, 2012), 223–47; 'Critical or Empiricist Interpretation of Modern Physics?: Remarks on Ernst Cassirer's "Einstein's Theory of Relativity"', in *Philosophical Papers. Volume 1: (1909–1922)*, ed. by Henk L. Mulder and Barbara F. B. Van De Velde-Schlick (Dordrecht; Boston: Kluwer, 1978), 322–34.

22 Cf. Moritz Schlick, *Allgemeine Erkenntnislehre* (Berlin: Julius Springer, 1918), 180 ff.; *General Theory of Knowledge*, ed. by Albert E. Blumberg and Herbert Feigl (Wien: Springer, 1974), 209 ff.

23 Schlick, *Allgemeine Erkenntnislehre*, 180; *General Theory of Knowledge*, 210.

24 Cf. Schlick, *Allgemeine Erkenntnislehre*, 181; *General Theory of Knowledge*, 212.

25 Schlick, *Allgemeine Erkenntnislehre*, 182; *General Theory of Knowledge*, 213.

26 Schlick, *Allgemeine Erkenntnislehre*, 195; *General Theory of Knowledge*, 228.

27 Cf. Schlick, *Allgemeine Erkenntnislehre*, 187–8; *General Theory of Knowledge*, 223.

28 Schlick, *Allgemeine Erkenntnislehre*, 190; *General Theory of Knowledge*, 222. Translation slightly modified.

29 Schlick, 'Die Philosophische Bedeutung des Relativitätsprinzips', 43; 'The Philosophical Significance of the Principle of Relativity', 178.

30 Schlick, 'Die Philosophische Bedeutung des Relativitätsprinzips', 44; 'The Philosophical Significance of the Principle of Relativity', 178–9.

31 Schlick, 'Die Philosophische Bedeutung des Relativitätsprinzips', 44; 'The Philosophical Significance of the Principle of Relativity', 179.

32 Ibid.

33 In particular, Schlick criticizes Mach's interpretation of the Newton's bucket thought experiment. While Mach is correct that the experiment does not demonstrate the existence of absolute motion, he errs in his assertion that the emergence of centrifugal forces as a result of the rotation *of the bucket* and the emergence of centrifugal forces as a result of the rotation of *the fixed stars* are not only indistinguishable but are *the same case*. Indeed, although the two cases might coincide from a kinematic point of view, they do not necessarily coincide from a dynamic one. Moreover, it is quite possible to imagine experiments able to prove the truth of one or the other of the cases without having to immobilize the fixed stars (cf. Schlick, 'Die Philosophische Bedeutung Des Relativitätsprinzips', 47 ff.; 'The Philosophical Significance of the Principle of Relativity', 180 ff.).

34 Schlick, 'Die Philosophische Bedeutung Des Relativitätsprinzips', 52; 'The Philosophical Significance of the Principle of Relativity', 184.

35 Schlick, 'Kritizistische oder empiristische Deutung der neuen Physik?', 224; 'Critical or Empiricist Interpretation of Modern Physics?', 322.

36 Schlick, 'Kritizistische oder empiristische Deutung der neuen Physik?', 225; 'Critical or Empiricist Interpretation of Modern Physics?', 323. My italics.

37 Ibid. My italics.

38 Cf. Schlick, 'Kritizistische oder empiristische Deutung der neuen Physik?', 226–7; 'Critical or Empiricist Interpretation of Modern Physics?', 324.

39 Moritz Schlick, 'Die Relativitätstheorie in der Philosophie', in *Rostock, Kiel, Wien: Aufsatze, Beitrage, Rezensionen 1919–1925: Aufsätze, Beiträge, Rezensionen 1919–1925*, ed. by Edwin Glassner, Heidi Konig-Porstner, and Karsten Boger (Wien;

New York: Springer Nature, 2012), 540; 'The Theory of Relativity in Philosophy', in *Philosophical Papers. Volume 1: (1909–1922)*, ed. by Henk L. Mulder and Barbara F. B. Van De Velde-Schlick (Dordrecht, Boston: Kluwer, 1978), 349.

40 Schlick, 'Die Relativitätstheorie in der Philosophie', 538; 'The Theory of Relativity in Philosophy', 348. My italics.
41 Ibid. My italics.
42 Schlick, 'Die Relativitätstheorie in der Philosophie', 539; 'The Theory of Relativity in Philosophy', 349.
43 Schlick, 'Die Relativitätstheorie in der Philosophie', 540–1; 'The Theory of Relativity in Philosophy', 349.
44 Schlick, 'Die Relativitätstheorie in der Philosophie', 541; 'The Theory of Relativity in Philosophy', 349.
45 Schlick, 'Die Relativitätstheorie in der Philosophie', 537; 'The Theory of Relativity in Philosophy', 348.
46 It should, however, be noted that Petzoldt expresses himself quite explicitly on this point in the passage in the *Einführung in die Philosophie der reinen Erfahrung*, where he reaffirms the existence of objects in the world, attacking the 'idealism of sensations' (cf. Einführung1904, pp. 307–308).
47 Weltproblem1921, 189, footnote.
48 Ibid.
49 Ibid. My italics.
50 Ibid.
51 Schlick, *Allgemeine Erkenntnislehre*, 206; *General Theory of Knowledge*, 241, cited in Weltproblem1921, 190, footnote.
52 Weltproblem1921, 190, footnote.
53 Ibid.
54 Stellung1923, 62–3.
55 Stellung1923, 66.
56 For an outline of Reichenbach's academic career, see Tilitzki, *Die deutsche Universitätsphilosophie*, II, 233 ff.; Harmut Hecht and Dieter Hoffmann, 'Die Berufung Hans Reichenbachs an die Berliner Universität. Zur Einheit von Naturwissenschaft, Philosophie und Politik', *Deutsche Zeitschrift für Philosophie*, 30/5 (1982), 651–62. For a brief profile of Reichenbach, see Wesley C. Salmon, 'The Philosophy of Hans Reichenbach', *Synthese*, 34/1 (1977), 5–88. On his interpretation of the theory of relativity, see Marco Giovanelli, '"… But I Still CanT Get Rid of a Sense of Artificiality": The Reichenbach–Einstein Debate on the Geometrization of the Electromagnetic Field', *Studies in History and Philosophy of Science Part B: Studies in History and Philosophy of Modern Physics*, 54 (2016), 35–51; Max Jammer, 'Hans Reichenbach und der Begriff der Gleichzeitigkeit', in *Hans Reichenbach, Philosophie im Umkreis der Physik*, ed. by Hans Poser and Ulrich Dirks (Berlin: Akademie Verlag, 2015), 11–24; Flavia Padovani, 'Relativizing the Relativized a Priori: Reichenbach's Axioms of Coordination Divided', *Synthese*, 181/1 (2011), 41–62. On Reichenbach in relation to Petzoldt, see also Hentschel, *Die Korrespondenz Petzoldt-Reichenbach*. On Reichenbach's interpretation of the theory of relativity see Hentschel, *Interpretationen und Fehlinterpretationen*, 178–95, 367ff. and Hentschel, 'Zur Rolle Hans Reichenbachs in den Debatten um die Relativitätstheorie (mit der vollständigen Korrespondenz Reichenbach – Friedrich Adler im Anhang)', in Lutz Danneberg, Andreas Kamlah and Lothar Schäfer (eds.), *Hans Reichenbach und die Berliner Gruppe* (Braunschweig: Vieweg, 1994), 295–324.

57 Hans Reichenbach, 'Bericht über eine Axiomatik der Einsteinschen Raum-Zeit Lehre', *Physikalische Zeitschrift*, 22 (1921), 683–7; Hans Reichenbach, *Axiomatik der relativistischen Raum-Zeit Lehre* (Braunschweig: Vieweg, 1924).
58 Cf. Hans Reichenbach, *Relativitätstheorie und Erkenntnis Apriori* (Berlin: Springer, 1920), 4 ff.; *The Theory of Relativity and a Priori Knowledge* (Berkeley; Los Angeles: University of California Press, 1965), 2 ff. It should be noted that this work was published at almost the same time as Cassirer's work on relativity, so Reichenbach was above all addressing the positions of other neo-Kantians, not having yet had the time to read Cassirer.
59 Cf. Reichenbach, *Relativitätstheorie und Erkenntnis Apriori*, 46 ff.; *The Theory of Relativity and a Priori Knowledge*, 48 ff.
60 Letter from Reichenbach to Schlick dated 29 November 1920, in the Hans Reichenbach Collection, University of Pittsburgh, HR 015–63–21.
61 Hans Reichenbach, 'Der gegenwärtige Stand der Relativitätsdiskussion', *Logos*, 10 (1922), 328 ff.; 'The Present State of the Discussion on Relativity', in *Selected Writings 1909–1953*, ed. by Maria Reichenbach and Robert S. Cohen, 2 vols (Dordrecht, Boston: Springer, 1978), II, 13 ff.
62 Reichenbach, 'Der gegenwärtige Stand der Relativitätsdiskussion', 330; 'The Present State of the Discussion on Relativity', II, 14.
63 Cf. Reichenbach, 'Der gegenwärtige Stand der Relativitätsdiskussion', 332; 'The Present State of the Discussion on Relativity', II, 16.
64 Reichenbach, 'Der gegenwärtige Stand der Relativitätsdiskussion', 333; 'The Present State of the Discussion on Relativity', II, 16.
65 Ibid.
66 Reichenbach, 'Der gegenwärtige Stand der Relativitätsdiskussion', 336; 'The Present State of the Discussion on Relativity', II, 18. My italics.
67 Reichenbach, 'Der gegenwärtige Stand der Relativitätsdiskussion', 334; 'The Present State of the Discussion on Relativity', II, 17. My italics.
68 Reichenbach, 'Der gegenwärtige Stand der Relativitätsdiskussion', 335; 'The Present State of the Discussion on Relativity', II, 18.
69 Reichenbach, 'Der gegenwärtige Stand der Relativitätsdiskussion', 333, 337; 'The Present State of the Discussion on Relativity', II, 17, 19.
70 Letter from Petzoldt to Reichenbach dated 12 May 1922, in Hentschel, *Die Korrespondenz Petzoldt-Reichenbach*, 40–1.
71 Letter from Reichenbach to Petzoldt dated 16 May 1922, in Hentschel, *Die Korrespondenz Petzoldt-Reichenbach*, 42.
72 Letter from Einstein to Reichenbach dated 27 March 1922, in *The Collected Papers of Albert Einstein. Volume 13: The Berlin Years: Writings & Correspondence January 1922-March 1923 (English translation supplement)* (Princeton: Princeton University Press, 2012), 122–3.
73 Letter from Reichenbach to Petzoldt dated 16 May 1922, Hentschel, *Die Korrespondenz Petzoldt-Reichenbach*, 40–4.
74 Ibid.
75 Ibid.
76 Letter from Petzoldt to Reichenbach dated 24 May 1922, in Hentschel, *Die Korrespondenz Petzoldt-Reichenbach*, 40–7.
77 Letter from Petzoldt to Reichenbach dated 24 May 1922, Hentschel, *Die Korrespondenz Petzoldt-Reichenbach*, 40–7.

78 Hans Reichenbach, 'La signification philosophique de la théorie de la relativité', *Revue Philosophique de la France et de l'Etranger*, 94 (1922), 58; 'The Philosophical Significance of the Theory of Relativity', in *Defending Einstein: Hans Reichenbach's Writings on Space, Time and Motion*, ed. by Steven Gimbel and Anke Walz (New York: Cambridge University Press, 2006), 156.
79 Letter from Petzoldt to Reichenbach dated 9 April 1923, in Hentschel, *Die Korrespondenz Petzoldt-Reichenbach*, 56.
80 Cf. Hentschel, *Die Korrespondenz Petzoldt-Reichenbach*, 56-7.
81 Letter from Petzoldt to Reichenbach dated 17 April 1923, in Hentschel, *Die Korrespondenz Petzoldt-Reichenbach*, 64.
82 Letter from Reichenbach to Petzoldt dated 11 April 1923, in Hentschel, *Die Korrespondenz Petzoldt-Reichenbach*, 60.
83 Ibid.
84 Joseph Petzoldt, 'Die physikalische Wirklichkeit', *Die Naturwissenschaften*, 11 (1923), 828.
85 Joseph Petzoldt, 'Die Elemente der Welt', *Monistische Monatshefte*, 8 (1923), 102.

Chapter 6

1 Gilles Deleuze and Félix Guattari, *What Is Philosophy?* (New York: Columbia University Press, 1996), 130.
2 Cf. James Conant, 'The Dialectic of Perspectivism', *Nordic Journal of Philosophy*, 6-7 (2005), I, 6-7.
3 Cf. Conant, 'The Dialectic of Perspectivism', I, 8-34.
4 Conant, 'The Dialectic of Perspectivism', I, 15.
5 Ibid.
6 Conant, 'The Dialectic of Perspectivism', II, 47. On Nietzsche's perspectivism see also Brian Leiter, 'Knowledge and affect. Perspectivism reconsidered', in *The Emergence of Relativism*, ed. by Martin Kusch and others (New York: Routledge, 2019), 133-50.
7 Weltproblem1906, 63-4. My italics.
8 Weltproblem1906, 99.
9 Joseph Petzoldt, *Das Weltproblem vom Standpunkte des relativistischen Positivismus aus, historisch-kritisch Dargestellt*, 4th edn (Leipzig: Teubner, 1924), 131.
10 See M. Heidelberger, *Nature from Within. Gustav Theodor Fechner and His Psychophysical Worldview*, University of Pittsburgh Press, Pittsburgh (PA), 2004, 126-7.
11 See in particular Stellung1923.
12 For an account that covers until the end of the nineteenth century see Volker Peckhaus, *Logik, Mathesis universalis und allgemeine Wissenschaft: Leibniz und die Wiederentdeckung der formalen Logik im 19. Jahrhundert* (Berlin: Akademie Verlag, 1997) or, for a broader account, Ralf Kromer and Yannick Chin-drian, *New Essays on Leibniz Reception: In Science and Philosophy of Science 1800–2000* (Basel; New York; London: Birkhäuser, 2012).
13 Stellung1923, 30. The links between Mach and Nietzsche on the topic of knowledge has become an established line of research, see Pietro Gori, *Nietzsche's Pragmatism: Essays on Perspectival Thought*, 1° edizione (Berlin; Boston: De Gruyter, 2019), and

Nadeem J. Z. Hussain, 'Reading Nietzsche Through Ernst Mach', in *Nietzche and Science*, ed. by Gregory Moore and Thomas H. Brobjer (Aldershot: Ashgate, 2004), 11–129.

14 See Lydia Patton, 'Perspectivalism in the Development of Scientific Observer-relativity', in *The Emergence of Relativism*, 63–78, 74.
15 See Michela Massimi and Casey D. McCoy, 'Introduction', in *Understanding Perspectivism: Scientific Challenges and Methodological Prospects* (Abingdon: Routledge, 2019).
16 Michela Massimi, 'Perspectivism', in *The Routledge Handbook of Scientific Realism*, ed. by Juha Saatsi (Abingdon: Routledge, 2017).
17 Ronald N. Giere, *Scientific Perspectivism* (Chicago: University of Chicago Press, 2010).
18 Luca Guzzardi, 'Holding the Hand of History. Mach on the History of Science, the Analysis of Sensations, and the Economy of Thought', in *Interpreting Mach: Critical Essays*, ed. by John Preston (Cambridge: Cambridge University Press, 2021), 164–83, 170.
19 Of the writers that have recently devoted themselves to countering the perception of Mach as a phenomenalist, it is worth noting two works by Erik C. Banks, *Ernst Mach's World Elements: A Study in Natural Philosophy* (Dordrecht: Springer, 2013); Banks, *The Realistic Empiricism of Mach, James, and Russell*.

Bibliography

Works by Joseph Petzoldt

'Zu Richard Avenarius' Prinzip des kleinsten Kraftmasses und zum Begriff der Philosophie', *Vierteljahrsschrift für wissenschaftliche Philosophie*, 11 (1887), 177–203.

'Anzeige von R. Avenarius "Kritik der reinen Erfahrung"', *Das Magazin für die Litteratur des In- und Auslandes*, 58 (1889), 8, 120–36.

'Vitaldifferenz und Erhaltungswert', *Vierteljahrsschrift für wissenschaftliche Philosophie*, 14 (1890), 85–7.

'Maxima, Minima und Ökonomie', *Vierteljahrsschrift für wissenschaftliche Philosophie*, 14 (1890), 206–39, 354–66, 417–42.

'Einiges zur Grundlegung der Sittenlehre', *Vierteljahrsschrift für wissenschaftliche Philosophie*, I: 17 (1893), 145–77; II: 18 (1894), 32–76; III: 18 (1894), 196–248.

'Über den Begriff der Entwickelung und einige Anwendungen desselben', *Naturwissenschaftliche Wochenschrift*, 9 (1894), 7, 77–81, 8, 89–93.

'Das Gesetz der Eindeutigkeit', *Vierteljahrsschrift für wissenschaftliche Philosophie*, 19 (1895), 146–203.

Einführung in die Philosophie der reinen Erfahrung, 2 vols (Leipzig: Teubner, 1900–1904).

'Solipsismus auf praktischem Gebiet', *Vierteljahrsschrift für wissenschaftliche Philosophie*, XXV (1901), 339–62.

'Die Notwendigkeit und Allgemeinheit des psychophysischen Parallelismus', *Archiv für systematische Philosophie*, 8 (1902), 281–37.

'Metaphysikfreie Naturwissenschaft', *Naturwissenschaftliche Wochenschrift*, 17 (1902), 31, 361–4.

'Sonderschulen für hervorragend Befähigte', *Neue Jahrbücher für Pädagogik*, 7 (1904), 425–56.

'Vorbemerkung zur zweiten Auflage', in Richard Avenarius, *Der menschliche Weltbegriff*. Zweite nach dem Tode des Verfassers herausgegebene Auflage (Leipzig: Reisland, 1905), XIII–XIV.

'Vorwort zur zweiten Auflage des ersten Bandes', in Richard Avenarius, *Kritik der reinen Erfahrung*. Zweite namentlich nach hinterlassenen Aufzeichnungen des Verfassers verbesserte Auflage, 2 vols (Leipzig: Reisland, 1905), I, VII–XII.

Das Weltproblem vom positivistischen Standpunkt aus (Leipzig: Teubner, 1906).

'Die mechanische Naturansicht und das Weltproblem. Vorlesungen bei den Hochschulkursen in Salzburg. 1906', *Das Wissen für alle*, 1907.

'Vorwort des Herausgebers', in Richard Avenarius, *Kritik der reinen Erfahrung*. Zweite namentlich nach hinterlassenen Aufzeichnungen des Verfassers verbesserte Auflage, 2 vols (Leipzig: Reisland, 1908), II, V–IX.

'Die Gebiete der absoluten und der relativen Bewegung', *Annalen der Naturphilosophie*, 7 (1908), 29–62.

'Die vitalistische Reaktion auf die Unzugänglichkeit der mechanischen Naturansicht', *Zeitschrift für allgemeine Physiologie*, 10 (1910), 69–119.

Das Weltproblem vom Standpunkte des relativistischen Positivismus aus. Historisch-kritisch dargestellt. Zweite vermehrte Auflage (Leipzig, Berlin: Teubner, 1911).
'Die Einwände gegen Sonderschulen für hervorragend Befähigte', *Neue Jahrbücher für Pädagogik*, 14 (1911), 1–24.
'Zwei Stunden für körperliche Ausbildung täglich ohne Rückgang der Geistigen', *Neue Jahrbücher für Pädagogik*, 14 (1911), 552–65.
'Die Relativitätstheorie im erkenntnistheoretischen Zusammenhange des relativistischen Positivismus', *Berichte der Deutschen physikalischen Gesellschaft*, 23 (1912), 1055–64.
'Naturwissenschaft', in Eugen Korschelt (ed.), *Handwörterbuch der Naturwissenschaften*, 10 vols (Jena: Fischer, 1912), VII, 50–94.
'Sonderschulen für Begabte', in Ludwig Fulda (ed.), *Die Schule der Zukunft. 8 Vorträge gehalten auf der Versammlung des Goethebundes in Berlin am 3. Dezember 1911* (Berlin: Fortschritt, 1912).
'Zur Positivistische Philosophie', *Zeitschrift für positivistische Philosophie*, 1 (1913), 1–16.
'Die Relativitätstheorie der Physik', *Zeitschrift für positivistische Philosophie*, 2 (1914), 1–56.
'Die biologischen Grundlagen der Psychologie', *Zeitschrift für positivistische Philosophie*, 2 (1914), 161–90.
'Die sittliche Rechtfertigung unseres Sieges', *Illustrierte Zeitung*, 3762 (06/06/1915).
'Ertüchtigung der Jugend', *Jugendfürsorge*, 16 (1915), 350–2.
'Die Existenz der Atome', *Chemiker-Zeitung*, 120 (1916), 846–7.
'Ernst Mach', *Deutscher Wille des Kunstwarts*, 29 (1916), 232–3.
'Philosophie auf naturwissenschaftlicher Grundlage', *Literarische Ratgeber und Jahresbericht des Dürerbundes*, 1916, 123–8.
'Verbietet die Relativitätstheorie Raum und Zeit als etwas wirkliches zu denken?', *Verhandlungen der Deutsche Physikalische Gesellschaft*, 20 (1918), 189–201.
'Hochschule für die Begabtenschulen', *Die neue Erziehung*, 1 (1919), 161–8.
'Die Unmöglichkeit mechanischer Modelle zur Veranschaulichung der Relativitätstheorie', *Berichte der Deutsche Physikalische Gesellschaft*, 13–14 (1919), 495–503.
'Neue Grundlegung der philosophischen Propädeutik', *Monatsschrift für höhere Schulen*, 19 (1920), 142–64.
'Biologischen Grundlage des Strafrechts', in Fritz Dehnow (ed.), *Die Zukunft des Strafrechts* (Berlin, Leipzig: Reisland, 1920), 5–23.
'Kausalität und Relativitätstheorie', *Zeitschrift für Physik*, 1 (1920), 467–74.
Das Weltproblem vom Standpunkte des relativistischen Positivismus aus. Dritte, neubearbeitete Auflage unter besonderer Berücksichtigung der Relativitätstheorie (Leipzig, Berlin: Teubner, 1921).
Die Stellung der Relativitätstheorie in der geistigen Entwicklung der Menschheit (Leipzig: Barth, 1921).
'Das Verhältnis der Machschen Gedankenwelt zur Relativitätstheorie', in Ernst Mach, *Die Mechanik in ihrer Entwicklung historisch-kritisch dargestellt*, 8. Auflage (Leipzig: Brockhaus, 1921), 490–517.
'Mechanistische Naturauffassung und Relativitätstheorie', *Annalen der Philosophie*, 2 (1921), 447–62.
'Mach und die Atomistik', *Die Wissenschaft*, 10 (1922), 230–1.
'Zur Krisis des Kausalitätsbegriffes', *Die Wissenschaft*, 32 (1922), 693–5.
'Der Fortschritt der Moral', in Fritz Dehnow (ed.), *Ethik der Zukunft* (Berlin, Leipzig: Reisland, 1922).
'An der Pforte der Relativitätstheorie', *Velhagen und Klasings Monatshefte*, 36 (1922-1923), 468–72.

Die Stellung der Relativitätstheorie in der geistigen Entwicklung der Menschheit, Zweite verbesserte und vermehrte Auflage (Leipzig: Barth, 1923).
Das allgemeinste Entwicklungsgesetz (München: Rösl, 1923).
'Hebung des Rechtsbewusstseins', *Die Tat*, 15 (1923-1924), 115-19.
'Die Elemente der Welt', *Monistische Monatshefte*, 8 (1923), 97-102, 129-33, 289-95, 363-71.
'Die physikalische Wirklichkeit', *Die Naturwissenschaften*, 2 (1923), 828.
Das Weltproblem vom Standpunkte des relativistischen Positivismus aus. Vierte Auflage, wie die dritte unter besonderer Berücksichtigung der Relativitätstheorie (Leipzig, Berlin: Teubner, 1924).
'Entwicklung selbst der Kirche', *Monistische Monatshefte*, 9 (1924), 69-71.
'Dämmerung der mechanische Naturansicht', *Monatsschrift für höhere Schulen*, 23 (1924), 83-7.
'Postulat der absoluten und relativen Welt', *Zeitschrift für Physik*, 21 (1924), 143-50.
'Beseitigung der mengentheoretischen Paradoxa durch logisch einwandfreie Definition des Mengenbegriffs', *Kant-Studien*, 30 (1925), 346-56.
'Naturwissenschaftliche Denkpsychologie und Gestalttheorie', *Die Naturwissenschaften*, 13 (1925), 801-2.
'Philosophie der Technik', *Monatsblätter des Berliner Bezirksverein Deutscher Ingenieure*, 6-7 (1926).
'Falsche Begabtenauslese', *Deutsche Philologenblatt*, 34 (1926), 587-8.
'Komplex und Begriff', *Zeitschrift für Psychologie*, 99 (1926), 74-103; 102 (1927), 265-306; 108 (1928), 336-70; 113 (1929), 287-344.
Das natürliche Höhenziel der menschheitlichen Entwicklung (Berlin, Leipzig: Paetel, 1927)
'Rationales und empirisches Denken', *Annalen der Philosophie und philosophischen Kritik*, 6 (1927) 145-60.
'Vorfragen zur Frage der Telepathie', *Annalen der Philosophie und philosophischen Kritik*, 7 (1928), 200-4.
'Kausalität und Wahrscheinlichkeit', *Naturwissenschaften*, 17 (1929), 51-2.

Works cited

Amrhein, Hans, *Kants Lehre vom 'Bewusstsein überhaupt' und ihre Weiterbildung bis auf die Gegenwart* (Berlin: Reuther und Reichard, 1909).
Araujo, Saulo De Freitas, *Wundt and the Philosophical Foundations of Psychology: A Reappraisal* (Cham: Springer, 2016).
'Aufruf', *Archiv Für Geschichte Der Philosophie*, 25 (1912), 502.
Avenarius, Ludwig, *Avenarianische Chronik: Blätter aus drei Jahrhunderten einer deutschen Bürgerfamilie* (Leipzig: Reisland, 1912).
Avenarius, Richard, 'Anmerkung zu der vorstehenden Abhandlung' [R. Willy, 'Das Erkenntnisstheoretische Ich und der natürliche Weltbegriff'], *Vierteljahrsschrift für wissenschaftliche Philosophie*, 18 (1894), 29-31.
Avenarius, Richard, 'Bemerkungen zum Begriff des Gegenstandes der Psychologie', *Vierteljahrsschrift für wissenschaftliche Philosophie*, 18-19 (1894-1895), 137-61, 400-20, 1-18, 129-45.
Avenarius, Richard, *Der menschliche Weltbegriff* [1891], 2nd edn (Leipzig: Reisland, 1905).
Avenarius, Richard, *Kritik der reinen Erfahrung*, 2 vols (Leipzig: Fues, 1888-1890).

Avenarius, Richard, *Philosophie als Denken der Welt gemäss dem Princip des kleinsten Kraftmasses: Prolegomena zu einer Kritik der reinen Erfahrung* (Leipzig: Fues, 1876).
Avenarius, Richard, 'Über die Stellung der Psychologie zur Philosophie', *Vierteljahrsschrift für wissenschaftliche Philosophie*, 1 (1877), 471–88.
Avenarius, Richard, and Ernst Mach, 'Briefe von Richard Avenarius und Ernst Mach an Wilhelm Schuppe', *Erkenntnis*, 6 (1936), 73–80.
Banks, Erik C., *Ernst Mach's World Elements: A Study in Natural Philosophy* (Dordrecht: Springer, 2013).
Banks, Erik C., *The Realistic Empiricism of Mach, James, and Russell. Neutral Monism Reconceived* (Cambridge: Cambridge University Press, 2014).
Barbour, Julian B., 'Einstein and Mach's Principle', in *The Genesis of General Relativity*, ed. by Michel Janssen, John D. Norton, Jürgen Renn, Tilman Sauer, and John Stachel, Boston Studies in the Philosophy of Science (Dordrecht: Springer Netherlands, 2007), 1492–1527. https://doi.org/10.1007/978-1-4020-4000-9_32 [accessed 29 October 2021].
Barbour, Julian B., *The Discovery of Dynamics: A study from a Machian point of view of the discovery and the structure of dynamical theories.* (Oxford: Oxford University Press, 2001).
Barbour, Julian B., and Herbert Pfister, eds., *Mach's Principle: From Newton's Bucket to Quantum Gravity* (Boston: Birkhäuser, 1995).
Bartels, Andreas, 'Die Auflösung der Dinge. Schlick und Cassirer über wissenschaftliche Erkenntnis und Relativitätstheorie', in *Philosophie und Wissenschaften. Formen und Prozesse ihrer Interaktion*, ed. by Hans Jörg Sandkühler (Frankfurt am Main: Peter Lang, 1997), 193–210.
Bayertz, Kurt, Myriam Gerhard, and Walter Jaeschke, *Der Darwinismus-Streit: Texte von L. Büchner, B. von Carneri, F. Fabri. G. von Gyzicki, E. Haeckel, E. von Hartmann, F. A. Lange, R. Stoeckl und K. Zittel* (Hamburg: Meiner, 2012).
Bayertz, Kurt, Myriam Gerhard, and Walter Jaeschke, eds., *Der Materialismus-Streit, Weltanschauung, Philosophie und Naturwissenschaft im 19. Jahrhundert* (Hamburg: Meiner, 2007), I.
Bayertz, Kurt, Myriam Gerhard, and Walter Jaeschke, eds., *Der Materialismus-Streit: Texte von L. Büchner, H. Czolbe, L. Feuerbach, I. H. Fichte, J. Frauenstädt, J. Froschammer, J. Henle, J. Moleschott, M. J. Schleiden, C. Vogt und R. Wagner* (Hamburg: Meiner, 2012).
Beiser, Frederick, 'Herbart's Monadology', *British Journal for the History of Philosophy*, 23/6 (2015), 1056–73.
Beiser, Frederick, *The Genesis of Neo-Kantianism, 1796–1880* (Oxford: Oxford University Press, 2014).
Beiser, Frederick, 'Two Traditions of Idealism', in *From Hegel to Windelband: Historiography of Philosophy in the 19th Century*, ed. by Gerald Hartung and Valentin Pluder (Berlin: De Gruyter, 2015), 81–98.
Berg, Otto, 'Das Relativitätsprinzip Der Elektrodynamik', *Abhandlungen Der Frieschen Schule*, 3 (1910), 333–82.
Blackmore, John, ed., *Ernst Mach – A Deeper Look: Documents and New Perspectives* (Dordrecht; Boston: Kluwer, 1992).
Blackmore, John, and Klaus Hentschel, *Ernst Mach als Aussenseiter: Machs Briefwechsel über Philosophie und Relativitätstheorie mit Persönlichkeiten seiner Zeit* (Wien: Braumüller, 1985).
Blackmore, John, Ryōichi Itagaki, and S. Tanaka, eds., *Ernst Mach's Science: Its Character and Influence on Einstein and Others* (Minamiyana, Hadano-shi, Kanagawa: Tokai University Press, 2006).

Bowler, Peter J., *The Non-Darwinian Revolution: Reinterpreting a Historical Myth* (Baltimore: Johns Hopkins University Press, 1988).
Bradley, J., *Mach's Philosophy of Science* (London: Bloomsbury, 2014).
Büchner, Ludwig, *Sechs Vorlesungen über die Darwin'sche Theorie von der Verwandlung der Arten und die erste Entstehung der Organismenwelt*, 3rd edn (Leipzig: Thomas, 1872).
Calaprice, Alice, *The Einstein Almanac* (Baltimore: Johns Hopkins University Press, 2004).
Carstanjen, Friedrich, 'Der Empiriokritizismus: Zugleich eine Erwiderung auf W. Wundts Aufsätze "Über naiven und kritischen Realismus"', *Vierteljahrsschrift für wissenschaftliche Philosophie*, 22 (1898), 45–95, 190–214, 267–93.
Cassini, Alejandro, and Marcelo Leonardo Levinas, 'Einstein's Reinterpretation of the Fizeau Experiment: How It Turned Out to Be Crucial for Special Relativity', *Studies in History and Philosophy of Science Part B: Studies in History and Philosophy of Modern Physics*, 65 (2019), 55–72.
Cassirer, Ernst, *Determinism and Indeterminism in Modern Physics. Historical and Systematic Studies of the Problem of Causality*, trans. by Theodor Benfey (New Haven: Yale University Press, 1956).
Cassirer, Ernst, *Determinismus und Indeterminismus in der modernen Physik. Historische und systematische Studien zum Kausalproblem* (Göteborg: Wettergren & Kerber, 1936).
Cassirer, Ernst, 'Die Philosophischen Probleme Der Relativitätstheorie', in *Vorlesungen Und Vorträge Zu Philosophischen Problemen Der Wissenschaften. 1907–1945*, ed. by Jörg Fingerhut, Gerald Hartung, and Rüdiger Kramme, Nachgelassene Manuskripte Und Texte, 8 (Hamburg: Meiner, 2010), 29–116.
Cassirer, Ernst, 'Einstein's Theory of Relativity', in *Substance and Function and Einstein's Theory of Relativity*, trans. by William Curtis Swabey and Marie Taylor Swabey (Chicago, London: Open Court, 1923), 351–460.
Cassirer, Ernst, 'Erkenntnistheorie Nebst Den Grenzfragen Der Logik', in *Aufsätze Und Kleine Schriften. 1902–1921*, ed. by Birgit Recki, Gesammelte Werke, 9 (Hamburg: Meiner, 2001), 139–200.
Cassirer, Ernst, 'Substance and Function', in *Substance and Function and Einstein's Theory of Relativity*, trans. by William Curtis Swabey and Marie Taylor Swabey (Chicago, London: Open Court, 1923), 1–346.
Cassirer, Ernst, *Zur Einsteinschen Relativitätstheorie. Erkenntnistheoretische Betrachtungen* (Berlin: Bruno Cassirer, 1921).
Classen, Johannes, 'Über das Relativitätsprinzip in der modernen Physik', *Zeitschrift für Physikalischen und Chemischen Unterricht*, 23 (1910), 257–67.
Clausius, Rudolf, 'Zur Geschichte der Mechanischen Wärmetheorie', *Annalen der Physik und Chemie*, 145 (1872), 132–46.
Cohn, Emil, *Physikalisches über Raum und Zeit* (Leipzig: Teubner, 1913).
Comte, Auguste, *Discours Sur l'esprit Positif* (Paris: Vrin, 1844).
Conant, James, 'The Dialectic of Perspectivism', *Nordic Journal of Philosophy*, 6–7 (2005), 5–50, 6–57.
Danneberg, Lutz, Andreas Kamlah and Lothar Schäfer, eds., *Hans Reichenbach und die Berliner Gruppe* (Braunschweig: Vieweg, 1994).
Darwin, Charles, *On the Origin of Species*, ed. by G. Beer (Oxford: Oxford University Press, 2008).
Deleuze, Gilles, and Félix Guattari, *What Is Philosophy?* (New York: Columbia University Press, 1996).
DiSalle, Robert, 'Carl Gottfried Neumann', *Science in Context*, 6/1 (1993), 345–53.

DiSalle, Robert, 'Space and Time: Inertial Frames', in *The Stanford Encyclopedia of Philosophy*, ed. by Edward N. Zalta, Winter 2020 (Metaphysics Research Lab, Stanford University, 2020). https://plato.stanford.edu/archives/win2020/entries/spacetime-iframes/ [accessed 29 October 2021].

van Dongen, Jeroen, 'Reactionaries and Einstein's Fame: "German Scientists for the Preservation of Pure Science," Relativity, and the Bad Nauheim Meeting', *Physics in Perspective*, 9/2 (2007), 212–30.

Du Bois-Reymond, Emil, 'The Limits of Our Knowledge of Nature', *The Popular Science Monthly*, 5 (1874), 17–32.

Du Bois-Reymond, Emil, *Über die Grenzen des Naturerkennens* (Leipzig: Veit & Comp., 1872).

Du Bois-Reymond, Emil, *Untersuchungen über thierische Elektricität* (Berlin: Reimer, 1848).

Dubislav, Walter, 'Joseph Petzoldt in Memoriam', *Annalen Der Philosophie Und Philosophischen Kritik*, 8 (1929), 289–95.

Edgar, Scott, 'The Physiology of the Sense Organs and Early Neo-Kantian Conceptions of Objectivity: Helmholtz, Lange, Liebmann', in *Objectivity in Science: New Perspectives from Science and Technology Studies*, ed. by Flavia Padovani, Alan Richardson, and Jonathan Y. Tsou, Boston Studies in the Philosophy and History of Science (Cham: Springer International Publishing, 2015), 101–22.

Einstein, Albert, 'Die Grundlage der allgemeinen Relativitätstheorie', *Annalen der Physik*, 49 (769), 1916.

Einstein, Albert, 'Ernst Mach', in *The Collected Papers of Albert Einstein. Volume 6. The Berlin Years: Writings, 1914–1917 (English translation supplement)* (Princeton: Princeton University Press, 1997), 141–5.

Einstein, Albert, *Geometrie und Erfahrung. Erweiterte Fassung des Festvortrages gehalten an der Preußischen Akademie der Wissenschaften zu Berlin am 27. Januar 1921* (Berlin: Springer, 1921).

Einstein, Albert, 'Geometry and Experience', in *The Collected Papers of Albert Einstein. Volume 7: The Berlin Years: Writings, 1918–1921 (English translation supplement)* (Princeton: Princeton University Press, 2002), 208–22.

Einstein, Albert, 'On the Electrodynamics of Moving Bodies', in *The Collected Papers of Albert Einstein. Volume 2: The Swiss Years: Writings, 1900–1909 (English translation supplement)* (Princeton: Princeton University Press, 1989), 140–71.

Einstein, Albert, 'On the Foundations of the General Theory of Relativity', in *The Collected Papers of Albert Einstein. Volume 7. The Berlin Years: Writings, 1918–1921 (English translation supplement)*, trans. by Alfred Engel (Princeton: Princeton University Press, 2002), 33–5.

Einstein, Albert, 'On the Principle of Relativity', in *The Collected Papers of Albert Einstein. Volume 6. The Berlin Years: Writings, 1914–1917 (English translation supplement)* (Princeton: Princeton University Press, 1997), 3–5.

Einstein, Albert, 'Prinzipielles zur allgemeinen Relativitätstheorie', *Annalen der Physik*, 55 (1918), 241–4.

Einstein, Albert, *The Collected Papers of Albert Einstein. Volume 5. The Swiss Years: Correspondence, 1902–1914 (English translation supplement)* (Princeton: Princeton University Press, 1995).

Einstein, Albert, *The Collected Papers of Albert Einstein. Volume 8. The Berlin Years: Correspondence, 1914–1918 (English translation supplement)* (Princeton: Princeton University Press, 1998).

Einstein, Albert, *The Collected Papers of Albert Einstein. Volume 9. The Berlin Years: Correspondence, January 1919–April 1920 (English translation supplement)* (Princeton: Princeton University Press, 2004).

Einstein, Albert, *The Collected Papers of Albert Einstein. Volume 10. The Berlin Years: Correspondence, May–December 1920, and Supplementary Correspondence, 1909–1920 (English translation supplement)* (Princeton: Princeton University Press, 2006).

Einstein, Albert, *The Collected Papers of Albert Einstein. Volume 13: The Berlin Years: Writings & Correspondence January 1922–March 1923 (English translation supplement)* (Princeton: Princeton University Press, 2012).

Einstein, Albert, 'The Foundation of the General Theory of Relativity', in *The Collected Papers of Albert Einstein. Volume 6. The Berlin Years: Writings, 1914–1917 (English translation supplement)* (Princeton: Princeton University Press, 1997), 146–99.

Einstein, Albert, 'The Theory of Relativity. Discussion Remarks at a Meeting of the Société française de Philosophie', in *The Collected Papers of Albert Einstein. Volume 13: The Berlin Years: Writings & Correspondence January 1922-March 1923 (English translation supplement)*, ed. by Diana Kormos Buchwald, József Illy, Ze'ev Rosenkranz, and Tilman Sauer (Princeton: Princeton University Press, 2012).

Einstein, Albert, 'Vom Relativitäts-Prinzip', *Vossische Zeitung*, 209 (1914), 33–4.

Einstein, Albert, 'Zur Elektrodynamic bewegter Körper', *Annalen der Physik*, 17 (1905), 891–921.

Eisler, Rudolf, 'Energie', *Handwörterbuch Der Philosophie* (Berlin: Mittler, 1922).

Engler, Fynn Ole, 'Über das erkenntnistheoretische Raumproblem bei Moritz Schlick, Wilhelm Wundt und Albert Einstein', in *Stationen. Dem Philosophen Und Physiker Moritz Schlick Zum 125. Geburtstag*, ed. by Friedrich Stadler, Hans Jürgen Wendel, Edwin Glassner, Fynn Ole Engler, and Massimo Ferrari (Wien: Springer Nature, 2008), 107–45.

Euler, Leonhard, *Mechanica sive motus scientia analytice exposita* (Petropoli: Typographia Acadamiae Scientiarum, 1736).

Euler, Leonhard, 'Réflexions sur l'espace et le temps', *Histoire de l'Académie des Sciences et Belles Lettres de Prusse*, 4 (1748), 324–33.

Fechner, Gustav Theodor, *Die Tagesansicht gegenüber der Nachtansicht* (Leipzig: Breitkopf und Härtel, 1879).

Fechner, Gustav Theodor, *Einige Ideen zur Schöpfungs- und Entwickelungsgeschichte der Organismen* (Leipzig: Breitkopf und Härtel, 1873).

Fechner, Gustav Theodor, *Elemente der Psychophysik*, 2 vols (Leipzig: Breitkopf und Härtel, 1860), I.

Fechner, Gustav Theodor, 'Über das Lustprincip des Handelns', *Zeitschrift für Philosophie und philosophische Kritik*, 19 (1848), 1–30, 163–94.

Feigl, Herbert, 'Positivism and Logical Empiricism', *The New Encyclopaedia Britannica* (Chicago et al.: Encyclopaedia Britannica, 1978), 877–83.

Finger, Stanley, *Minds Behind the Brain: A History of the Pioneers and Their Discoveries* (Oxford: Oxford University Press, 1999).

Frank, Philipp, 'Das Relativitätsprinzip und die Darstellung der physikalischen Erscheinungen im vierdimensionalen Raum', *Zeitschrift für physikalische Chemie, Stöchiometrie und Verwandtschaftslehre*, 74 (1910), 466–95.

Fuhrmann, Arwed, *Aufgaben aus der analytischen Mechanik: ein Übungsbuch für Studierende der Mathematik, Physik, Technik* (Leipzig: Teubner, 1867).

Gauss, Carl Friedrich, 'Ueber ein allgemeines Grundgesetz der Mechanik', *Journal für die reine und angewandte Mathematik*, 4 (1829), 232–5.

'Gesellschäftliche Mitteilungen', *Zeitschrift Für Positivistische Philosophie*, 1 (1913), 1.
Giere, Ronald N., *Scientific Perspectivism* (Chicago: University Of Chicago Press, 2010).
Giovanelli, Marco, '"… But I Still Can't Get Rid of a Sense of Artificiality": The Reichenbach–Einstein Debate on the Geometrization of the Electromagnetic Field', *Studies in History and Philosophy of Science Part B: Studies in History and Philosophy of Modern Physics*, 54 (2016), 35–51.
Giovanelli, Marco, 'Nothing but Coincidences: The Point-Coincidence and Einstein's Struggle with the Meaning of Coordinates in Physics', *European Journal for Philosophy of Science*, 11/2 (2021), 45.
Glassner, Edwin, 'Was heißt Koinzidenz bei Schlick?', in *Stationen. Dem Philosophen Und Physiker Moritz Schlick Zum 125. Geburtstag*, ed. by Friedrich Stadler, Hans Jürgen Wendel, Edwin Glassner, Fynn Ole Engler, and Massimo Ferrari (Wien: Springer Nature, 2008), 146–66.
Glick, Thomas F., ed., *The Comparative Reception of Relativity*, Boston Studies in the Philosophy of Science, 103 (Dordrecht: Reidel, 1987).
Goenner, Hubert, 'The Reception of the Theory of Relativity in Germany as Reflected by Books Published between 1908 and 1945', in *Studies in the History of General Relativity*, ed. by Jean Eisenstaedt and A. J. Kox (Boston, Basel, Berlin: Birkhäuser, 1992), 15–38.
Goenner, Hubert, 'The Reaction to Relativity Theory I: The Anti-Einstein Campaign in Germany in 1920', *Science in Context*, 6/1 (1993), 107–33.
Gomperz, Theodor, *Griechische Denker: eine Geschichte der antiken Philosophie*, 3 vols (Leipzig: Veit, 1896).
Gori, Pietro, *Nietzsche´s Pragmatism: Essays on Perspectival Thought* (Berlin, Boston: De Gruyter, 2019).
Grünbaum, Heinrich, *Zur Kritik der modernen Causalanschauungen* (Würzburg: Stürtz, 1899).
Gruner, Paul, 'Elementare Darlegung Der Relativitätstheorie', *Mitteilungen Der Naturforschenden Gesellschaft in Bern*, 1910.
Guzzardi, Luca, 'Holding the Hand of History. Mach on the History of Science, the Analysis of Sensations, and the Economy of Thought', in *Interpreting Mach: Critical Essays*, ed. by John Preston (Cambridge: Cambridge University Press, 2021), 164–83.
Hecht, Harmut, and Dieter Hoffmann, 'Die Berufung Hans Reichenbachs an die Berliner Universität. Zur Einheit von Naturwissenschaft, Philosophie und Politik', *Deutsche Zeitschrift für Philosophie*, 30/5 (1982), 651–62.
Hecht, Harmut, and Dieter Hoffmann, 'The Berlin "Society for Scientific Philosophy" as Organizational Form of Philosophizing in the Medium of Natural Science', in *World Views and Scientific Discipline Formation*, ed. by William R. Woodward and Robert S. Cohen (Dordrecht et al.: Kluwer, 1991), 75–87.
Hegselmann, Reiner, and Geo Siegwart, 'Zur Geschichte der "Erkenntnis"', *Erkenntnis*, 35 (1991), 461–71.
Heidelberger, Michael, 'Functional Relations and Causality in Fechner and Mach', *Philosophical Psychology*, 23/2 (2010), 163–72.
Heidelberger, Michael, *Nature from Within: Gustav Theodor Fechner and His Psychophysical Worldview* (Pittsburgh: University of Pittsburgh Press, 2004).
Helmholtz, Hermann von, 'Ueber die Erhaltung der Kraft. Zusätze', in *Wissenschaftliche Abhandlungen*, by Hermann von Helmholtz (Leipzig: Barth), I, 68–75.
Hentschel, Klaus, *Die Korrespondenz Petzoldt-Reichenbach: Zur Entwicklung der „wissenschaftliche Philosophie" in Berlin* (Berlin: Sigma, 1990).

Hentschel, Klaus, *Interpretationen und Fehlinterpretationen der speziellen und der allgemeinen Relativitätstheorie durch Zeitgenossen Albert Einsteins*, Science Networks. Historical Studies (Basel: Birkhäuser, 1990).

Hentschel, Klaus, 'Philosophical Interpretations of Einstein's Theories of Relativity', *PSA (Philosophy of Science Association)*, 2 (1990), 169–79.

Hentschel, Klaus, 'Zur Rolle Hans Reichenbachs in den Debatten um die Relativitätstheorie (mit der vollständigen Korrespondenz Reichenbach – Friedrich Adler im Anhang)', in *Hans Reichenbach und die Berliner Gruppe*, ed. by Lutz Danneberg, Andreas Kamlah and Lothar Schäfer (Braunschweig: Vieweg, 1994), 295–324.

Hentschel, Klaus, 'Zur Rezeption von Vaihingers Philosophie des Als Ob in der Physik', in *Fiktion und Fiktionalismus. Beiträge zu Hans Vaihingers 'Philosophie des Als-Ob'* ed. by Matthias Neuber (Würzburg: Königshausen & Neumann, 2014), 161–86.

Hentschel, Klaus, '(Mis-)Interpretations of the Theory of Relativity – Considerations on how they arise and how to analyze them', in *Philosophers and Einstein's Relativity*, ed. by Chiara Russo Krauss and Luigi Laino (Cham: Springer, forthcoming).

Herneck, Friedrich, 'Ernst Mach und Albert Einstein', in *Symposium aus Anlass des 50. Todestages von Ernst Mach*, ed. by Wolfgang Merzkirch (Freiburg: Ernst-Mach-Institut, 1966), 45–61.

Herrmann, Christian, 'Joseph Petzoldt', *Kant-Studien*, 34 (1929), 508–10.

Herzberg, Lily, 'Joseph Petzoldt Tot', *Monistische Monatshefte*, 14 (1929), 223–4.

Hiebert, Erwin N., 'Mach's Philosophical Use of the History of Science', in *Historical and Philosophical Perspectives of Science*, ed. by Roger H. Stuewuer, Minnesota Studies in the Philosophy of Science, 5 (Minneapolis: University of Minnesota Press, 1970), 184–203.

Hiebert, Erwin N., 'The Influence of Mach's Thought on Science', *Philosophia Naturalis*, 21 (1984), 598–615.

Hoffmann, Dieter, 'The Society for Empirical/Scientific Philosophy', in *The Cambridge Companion to Logical Empiricism*, ed. by Alan Richardson and Thomas Uebel (Cambridge: Cambridge University Press, 2007), 41–57.

Holmes, S. J., 'The Principle of Stability as a Cause of Evolution. A Review of Some Theories', *The Quarterly Review of Biology*, 23 (1948), 324–32.

Holst, Helge, 'Die kausale Relativitätsforderung und Einsteins Relativitätstheorie', *Det Kgl. Danske Videnskabernes Selskab, Matematisk-fysiske Meddelelser*, 2 (1919).

Holst, Helge, 'Wirft die Relativitätstheorie den Ursachsbegriff über Bord?', *Zeitschrift für Physik*, 1 (1920), 32–9.

Holton, Gerald J., *Science and Anti-Science* (Cambridge: Harvard University Press, 1993).

Howard, Don, 'Einstein and Eindeutigkeit: A Neglected Theme in the Philosophical Background to General Relativity', in *Studies in the History of General Relativity*, ed. by Jean Eisenstaedt and A. J. Kox (Boston, Basel, Berlin: Birkhäuser, 1992), 154–243.

Howard, Don, 'Einstein, Kant and the Origins of Logical Positivism', in *Logic, Language, and the Structure of Scientific Theories: Proceedings of the Carnap-Reichenbach Centennial*, ed. by Wesley C. Salmon and Gereon Wolters (Pittsburgh: University of Pittsburgh Press, 1994), 45–105.

Howard, Don, 'Relativity, Eindeutigkeit, and Monomorphism: Rudolf Carnap and the Development of the Categoricity Concept in Formal Semantics', in *Origins of Logical Empiricism*, ed. by Ronald N. Giere and Alan W. Richardson (Minneapolis and London: University of Minnesota Press, 1996), 115–64.

Hume, David, *A Treatise of Human Nature* [*1739*], ed. by David Norton and Mary Norton (Oxford: Clarendon, 2007).

Hussain, Nadeem J. Z., 'Reading Nietzsche through Ernst Mach', in *Nietzche and Science*, ed. by Gregory Moore and Thomas H. Brobjer (Aldershot: Ashgate, 2004), 11–129.
Jakob, Max, 'Bemerkung zu dem Aufsatz von J. Petzoldt: 'Verbietet die Relativitätstheorie Raum und Zeit als etwas Wirkliches zu denken?', *Verhandlungen der physikalischen Gesellschaft zu Berlin*, 21 (1919), 159–61.
Jakob, Max, 'Bemerkungen zu dem Aufsatz von J. Petzoldt: 'Über die Unmöglichkeit mechanischer Modelle zur Veranschaulichung der Relativitätstheorie', *Verhandlungen der physikalischen Gesellschaft zu Berlin*, 21 (1919), 501–3.
Jammer, Max, *Concepts of Space. The History of Theories of Space in Physics*, 3rd enlarged edn (New York: Dover, 1993).
Jammer, Max, 'Hans Reichenbach und der Begriff der Gleichzeitigkeit', in *Hans Reichenbach, Philosophie im Umkreis der Physik*, ed. by Hans Poser and Ulrich Dirks (Berlin: Akademie Verlag, 2015), 11–24.
Jungnickel, Christa, and Russell McCormmach, *Intellectual Mastery of Nature. Theoretical Physics from Ohm to Einstein, Volume 2: The Now Mighty Theoretical Physics, 1870 to 1925* (Chicago: University of Chicago Press, 1990).
Kant, Immanuel, *Critique of Pure Reason* [1787], ed. by Allen Wood and Paul Guyer (Cambridge: Cambridge University Press, 1999).
Kant, Immanuel, *Handschriftlicher Nachlass: Logik*, Gesammelten Werken, 23 vols (Akademie Ausgabe, 1914), xvi.
Kant, Immanuel, *Metaphysical Foundations of Natural Science* [1786], trans. by Michael Friedman (Cambridge: Cambridge University Press, 2004).
Kelly, Alfred, *The Descent of Darwin: The Popularization of Darwinism in Germany, 1860–1914* (Chapel Hill: The University of North Carolina Press, 2012).
Kirchhoff, Gustav Robert, *Vorlesungen über mathematische Physik* (Leipzig: Teubner, 1874).
Kirchner, Friedrich, 'Energy', *Wörterbuch Der Philosophischen Grundbegriffe* (Leipzig: Dürr, 1907), 177–8.
Kline, Morris, *Mathematical Thought from Ancient to Modern Times* (Oxford: Oxford University Press, 1990), ii.
Köhnke, Klaus Christian, *The Rise of Neo-Kantianism: German Academic Philosophy between Idealism and Positivism* (Cambridge: Cambridge University Press, 1991).
Kromer, Ralf, and Yannick Chin-drian, *New Essays on Leibniz Reception: In Science and Philosophy of Science 1800–2000* (Basel, New York and London: Birkhäuser, 2012).
Kuntze, Johannes E., *Gustav Theodor Fechner (Dr. Mises). Ein deutsches Gelehrtenleben* (Leipzig: Breitkopf und Härtel, 1892).
Laas, Ernst, *Idealismus und positivismus. Eine kritische Auseinandersetzung*, 3 vols (Berlin: Weidmann, 1879).
Laas, Ernst, *Kants Analogien der Erfahrung. Eine kritische Studie über die Gründlagen der theoretisehen Philosophie* (Berlin: Weidmann, 1876).
Laino, Luigi, 'The Conditions of Possibility of Scientific Experience: Cassirer's Interpretation of the Theory of Relativity', in *The Changing Faces of Space*, ed. by Maria Teresa Catena and Felice Masi, Studies in Applied Philosophy, Epistemology and Rational Ethics, 39 (Cham: Springer, 2017), 235–54.
Lampa, Anton, 'Über die Tendenz zur Stabilität', in *Festschrift für Wilhelm Jerusalem zu seinem 60. Geburtstag* (Vienna: Braumüller, 1915), 147–53.
Lange, Friedrich Albert, *History of Materialism and Criticism of Its Present Importance* [1873–5], trans. by Ernest Chester Thomas, 3 vols (London: Trübner & Company, 1879).

Lange, Ludwig, *Die geschichtliche Entwickelung des Bewegungsbegriffes und ihr voraussichtliches Endergebniss. Ein Beitrag zur historischen Kritik der mechanischen Principien* (Leipzig: Engelmann, 1886).
Lange, Ludwig, 'Nochmals über das Beharrungsgesetz', *Philosophische Studien*, 2 (1885), 539–45.
Lange, Ludwig, 'On the Law of Inertia', trans. by Herbert Pfister, *European Physical Journal H*, 39 (2014), 251–62.
Lange, Ludwig, 'Über das Beharrungsgesetz', *Berichte über Verhandlungen der Königliche Sächsischen Gesellschaft der Wissenschaften. Mathematische-physikalische Klasse*, 1885, 333–51.
Lange, Ludwig, 'Über die wissenschaftliche Fassung des Galilei'schen Beharrungsgesetzes', *Philosophische Studien*, 2 (1885), 266–97.
Langevin, Paul, 'L'Évolution de l'espace et du temps', *Scientia*, 10 (1911), 31–54.
Laplace, Pierre Simon, *A Philosophical Essay on Probabilities*, trans. by Frederick Wilson Truscott and Frederick Lincoln Emory (New York: Dover, 1951).
Laue, Max von, 'Dr. Ludwig Lange (1863–1936). Ein zu Unrecht Vergessener', *Die Naturwissenschaften*, 35/7 (1948), 193–6.
Leiter, Brian, 'Knowledge and affect. Perspectivism reconsidered', in *The Emergence of Relativism*, ed. by Martin Kusch, Katherina Kinzel, Johannes Steizinger, and Niels Jacob Wildschut (New York: Routledge, 2019), 133–50.
Lenin, Vladimir Il'ich, *Materialism and Empirio-Criticism. Critical Notes Concerning a Reactionary Philosophy*, Collected Works of V. I. Lenin (Moscow: Progress Publisher, 1970), XIII.
Lenoir, Timothy, *The Strategy of Life: Teleology and Mechanics in Nineteenth Century German Biology* (Dordrecht: Reidel, 2011).
Liebmann, Otto, *Gedanken und Tatsachen. Philosophische Abhandlungen, Aphorismen und Studien* (Strasburg: Trübner, 1882).
Liebmann, Otto, *Kant und die Epigonen. Eine kritische Abhandlung* (Stuttgart: Schober, 1865).
Liebmann, Otto, *Zur Analysis der Wirklichkeit: philosophische Untersuchungen* (Strasburg: Trübner, 1876).
Longuenesse, Beatrice, *Kant and the Capacity to Judge: Sensibility and Discursivity in the Transcendental Analytic of the 'Critique of Pure Reason'* (Princeton: Princeton University Press, 1998).
Lovrenov, Maja, 'The Role of Invariance in Cassirer's Interpretation of the Theory of Relativity', *Synthesis Philosophica*, 21 (2006).
Mach, Ernst, *Beiträge zur Analyse der Empfindungen* (Jena: Gustav Fischer, 1886).
Mach, Ernst, *Contributions to the Analysis of the Sensations*, trans. by C. M. Williams (Chicago: Open Court, 1897).
Mach, Ernst, *Die Analyse der Empfindungen und das Verhältniss des Physischen zum Psychischen*, Zweite vermehrte Auflage der Beiträge zur Analyse der Empfindungen (Jena: Fischer, 1900).
Mach, Ernst, *Die Analyse der Empfindungen und das Verhältniss des Physischen zum Psychischen*, Vierte vermehrte Auflage (Jena: Fischer, 1903).
Mach, Ernst, *Die Geschichte und die Wurzel des Satzes von der Erhaltung der Arbeit* (Prague: J. G. Calve, 1872).
Mach, Ernst, *Die Geschichte und die Wurzel des Satzes von der Erhaltung der Arbeit*, 2nd edn (Leipzig: Barth, 1909).

Mach, Ernst, 'Die Leitgedanken meiner wissenschaftlichen Erkenntnislehre und ihre Aufnahme durch die Zeitgenossen', *Scientia*, 7 (1910), 225–40.

Mach, Ernst, *Die Mechanik in ihrer Entwickelung; historisch-kritisch dargestellt*, 1st edn (Leipzig: Brockhaus, 1883).

Mach, Ernst, *Die Mechanik in ihrer Entwickelung; historisch-kritisch dargestellt*, Dritte verbesserte und vermehrte Auflage (Leipzig: Brockhaus, 1897).

Mach, Ernst, *Die Mechanik in ihrer Entwickelung; historisch-kritisch dargestellt*, Fünfte vermehrte Auflage (Leipzig: Brockhaus, 1904).

Mach, Ernst, *Die Mechanik in ihrer Entwickelung; historisch-kritisch dargestellt*, Vierte verbesserte und vermehrte Auflage (Leipzig: Brockhaus, 1901).

Mach, Ernst, *Die Mechanik in ihrer Entwickelung; historisch-kritisch dargestellt*, Zweite verbesserte Auflage (Leipzig: Brockhaus, 1889).

Mach, Ernst, *Die Ökonomische Natur der physikalischen Forschung. Vortrag gehalten in der feierlichen Sitzung der kaiserlichen Akademie der Wissenschaften am 25 Mai 1882* (Wien: Staatdruckerei, 1882).

Mach, Ernst, *Die Prinzipien der physikalischen Optik. Historisch und erkenntnispsychologisch entwickelt* (Leipzig: Barth, 1921).

Mach, Ernst, *Die Prinzipien der Wärmelehre historisch-kritisch entwickelt* (Leipzig: Barth, 1896).

Mach, Ernst, *Erkenntnis und Irrtum. Skizzen zur Psychologie der Forschung* (Leipzig: Barth, 1905).

Mach, Ernst, *History and Root of the Principle of the Conservation of Energy*, trans. by Philip E. B. Jourdain (Chicago: Open Court, 1911).

Mach, Ernst, *Knowledge and Error: Sketches on the Psychology of Enquiry*, trans. by Thomas J. McCormack, Vienna Circle Collection (Dordrecht, Boston: Reidel, 1976).

Mach, Ernst, *Principles of the Theory of Heat. Historically and Critically Elucidated*, ed. by Brian McGuinness (Dordrecht: Reidel, 1986).

Mach, Ernst, *The Analysis of Sensations, and the Relation of the Physical to the Psychical*, trans. by C. M. Williams (Chicago, London: Open Court, 1914).

Mach, Ernst, 'The Economical Nature of Physical Inquiry', in *Popular Scientific Lectures*, by Ernst Mach, trans. by Thomas J. McCormack (Cambridge: Cambridge University Press, 2014), 186–213.

Mach, Ernst, *The Principles of Physical Optics*, trans. by John S. Anderson and A. F. A. Young (London: Methuen, 1926).

Mach, Ernst, *The Science of Mechanics: A Critical and Historical Account of Its Development*, trans. by Thomas J. McCormack (Chicago, London: Open Court, 1919).

Mach, Ernst, 'Über Das Princip der Vergleichung in der Physik', *Verhandlungen Deutscher Naturforscher*, 1894, 44–56.

Massimi, Michela, 'Perspectivism', in *The Routledge Handbook of Scientific Realism*, ed. by Juha Saatsi (Abingdon: Routledge, 2017), 164–75.

Massimi, Michela, and Casey D. McCoy, eds., *Understanding Perspectivism: Scientific Challenges and Methodological Prospects* (Abingdon: Routledge, 2019).

Maxwell, James Clerk, *A Treatise on Electricity and Magnetism* (Oxford: Clarendon, 1873).

Maxwell, James Clerk, *Lehrbuch der Electricität und des Magnetismus*, trans. by Max Bernhard Weinstein (Berlin: Julius Springer, 1883).

Mayer, Julius Robert, 'Bemerkungen über die Kräfte der unbelebten Natur', *Annalen der Chemie und Pharmazie*, 47 (1842), 233–40.

Mayr, Ernst, 'The Idea of Teleology', *Journal of the History of Ideas*, 53/1 (1992), 117–35.

Meli, Domenico Bertoloni, 'The Relativization of Centrifugal Force', *Isis*, 81/1 (1990), 23–43.

Milkov, Nikolay, ed., *Die Berliner Gruppe. Texte zum Logischen Empirismus* (Hamburg: Meiner, 2015).

Milkov, Nikolay and Volker Peckhaus, eds., *The Berlin Group and the Philosophy of Logical Empiricism*, Boston Studies in the Philosophy and History of Science (Cham: Springer, 2013).

Miller, Arthur I., *Albert Einstein's Special Theory of Relativity: Emergence (1905) and Early Interpretation (1905–1911), Addison-Wesley, Massachusetts, 1981* (Reading: Addison-Wesley, 1981).

Minkowski, Hermann, *Raum und Zeit: Vortrag, gehalten auf der 80. Naturforscher-Versammlung zu Köln am 21. September 1908*, 1909th edn (Leipzig: Teubner, 1909).

Minkowski, Hermann, 'Space and Time', in *The Principle of Relativity. A Collection of Original Memoirs on the Special and General Theory of Relativity*, ed. by Wilfrid Perrett and George B. Jeffrey, Fundamental Theories of Physics (New York: Dover, 1923), 73–91.

Müller, Aloys, *Die philosophischen Probleme der Einsteinschen Relativitätstheorie* (Braunschweig: Vieweg, 1922).

Müller, Aloys, 'Probleme der speziellen Relativitätstheorie', *Zeitschrift für Physik*, 17 (1923), 409–20.

Müller, Horst, 'Joseph Petzoldt', *Humanismus Und Technik. Technische Universität Berlin*, 11 (1966), 33–6.

Müller, Ingo, *A History of Thermodynamics: The Doctrine of Energy and Entropy* (Berlin: Springer Nature, 2010).

Murray, David J., *The Creation of Scientific Psychology* (Routledge, 2020).

Natorp, Paul, *Die logischen Grundlagen der exakten Wissenschaften* (Leipzig: Teubner, 1910).

Needham, Joseph, 'Evolution and Thermodynamics: A Paradox with Social Significance', *Science & Society*, 6/4 (1942), 352–75.

Neuber, Matthias, *Die Grenzen des Revisionismus. Schlick, Cassirer und das Raumproblem* (Vienna and New York: Springer, 2012).

Neumann, Carl Gottfried, 'On the Principles of the Galilean-Newtonian Theory', trans. by Robert DiSalle, *Science in Context*, 6/1 (1993), 355–68.

Neumann, Carl Gottfried, 'Über den Körper Alpha', *Berichte über die Verhandlungen der Königlich Sächsischen Gesellschaft der Wissenschaften. Physische Klasse*, 62 (1910), 69–86.

Neumann, Carl Gottfried, *Ueber die Prinzipien der Galilei-Newtonschen Theorie* (Leipzig: Teubner, 1870).

Newton, Isaac, *The Principia: Mathematical Principles of Natural Philosophy: The Authoritative Translation*, trans. by I. Bernard Cohen, Anne Whitman, and Julia Budenz (Berkeley, Los Angeles, London: University of California Press, 1999).

Norton, John D., 'Einstein's Investigations of Galilean Covariant Electrodynamics Prior to 1905', *Archive for History of Exact Sciences*, 59/1 (2004), 45–105.

Norton, John D., 'Einstein's Special Theory of Relativity and the Problems in the Electrodynamics of Moving Bodies That Led Him to It', in *The Cambridge Companion to Einstein*, ed. by Christoph Lehner and Michel Janssen (Cambridge: Cambridge University Press, 2014), 72–102.

Oberdan, Thomas, 'Geometry, Convention, and the Relativized Apriori: The Schlick-Reichenbach Correspondence', in *Stationen. Dem Philosophen Und Physiker Moritz*

Schlick Zum 125. Geburtstag, ed. by Friedrich Stadler, Hans Jürgen Wendel, Edwin Glassner, Fynn Ole Engler, and Massimo Ferrari (Wien: Springer Nature, 2008), 186–211.

Ostwald, Wilhelm, 'Ueber das Prinzip des ausgezeichneten Falles', *Königlich sächsische Gesellschaft der Wissenschaften. Mathematisch-physikalische Classe. Berichte über die Verhandlungen*, 1893, 599–603.

Padovani, Flavia, 'Relativizing the Relativized a Priori: Reichenbach's Axioms of Coordination Divided', *Synthese*, 181/1 (2011), 41–62.

Patton, Lydia, 'Perspectivalism in the Development of Scientific Observer-Relativity', in *The Emergence of Relativism*, ed. by Martin Kusch, Katherina Kinzel, Johannes Steizinger, and Niels Jacob Wildschut (New York: Routledge, 2019), 63–78.

Peckhaus, Volker, *Logik, Mathesis universalis und allgemeine Wissenschaft: Leibniz und die Wiederentdeckung der formalen Logik im 19. Jahrhundert* (Berlin: Akademie Verlag, 1997).

Pellegrino, Emilio Marco, Elena Ghibaudi, and Luigi Cerruti, 'Clausius' Disgregation: A Conceptual Relic That Sheds Light on the Second Law', *Entropy*, 17/7 (2015), 4500–18.

Petzoldt, Joseph, 'Anzeige von R. Avenarius "Kritik der reinen Erfahrung"', *Das Magazin für die Litteratur des In- und Auslandes*, 58/8 (1889), 120–36.

Petzoldt, Joseph, *Das allgemeinste Entwicklungsgesetz* (München: Rösl, 1923).

Petzoldt, Joseph, *Das natürliche Höhenziel der menschheitlichen Entwicklung* (Berlin, Leipzig: Paetel, 1927).

Petzoldt, Joseph, 'Das Verhältnis der Machschen Gedankenwelt zur Relativitätstheorie', in *Die Mechanik in ihrer Entwicklung historisch-kritisch dargestellt*, ed. by Ernst Mach, 8th edn (Leipzig: Brockhaus, 1921), 490–517.

Petzoldt, Joseph, *Das Weltproblem vom Standpunkte des relativistischen Positivismus aus, historisch-kritisch dargestellt*, 4th edn (Leipzig: Teubner, 1924).

Petzoldt, Joseph, 'Die Elemente der Welt', *Monistische Monatshefte*, 8 (1923), 97–102, 129–33, 189–95, 363–71.

Petzoldt, Joseph, 'Die physikalische Wirklichkeit', *Die Naturwissenschaften*, 11 (1923), 828.

Petzoldt, Joseph, 'Die Relativitätstheorie im erkenntnistheoretischen Zusammenhange des relativistischen Positivismus', *Berichte der Deutschen physikalischen Gesellschaft*, 23 (1912), 1055–64.

Petzoldt, Joseph, 'Die Unmöglichkeit mechanischer Modelle zur Veranschaulichung der Relativitätstheorie', *Berichte der Deutsche Physikalische Gesellschaft*, 13–14 (1919), 495–503.

Petzoldt, Joseph, 'Kausalität und Relativitätstheorie', *Zeitschrift für Physik*, 1/5 (1920), 467–74.

Petzoldt, Joseph, 'Mechanistische Naturauffassung und Relativitätstheorie', *Annalen der Philosophie*, 2/3 (1921), 447–62.

Petzoldt, Joseph, 'Postulat der absoluten und relativen Welt', *Zeitschrift für Physikalischen und Chemischen Unterricht*, 21 (1924), 143–50.

Petzoldt, Joseph, 'Solipsismus auf praktischem Gebiet', *Vierteljahrsschrift für wissenschaftliche Philosophie*, 25 (1901), 339–62.

Petzoldt, Joseph, 'Sonderschulen für hervorragend Befähigte', *Neue Jahrbücher für Pädagogik*, 7 (1904), 425–56.

Petzoldt, Joseph, 'Verbietet die Relativitätstheorie Raum und Zeit als etwas wirkliches zu denken?', *Verhandlungen der Deutsche Physikalische Gesellschaft*, 20 (1918), 189–201.

Petzoldt, Joseph, 'Vorwort des Herausgebers', in *Kritik der reinen Erfahrung*, ed. by Richard Avenarius, 2nd edn, 2 vols (Leipzig: Reisland, 1907), I, v–ix.

Petzoldt, Joseph, 'Zur positivistische Philosophie', *Zeitschrift für positivistische Philosophie*, 1 (1913), 1–16.
Pfister, Herbert, 'Ludwig Lange on the Law of Inertia', *The European Physical Journal H*, 39/2 (2014), 245–50.
Planck, Max, 'Das Prinzip der Relativität und die Grundgleichungen der Mechanik', *Verhandlungen der Deutschen Physikalischen Gesellschaft*, 8 (1906), 136–41.
Planck, Max, 'Die Einheit Des Physikalischen Weltbildes', *Physikalische Zeitschrift*, 10 (1909), 62–75.
Planck, Max, 'The Mach-Planck Polemics', in *Ernst Mach. A Deeper Look: Documents and New Perspectives*, ed. by John Blackmore, Boston Studies in the Philosophy of Science (Dordrecht: Springer, 1992), 127–50.
Popper, Karl R., 'A Note on Berkeley as Precursor of Mach', *The British Journal for the Philosophy of Science*, 4/13 (1953), 26–36.
Pross, Addy, 'Paradoxes of Stability: How Life Began and Why It Can't Rest', *Aeon*. https://aeon.co/essays/paradoxes-of-stability-how-life-began-and-why-it-can-t-rest [accessed 28 October 2021].
Pross, Addy, *What Is Life?: How Chemistry Becomes Biology* (Oxford: Oxford University Press, 2012).
Pyenson, Lewis, 'The Relativity Revolution in Germany', in *The Comparative Reception of Relativity*, ed. by Thomas F. Glick, Boston Studies in the Philosophy of Science, 103 (Dordrecht: Reidel, 1987), 59–111.
Pyenson, Lewis, *The Young Einstein. The Advent of Relativity* (Bristol, Boston: CRC Press, 1985).
Rehmke, Johannes, *Die Seele des Menschen* (Leipzig: Teubner, 1905).
Reichenbach, Hans, *Axiomatik der relativistischen Raum-Zeit Lehre* (Braunschweig: Vieweg, 1924).
Reichenbach, Hans, 'Bericht über eine Axiomatik der Einsteinschen Raum-Zeit Lehre', *Physikalische Zeitschrift*, 22 (1921), 683–7.
Reichenbach, Hans, 'Der gegenwärtige Stand der Relativitätsdiskussion', *Logos*, 10 (1922), 316–78.
Reichenbach, Hans, 'La signification philosophique de la théorie de la relativité', *Revue Philosophique de la France et de l'Etranger*, 94 (1922), 5–61.
Reichenbach, Hans, 'Lichtgeschwindigkeit und Gleichzeitigkeit', *Annalen der Philosophie und philosophische Kritik*, 6 (1927), 128–44.
Reichenbach, Hans, *Relativitätstheorie und Erkenntnis Apriori* (Berlin: Springer, 1920).
Reichenbach, Hans, 'The Philosophical Significance of the Theory of Relativity', in *Defending Einstein: Hans Reichenbach's Writings on Space, Time and Motion*, ed. by Steven Gimbel and Anke Walz (New York: Cambridge University Press, 2006), 95–160.
Reichenbach, Hans, 'The Present State of the Discussion on Relativity', in *Selected Writings 1909-1953*, ed. by Maria Reichenbach and Robert S. Cohen, 2 vols (Dordrecht, Boston: Springer, 1978), II, 3–47.
Reichenbach, Hans, *The Theory of Relativity and a Priori Knowledge* (Berkeley, Los Angeles: University of California Press, 1965).
Rescher, Nicolas, 'The Berlin School of Logical Empiricism and Its Legacy', *Erkenntnis* 64 (2006), 281–304.
Rickert, Heinrich, *Der Gegenstand der Erkenntnis. Einführung in die Transcendentalphilosophie* (Tübingen: Mohr, 1904).
Robinson, David K., 'Reaction-Time Experiments in Wundt's Institute and Beyond', in *Wilhelm Wundt in History*, ed. by Robert W. Rieber and David K. Robinson (Springer, Boston, MA, 2001), 161–204.

Roux, Wilhelm, *Der Kampf der Teile im Organismus. Ein Beitrag zur Vervollständigung der mechanischen Zweckmäßigkeitslehre* (Leipzig: Engelmann, 1881).

Rudolph, Enno, *Ernst Cassirer Im Kontext: Kulturphilosophie Zwischen Metaphysik Und Historismus* (Tübingen: Mohr Siebeck, 2003).

Russo Krauss, Chiara, *Wundt, Avenarius, and Scientific Psychology: A Debate at the Turn of the Twentieth Century* (New York: Palgrave Macmillan, 2019).

Ryckman, Thomas, 'Conditio Sine qua Non? Zuordnung in the Early Epistemologies of Cassirer and Schlick', *Synthese*, 88 (1991), 57–95.

Ryckman, Thomas, 'Einstein, Cassirer, and General Covariance. Then and Now', *Science in Context*, 12 (1999), 585–619.

Ryckman, Thomas, *The Reign of Relativity: Philosophy in Physics 1915–1925* (Oxford: Oxford University Press, 2004).

Rynasiewicz, Robert, 'The Construction of the Special Theory: Some Queries and Considerations', in *Einstein: The Formative Years. 1879–1909*, ed. by Don Howard and John Stachel (Cham: Springer, 2000), 159–201.

Sachs, Mendel, 'On Einstein's Later View of the Twin Paradox', *Foundations of Physics Volume*, 15 (1985), 977–80.

Salmon, Wesley C., 'The Philosophy of Hans Reichenbach', *Synthese*, 34/1 (1977), 5–88.

Scharff, Robert C., *Comte after Positivism*, Modern European Philosophy (Cambridge: Cambridge University Press, 1995).

Schlick, Moritz, *Allgemeine Erkenntnislehre* (Berlin: Julius Springer, 1918).

Schlick, Moritz, 'Critical or Empiricist Interpretation of Modern Physics?: Remarks on Ernst Cassirer's "Einstein's Theory of Relativity"', in *Philosophical Papers. Volume 1: (1909–1922)*, ed. by Henk L. Mulder and Barbara F. B. Van De Velde-Schlick (Dordrecht, Boston: Kluwer, 1978), 322–34.

Schlick, Moritz, 'Die Philosophische Bedeutung Des Relativitätsprinzips', *Zeitschrift Für Philosophie Und Philosophische Kritik*, 159 (1915), 129–75.

Schlick, Moritz, 'Die Relativitätstheorie in der Philosophie', in *Rostock, Kiel, Wien: Aufsatze, Beitrage, Rezensionen 1919–1925: Aufsätze, Beiträge, Rezensionen 1919–1925*, ed. by Edwin Glassner, Heidi Konig-Porstner, and Karsten Boger (Wien, New York: Springer Nature, 2012), 529–47.

Schlick, Moritz, *General Theory of Knowledge*, ed. by Albert E. Blumberg and Herbert Feigl (Wien: Springer, 1974).

Schlick, Moritz, 'Kritizistische oder empiristische Deutung der neuen Physik? Bemerkungen zu Ernst Cassirers Buch "Zur Einsteinschen Relativitätstheorie"', in *Rostock, Kiel, Wien: Aufsatze, Beitrage, Rezensionen 1919–1925: Aufsätze, Beiträge, Rezensionen 1919–1925*, ed. by Edwin Glassner, Heidi Konig-Porstner, and Karsten Boger (Wien, New York: Springer Nature, 2012), 223–47.

Schlick, Moritz, 'The Philosophical Significance of the Principle of Relativity', in *Philosophical Papers, Volume I (1909–1922)*, ed. by Henk L. Mulder and Barbara F. B. van de Velde-Schlick, Vienna Circle Collection (Dordrecht: Reidel, 1979), 153–89.

Schlick, Moritz, 'The Theory of Relativity in Philosophy', in *Philosophical Papers. Volume 1: (1909–1922)*, ed. by Henk L. Mulder and Barbara F. B. Van De Velde-Schlick (Dordrecht, Boston: Kluwer, 1978), 343–53.

Schubert-Soldern, Richard von, *Ueber Transcendenz des Objects und Subjects* (Leipzig: Fues, 1882).

Schuppe, Wilhelm, *Der Zusammenhang von Leib und Seele und das Grundproblem der Psychologie* (Wiesbaden: Bergmann, 1902).

Schuppe, Wilhelm, 'Die Bestätigung des naiven Realismus. Offener Brief an Herrn Prof. Dr. Richard Avenarius', *Vierteljahrsschrift für wissenschaftliche Philosophie*, 17 (1893), 365–88.

Schuppe, Wilhelm, *Grundriss der Erkenntnistheorie und Logik* (Berlin: Gaertners, 1894).

Société française de Philosophie, 'Comptes rendus des séances. Séance du 6 avril 1922. La théorie de la relativité', *Bulletin de la Société française de Philosophie*, 22 (1922), 91–113.

Sommer, Manfred, 'Das unrettbare Ich und die heitere Passivität des Ernst Mach', in *La metafisica del positivismo*, ed. by Luca Guidetti, Discipline Filosofiche (Macerata: Quodlibet), 149–59.

Spaulding, Edward Gleason, *Beiträge zur Kritik des psychophysischen Parallelismus vom Standpunkte der Energetik* [1900] (Hildesheim: Georg Olms, 1985).

Spencer, Herbert, *First Principles* (London: Williams & Norgate, 1862).

Stachel, John, *Einstein from 'B' to 'Z'* (Boston: Birkhäuser, 2001).

Staley, Richard, 'Mother's Milk and More: On the Role of Ernst Mach's Relational Physics in the Development of Einstein's Theory of Relativity', in *Interpreting Mach: Critical Essays*, ed. by John Preston (Cambridge: Cambridge University Press, 2021), 28–47.

Stein, Howard, 'Some Philosophical Prehistory of General Relativity', in *Foundations of Space-Time Theories*, Minnesota Studies in the Philosophy of Science, 8 (Minneapolis: University of Minnesota Press), 3–49.

Steinthal, Heymann, *Abriss der Sprachwissenschaften. Einleitung in die Psychologie und Sprachwissenschaft*, 2 vols (Berlin: Dümmler, 1871), I.

Steinthal, Heymann, *Der Ursprung der Sprache im Zusammenhange mit den letzten Fragen alles Wissens*, 4th edn (Berlin: Dümmler, 1888).

Stöltzner, Michael, 'Action Principles and Teleology', in *Inside versus Outside. Endo- and Exo-Concepts of Observation and Knowledge in Physics, Philosophy and Cognitive Science*, ed. by Harald Atmanspacher and Gerhard J. Dalenoort (Berlin, Heidelberg: Springer, 1994), 33–62.

Stubenberg, Leopold, 'Neutral Monism', in *The Stanford Encyclopedia of Philosophy*, ed. by Edward N. Zalta, Fall 2018 (Metaphysics Research Lab, Stanford University, 2018). https://plato.stanford.edu/archives/fall2018/entries/neutral-monism/ [accessed 28 October 2021].

Terletskii, Yakov P., *Paradoxes in the Theory of Relativity* (New York: Springer, 1978).

Thiele, Joachim, 'Briefe Albert Einsteins an Joseph Petzoldt', *NTM. Schriftenreihe Für Geschichte Der Naturwissenschaften, Technik Und Medizin*, 8 (1971), 70–4.

Thiele, Joachim, 'Briefe deutscher Philosophen an Ernst Mach', *Synthese*, 18/2/3 (1968), 285–301.

Tilitzki, Christian, *Die deutsche Universitätsphilosophie in der Weimarer Republik und im Dritten Reich*, 2 vols (Berlin: De Gruyter, 2002), II.

Ugarov, Vladimir A., *Special Theory of Relativity* (Moscow: Mir, 1979).

Ulrich, Georg, 'Bewusstsein und Ichheit', *Zeitschrift für Philosophie und philosophische Kritik*, 124 (1905), 58–79.

Vaihinger, Hans, *Die Philosophie des Als Ob. System der theoretischen, praktischen und religiösen Fiktionen der Menschheit auf Grund eines idealistischen Positivismus* (Berlin: Reuter und Richard, 1911).

Vaihinger, Hans, 'Geleitwort', in *Kants Lehre vom 'Bewusstsein überhaupt' und ihre Weiterbildung bis auf die Gegenwart*, ed. by Hans Amrhein (Berlin: Reuther und Reichard, 1909), iii–v.

Wagner, Hermann, 'Professor Joseph Petzoldt Zum Gedächtnis. Von Einem Ehemaligen Schüler', *Kai Nein Kai Grammata. 1853-1953*, 1953, 27–9.

Watkins, Eric, 'The Laws of Motion from Newton to Kant', *Perspectives on Science*, 5 (1997), 311–48.

Wazeck, Milena, *Einstein's Opponents: The Public Controversy about the Theory of Relativity in the 1920s*, trans. by Geoffrey S. Koby (Cambridge, New York: Cambridge University Press, 2014).

Weinstein, Galina, 'Einstein's Clocks and Langevin's Twins', 8. https://arxiv.org/ftp/arxiv/papers/1205/1205.0922.pdf.

Wiechert, Emil, 'The Relativity Concept in Physics', in *Ernst Mach – A Deeper Look: Documents and New Perspectives*, ed. by John Blackmore (Dordrecht, Boston: Kluwer, 1992), 165–70.

Willy, Rudolf, 'Das erkenntnistheoretische Ich und der natürliche Weltbegriff', *Vierteljahrsschrift für wissenschaftliche Philosophie*, 18 (1894), 1–29.

Wolters, Gereon, 'Globalized Parochialism: Consequences of English as Lingua Franca in Philosophy of Science', *International Studies in the Philosophy of Science*, 29 (2015), 189–200.

Wolters, Gereon, 'Mach and Einstein, or Clearing Troubled Waters in the History of Science', in *Einstein and the Changing World Views of Physics*, ed. by Christoph Lehner, Jürgen Renn, and Matthias Schemmel (New York: Springer, 2012), 39–57.

Wolters, Gereon, *Mach I, Mach II, Einstein und die Relativitätstheorie* (Berlin, New York: De Gruyter, 1987).

Wundt, Wilhelm, *Grundzüge der physiologischen Psychologie*, 3rd edn, 2 vols (Leipzig: Engelmann, 1887), ii.

Wundt, Wilhelm, *Grundzüge der physiologischen Psychologie*, 1st edn (Leipzig: Engelmann, 1874).

Wundt, Wilhelm, 'Über naiven und kritischen Realismus. II. Der Empiriokritizismus', *Philosophische Studien*, 13 (1898), 1–105, 323–433.

Wundt, Wilhelm, 'Über psychische Causalität und das Princip des psychophysischen Parallelismus', *Philosophische Studien*, 10 (1894), 1–124.

Zöllner, Johann Karl Friedrich, *Über die Natur der Cometen. Beiträge zur Geschichte und Theorie der Erkenntnis* (Leipzig: Engelmann, 1872)

Index

Anaxagoras 83
Aristotle 83
Avenarius, Richard
  Brain's evolution towards pure experience 62–4
  Conception of the physical and the psychical 48–52, 55, 60–2
  Debate with Schuppe 89–90
  Early psychological theory 20–7
  As forerunner of relativistic positivism 77–9, 81, 85, 88, 92–7, 99–100, 104, 107, 167, 169, 171–2
  Influence on Petzoldt 1–3, 5–9, 11
  Later psychological theory 27–8
  Overcoming of subjective idealism 67–75, 157
  Petzoldt criticism of 32–5
  Principle of the least amount of energy 13–14, 17
  Theory of introjection 49, 70, 79–80

Berg, Otto 132
Berkeley, George 72, 84, 92
Bernoulli, Daniel 28
Brain
  Dependence of mental activity on the brain 19, 54–62, 69, 75, 83, 86, 91–2, 97, 102, 107
  As frame of reference in physics 115–20, 150–1
  Its evolution as condition for the evolution of knowledge 62–5, 103–4, 156, 171
  Lawfulness and regularity of brain activity 52–3
  Psychical defined as what is dependent on the brain 47–51
  And relativistic positivism 12, 72, 100–1, 120, 143, 149–50, 160, 163, 167–8
  Self-preservation 27, 52
Büchner, Ludwig 11, 16

Carnap, Rudolf 5
Cassirer, Ernst 153–7, 159, 165, 167, 172
Classen, Johannes 131–2
Clausius, Rudolf 16, 40
Cohn, Emil 136
Comte, August 35
Conant, James 169
Conservation of energy, principle of the 7, 15, 21, 27, 35–6, 38–40, 42, 47, 53, 56–60, 99, 103, 107
Copernicus, Nicolaus and Copernican system 48, 109, 146
Cornelius, Hans 9
Czolbe, Heinrich 11

Darwin, Charles and Darwinism 1, 12–14, 16–17, 21, 23–4, 26, 28, 33, 63, 65
Deleuze, Gilles 169
Democritus 80
Descartes, René 83–4
Dilthey, Wilhelm 2
Dingler, Hugo 138, 162
Driesch, Hans 90
Du Bois-Reymond, Emil 1, 36, 74, 99
Dubislav, Walter 5

Ehrenfest, Paul 136–7
Eindeutigkeit, principle of
  Application to mental processes 47–56, 75–7, 98
  As argument against subjectivism and Kantianism 75–7
  As condition of possibility of experience (a priori) and as grounded on experience (a posteriori) 43–7, 86, 115
  As designed to overcome the metaphysical concept of causality 35–43, 91–2
  As grounded on the evolution of the brain 44, 64, 86, 115
  And law of inertia 113–15, 128

As philosophical interpretation of the variational principles 28–31, 34–5
And relativity theory 123, 133, 141–7, 149, 155–6, 158, 160–1, 163–4, 167–9, 173
Role in relativistic positivism 101–2, 120, 169
Einstein, Albert
  Cassirer's criticism of Petzoldt's interpretation of relativity theory 153–4, 156
  Formulation of Mach's principle 110, 118
  Formulation of the special relativity theory 7, 128, 130
  Grounding of relativity theory on sensory data 148–51
  Personal relationship with Petzoldt 3, 133, 135–8
  Petzoldt's interpretation of relativity theory as confirmation of Mach's philosophy and relativistic positivism 8, 77–9, 107–8, 122, 133–5, 140–4, 167–8, 170, 172–3
  Petzoldt's and Mach's letters about relativity theory 131–3, 135
  Petzoldt's objections to relativity theory 144–8
  Reichenbach's criticism of Petzoldt's interpretation of relativity theory 162–7
  Relationship with Mach 135, 137–9
  Schlick's criticism of Petzoldt's interpretation of relativity theory 157–8, 160–2
Empiricism 5–7, 36, 72, 80, 104, 154, 157, 159
Empiriocriticism 5–6, 8, 13, 64, 72–3, 75, 77–8, 89–90, 96, 100, 167, 172
Euclid 28
Euler, Leonhard and Euler's principle of least action 2, 28, 30, 113
Euripides 83

Fechner, Gustav Theodor
  Conception of the physical and the psychical 48, 100
  Influence on Petzoldt 9, 28, 169–71
  Principle of the tendency towards stability 13–17, 22–3, 25, 26, 32–3, 52, 100
  Psychophysics 19–22
  View of causality 36, 38–41, 43, 57, 81
Finalism 6, 11–12, 15–16, 18, 20, 23, 25, 26, 32, 33–5, 43, 67
Fizeau, Hippolyte 129–30, 135, 141, 145
Foucault, Jean Bernard Léon 129
Frank, Philipp 132
Fresnel, Augustin 129
Freud, Sigmund 3
Functional relations 35–6, 38–43, 50–2, 55–6, 81, 94, 97, 102, 122, 140, 142, 144–6, 150, 158, 168

Galilei, Galileo and Galilean relativity 108, 111, 121, 128–31
Gauss, Johann Friedrich Carl and Gauss' principle of least effort 2, 25–6, 28–9
Giere, Ronald N. 171
Grelling, Kurt 5
Gruner, Paul 132
Guattari, Felix 169

Haeckel, Ernst Heinrich 16
Hamilton, William Rowan and Hamilton's variational principle 28–9
Hegel, Georg Wilhelm Friedrich 74, 141
Helm, Georg 3, 136–7
Helmholtz, Hermann von 18, 36, 99, 171
Hentschel, Klaus 144
Heraclitus 80
Herbart, Johann Friedrich and Herbartian psychology 17–23, 26
Hering, Ewald 136
Hertz, Heinrich Rudolf 129
Hilbert, David 3
Hobbes, Thomas 39, 84
Hönigswald, Richard 105
Howard, Don 30–1
Hume, David 35–6, 83–5

Idealism 2, 12–13, 68–77, 84–5, 88–9, 92–5, 137, 167–9, 172
Indeterminism 38–41, 43, 57, 128, 173

Jerusalem, Wilhelm 3

Kant, Immanuel
  Petzoldt criticism of 6, 44, 67–8, 143–4
  Revival during the Nineteenth century 1, 12
  Role in the history of philosophy according to Petzoldt 85–7
  View of causality and finality 15, 18, 43, 113
Kantianism and neo-Kantianism
  As mainstream philosophical trend of the era 7–8, 12–13, 35, 69
  Neo-Kantian interpretations of relativity theory 134, 153–4, 156–7, 159, 162
  Petzoldt's early and latent Kantianism 1–2, 26, 99, 104, 161
  Petzoldt's rejection of 6, 13, 71, 85–8, 97, 100, 169, 172
  Schuppe as bridge between Kantianism and positivism 89, 93–4, 166
Kauffmann, Max Reinhard 90
Kirchhoff, Gustav Robert 36
Klein, Felix 3
Köhnke, Klaus C. 12
Kraus, Oskar 137

Laas, Emil 88
Lagrange, Joseph-Louis 28–9
Lagrangian mechanics, philosophical interpretation of 28–31
Lamarck, Jean-Baptiste 21
Lamprecht, Karl 3
Lange, Friedrich Albert 1, 61–2, 86, 99, 110
Lange, Ludwig 110, 112–13, 122, 125–8
Laplace, Pierre-Simon 44
Laue, Max von 134
Leibniz, Gottfried Wilhelm von 15, 31, 57, 158, 161–2, 169–70
Liebmann, Otto 85
Locke, John 84
Lorentz, Hendrik Antoon and Lorentz's transformations 129–32, 135, 137, 141–2, 149

Mach, Ernst
  Conception of the physical and the psychical 48–51
  Criticism of absolute space and time 108–13, 123–5, 127–8, 140
  Criticism of Petzoldt's Machian interpretation of relativity theory by other philosophers 153–4, 156–9, 162–3, 167
  Discussion with Petzoldt on human body as frame of reference in physics 116–22
  Discussion with Petzoldt on the law of inertia 113–16
  Discussion of relativity theory 131–3, 135, 138–9
  Indeterminism 39–40, 43, 57
  Influence on Einstein 135–7
  Influence on Petzoldt 1–3, 5–9, 11, 13, 169–73
  Petzoldt on relativity theory as confirmation of Mach's criticism of Newtonian physics 134–5, 141–6, 148
  Petzoldt's anti-subjectivist interpretation of Mach's philosophy 72–4, 87–8, 97, 100–1, 104
  Petzoldt's criticism of 32–4, 39–41
  Petzoldt's interpretation of relativity theory as consistent with Mach's physiology of senses 148–51, 161
  Petzoldt's turning Mach's thought into a philosophical system 105, 107
  Principle of economy of thought 2, 25, 28, 73–4
  Relationship with Schuppe 90–1
  Role in the history of philosophy according to Petzoldt 77–9, 81, 85, 92–7, 99–101
  View of causality 36–43
Mach, Ludwig 138–9
Mach's principle 110, 118
Massimi, Michela 171
Materialism 11–17, 55, 72, 77, 92, 95, 98–9, 140, 169, 172
Maupertius, Pierre Louis Moreau de 29
Maxwell, James Clerk 121, 128–9
Mayer, Julius Robert von 35
Michelson, Albert Abraham 129–30, 135, 141, 145
Minkowski, Hermann 132–3, 145, 147–8
Moleschott, Jakob 11
Morley, Edward 129–30, 135, 141, 145

Müller, Georg Elias 2–3
Müller, Johannes Peter 18, 69

Natorp, Paul 134
Naturphilosophie 6, 11, 15, 17, 25–6
Nervous system *See* Brain
Neumann, Carl 110–12, 115, 117–19, 124–5, 127
Newton, Isaac 28–9, 108–12, 115, 117, 125–9, 133–5, 144, 150, 156, 172
Nietzsche, Friedrich 169–71

Oken, Lorenz 17

Parmenides 80
Paulsen, Friedrich 2
Pericles 83
Perspectivism 167, 169–71, 173
Phenomenalism 6, 74, 100, 105, 156, 163, 165, 167–8, 171, 173
Phidias 83
Planck, Max 78, 132–3, 135, 157
Plato 83, 143–4, 154
Potonié, Henry 3
Pross, Addy 24
Protagoras 80–5, 96, 101, 133, 141, 154–5, 160, 169–70
Psychology 1, 4–5, 7, 17–24, 49–53, 58, 60, 69, 74, 105, 116, 165–8, 171–2
Psychophysical parallelism 35, 47, 51, 54–7, 60, 67, 91–2, 98, 100
Ptolemy and Ptolemaic system 48, 109, 146

Realism 6, 13, 68–70, 72, 77, 84, 87, 95, 104, 157, 168–9, 171, 173
Reichenbach, Hans 5, 153, 155, 162–7, 172
Relativistic positivism
    Criticism by other philosophers 153–5, 159–61, 163, 165–8
    Definition 75, 77–9, 99–105, 169–73
    Schuppe, Mach, and Avenarius as its forerunners 93–9
    And theory of relativity 107–8, 119–20, 122, 124, 127, 133–44, 147
Rickert, Heinrich 88
Roux, Wilhelm 3, 63

Scepticism 12–13, 65, 82–5, 100–1, 155, 159, 166–7, 171, 173
Schelling, Friedrich 17–18, 141
Schlick, Moritz 137, 153, 157–62, 167, 172
Schmidt, Raymund 5
Schubert Soldern, Richard von 9, 90
Schuppe, Wilhelm 3, 9, 77, 85, 88–95, 97–100, 107, 136, 157, 166
Society for Positivist Philosophy (*Gesellschaft für positivistische Philosophie*) and Society for Empirical Philosophy (*Gesellschaft für empirische Philosophie*) 3–5, 133, 138, 166, 172
Sophocles 83
Spinoza, Baruch 48, 84, 161
Staudinger, Franz 35
Steinthal, Heymann 9, 17, 19–22, 26, 28
Stumpf, Carl 2

Teleology *See* Finalism
Thales of Miletus 80–1, 83
Thomson, William (Lord Kelvin) 40
Tönnies, Ferdinand 3

Univocalness *See Eindeutigkeit*

Vaihinger, Hans 4–5, 137
Verworn, Max 3
Vogt, Carl 11

Weber, Ernst Heinrich 171
Wertheimer, Max 137
Willy, Rudolph 89
Wolters, Gereon 138
Wundt, Wilhelm 35, 38–40, 43, 55, 57–8, 61, 78

Young, Thomas 129

Zenodorus 28
Ziehen, Theodor 3, 9
Zöllner, Johann Karl Friedrich 13–14, 28, 32

www.ingramcontent.com/pod-product-compliance
Lightning Source LLC
Chambersburg PA
CBHW062139300426
44115CB00012BA/1984